Epigenetics
in
BIOLOGY and MEDICINE

Epigenetics
in
BIOLOGY and
MEDICINE

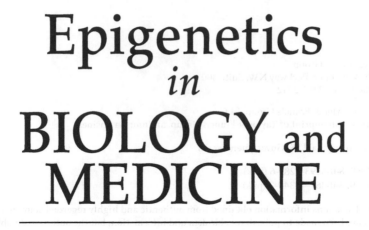

Edited by
Manel Esteller

CRC Press
Taylor & Francis Group
Boca Raton   London   New York

CRC Press is an imprint of the
Taylor & Francis Group, an **informa** business

CRC Press
Taylor & Francis Group
6000 Broken Sound Parkway NW, Suite 300
Boca Raton, FL 33487-2742

ISBN-13: 978-0-8493-7289-6 (hbk)
ISBN-13: 978-0-367-40349-2 (pbk)

Library of Congress Cataloging-in-Publication Data

Epigenetics in biology and medicine / editor, Manel Esteller.
p. ; cm.
Includes bibliographical references and index.
ISBN 978-0-8493-7289-6 (hardcover : alk. paper)
1. DNA--Methylation. 2. Histones. 3. Epigenesis. 4. Carcinogenesis. I. Esteller, Manel. II. Title.
[DNLM: 1. Epigenesis, Genetic. QU 475 E645 2008]

QP624.5.M46E648 2008
572.8'6--dc22   2008014629

Visit the Taylor & Francis Web site at
http://www.taylorandfrancis.com

and the CRC Press Web site at
http://www.crcpress.com

Editor

Manel Esteller, M.D., Ph.D. (Sant Boi de Llobregat, Barcelona, Catalonia, Spain), graduated in medicine with honors from the Universidad de Barcelona in 1992, where he also obtained his Ph.D., specializing in molecular genetics of endometrial carcinoma, in 1996. He was an invited researcher at the School of Biological and Medical Sciences at the University of St. Andrews (Scotland, UK), during which time his research interests focused on the molecular genetics of inherited breast cancer. From 1997 to 2001, Esteller was a postdoctoral fellow and a research associate at the Johns Hopkins University and School of Medicine (Baltimore, Maryland), where he studied DNA methylation and human cancer. His work was decisive in establishing promoter hypermethylation of tumor suppressor genes as a common hallmark of all human tumors.

From October 2001 to September 2008 Dr. Esteller was the leader of the Cancer Epigenetics Laboratory of CNIO (Centro Nacional de Investigaciones Oncológicas), where his principal areas of research were the alterations in DNA methylation, histone modifications, and chromatin in human cancer. Since October 2008, Dr. Esteller has been the director of the Cancer Epigenetics and Biology Program of the Catalan Institute of Oncology (ICO) in Barcelona, professor of genetics in the School of Medicine of the University of Barcelona, and an ICREA research professor.

Author of more than 190 original peer-reviewed manuscripts in biomedical sciences, Dr. Esteller is also a member of numerous international scientific societies and editorial boards and a reviewer for many journals and funding agencies. He is also associate editor for *Cancer Research*, *The Lancet Oncology and Carcinogenesis*, editor-in-chief of *Epigenetics*, advisor for the Human Epigenome Project, associate member of the Epigenome Network of Excellence, and president of the Epigenetics Society. His numerous awards include the Cancer Research Award from the European School of Medical Oncology (1999), first prize in basic research from Johns Hopkins University (1999), Investigator Award from the European Association for Cancer Research (2000), Carcinogenesis Award (2005), Beckman-Coulter Award (2006), Francisco Cobos Biomedical Research Award (2006), Swiss Bridge Award (2006), Innovation Award from the Commonwealth of Massachusetts (2007), and Human Frontier Science Program Award (2007).

His current research is devoted to the establishment of the epigenome maps of normal and transformed cells, the study of the interactions between epigenetic modifications and noncoding RNAs, and the development of new epigenetic drugs for cancer therapy.

Contributors

Ehab Atallah
Department of Leukemia
M.D. Anderson Cancer Center
University of Texas
Houston, Texas

Esteban Ballestar
Cancer Epigenetics Laboratory
Spanish National Cancer Research
  Centre (CNIO)
Madrid, Spain

Maria Berdasco
Cancer Epigenetics Laboratory
Spanish National Cancer Research
  Centre (CNIO)
Madrid, Spain

George A. Calin
Department of Molecular Virology
Immunology and Medical Genetics and
  Comprehensive Cancer Center
Ohio State University
Columbus, Ohio

Maria Jesús Cañal
Epiphysage Research Group
B.O.S. Department
University of Oviedo
Oviedo, Spain
Asturias Institute of Biotechnology
  (associated with CSIC)

Carlo M. Croce
Department of Molecular Virology
Immunology and Medical Genetics and
  Comprehensive Cancer Center
Ohio State University
Columbus, Ohio

Manel Esteller
Cancer Epigenetics Laboratory
Spanish National Cancer Research
  Centre (CNIO)
Madrid, Spain

Muller Fabbri
Department of Molecular Virology
Immunology and Medical Genetics and
  Comprehensive Cancer Center
Ohio State University
Columbus, Ohio

Isabel Feito
SERIDA, Asturias Service of
  Agricultural Research
Oviedo, Spain

Mario F. Fraga
Cancer Epigenetics Laboratory
Spanish National Cancer Research
  Centre (CNIO)
Madrid, Spain

Guillermo Garcia-Manero
Section of Myelodysplastic Syndromes
Department of Leukemia
M.D. Anderson Cancer Center
University of Texas
Houston, Texas

Richard J. Gibbons
MRC Molecular Haematology Unit
Weatherall Institute of Molecular
  Medicine
University of Oxford
John Radcliffe Hospital
Oxford, UK

Rodrigo Hasbún
Forest Biotechnology Laboratory
University of Concepción, Chile

Ricky W. Johnstone
Cancer Immunology Division
Peter MacCallum Cancer Centre
East Melbourne, Victoria, Australia

Xiangyi Lu
Institute for Environmental Health
  Sciences
Wayne State University
Detroit, Michigan

Gertrud Lund
Department of Genetic Engineering
Irapuato, Guanajuato, México

José Ignacio Martín-Subero
Institute of Human Genetics
Christian–Albrechts University
University Hospital Schleswig–Holstein
Kiel, Germany

Mónica Meijón
Epiphysage Research Group
B.O.S. Department
University of Oviedo
Oviedo, Spain
Asturias Institute of Biotechnology
  (associated with CSIC)

Gabriel Oh
The Krembil Family Epigenetics
  Laboratory
Centre for Addiction and Mental Health
Toronto, Ontario, Canada

Melissa Peart
Department of Biological Sciences
Columbia University
New York, New York

Arturas Petronis
The Krembil Family Epigenetics
  Laboratory
Centre for Addiction and Mental Health
Toronto, Ontario, Canada

Parsa Rasouli
Institute for Environmental Health
  Sciences
Wayne State University
Detroit, Michigan

Danny Reinberg
Department of Biochemistry
NYU Medical School
New York, New York

Bruce Richardson
University of Michigan and Ann Arbor
  Veterans Affairs Hospital
Ann Arbor, Michigan

Jose Luis Rodríguez
Epiphysage Research Group
B.O.S. Department
University of Oviedo
Oviedo, Spain
Asturias Institute of Biotechnology
  (associated with CSIC)

Roberto Rodríguez
Epiphysage Research Group
B.O.S. Department
University of Oviedo
Oviedo, Spain
Asturias Institute of Biotechnology
  (associated with CSIC)

Santiago Ropero
Cancer Epigenetics Laboratory
Spanish National Cancer Research
  Centre (CNIO)
Madrid, Spain

Douglas M. Ruden
Institute for Environmental Health
  Sciences
Wayne State University
Detroit, Michigan

Estrella Santamaría
Epiphysage Research Group
B.O.S. Department
University of Oviedo
Oviedo, Spain
Asturias Institute of Biotechnology
  (associated with CSIC)

Reiner Siebert
Institute of Human Genetics
Christian–Albrechts University
University Hospital Schleswig–Holstein
Kiel, Germany

Luis Valledor
Epiphysage Research Group
B.O.S. Department
University of Oviedo
Oviedo, Spain
Asturias Institute of Biotechnology
  (associated with CSIC)

Alejandro Vaquero
Cancer Epigenetics and Biology
  Program (PEBC)
Catalan Institute of Oncology
L'Hospitalet de Llobregat
Barcelona, Spain

Miguel Vidal
Department of Cell and Developmental
  Biology
Centro de Investigaciones Biológicas
CSIC
Madrid, Spain

Silvio Zaina
Institute for Medical Investigation
University of Guanajuato
León, Guanajuato, México

Contents

Editor v

Contributors vii

Chapter 1   An Introduction to Epigenetics 1
              Manel Esteller

Chapter 2   Epigenetics and Cancer: DNA Methylation 3
              Santiago Ropero and Manel Esteller

Chapter 3   Epigenetics and Cancer: Histone Modifications 17
              Mario F. Fraga and Manel Esteller

Chapter 4   Epigenetic Drugs: DNA Demethylating Agents 27
              Ehab Atallah and Guillermo Garcia-Manero

Chapter 5   Epigenetic Drugs: Histone Deacetylase Inhibitors 49
              Melissa Peart and Ricky W. Johnstone

Chapter 6   Sirtuins in Biology and Disease 73
              Alejandro Vaquero and Danny Reinberg

Chapter 7   microRNAs in Cell Biology and Diseases 105
              Muller Fabbri, Carlo M. Croce, and George A. Calin

Chapter 8   Chromatin Modifications by Polycomb Complexes 131
              Miguel Vidal

Chapter 9   Epigenetics and its Genetic Syndromes 155
              Richard J. Gibbons

Chapter 10 Epigenetics and Immunity 175
              Esteban Ballestar and Bruce Richardson

Chapter 11 Etiology of Major Psychosis: Why Do We Need Epigenetics? 189

Gabriel Oh and Arturas Petronis

Chapter 12 Epigenetics and Cardiovascular Disease 207

Gertrud Lund and Silvio Zaina

Chapter 13 Plant Epigenetics 225

*Mónica Meijón, Luis Valledor, Jose Luis Rodríguez,
Rodrigo Hasbún, Estrella Santamaría, Isabel Feito,
Maria Jesús Cañal, Maria Berdasco, Mario F. Fraga,
and Roberto Rodríguez*

Chapter 14 Epigenetics, Environment, and Evolution 241

Douglas M. Ruden, Parsa Rasouli, and Xiangyi Lu

Chapter 15 Epigenetics and Epigenomics 261

José Ignacio Martín-Subero and Reiner Siebert

Index 285

1 An Introduction to Epigenetics

Manel Esteller

Genetics alone cannot explain human variation and disease. Humans with the same DNA sequence, such as monozygotic twins, and cloned animals frequently present different phenotypes and degrees of sickness penetrance. The increasingly popular term *epigenetics* embodies a partial explanation of both phenomena. First introduced by C.H. Waddington in 1939 to name "the causal interactions between genes and their products, which bring the phenotype into being," it was subsequently defined as those heritable changes in gene expression that are not due to any alteration in DNA sequence. The best-known epigenetic marker is DNA methylation. From the initial characterization of global hypomethylation of human tumors and the first hyper-methylated tumor suppressor, to the DNA methylation silencing of microRNAs, the ongoing human epigenome projects, and the clinical approval of therapies with DNA demethylating agents and histone deacetylase inhibitors, epigenetics has seized the attention of biomedical researchers. The scenario is further enriched because DNA methylation occurs in a highly complex chromatin network mediated by histone modifications, which are also disrupted in human diseases. Aberrant epigenetic patterns go beyond oncology to touch a wide range of fields of biomedical (immunology, neurology, metabolism, imprinting, cardiovascular, etc.), scientific (machineries for transcriptional activation and repression, high-order organization of DNA, etc.), and industrial (animal and yeast models, agriculture, nutrition) knowledge that have an impact on our lives.

The current hype of epigenetic research also relates to the introduction of powerful and user-friendly techniques for the study of DNA methylation, such as sodium bisulfite modification associated with polymerase chain reaction procedures. More recently, the advent of comprehensive epigenomic technologies has given rise to the first preliminary descriptions of the epigenomes of human cells. This book discusses the new developments, the overall main features, and the translational applications for disease in the still-young field of epigenetics, particularly regarding DNA methylation, histone modifications, noncoding RNAs, and chromatin remodeling. Enjoy!

1

2 Epigenetics and Cancer
DNA Methylation

Santiago Ropero and Manel Esteller

CONTENTS

2.1 Introduction 3
2.2 DNA Methylation in Healthy Cells 4
2.3 DNA Methylation Patterns Change in Cancer Cells 5
    2.3.1 Global DNA Hypomethylation in Human Cancer 5
    2.3.2 Aberrant Gene Hypermethylation in Human Cancer 6
2.4 miRNA Silencing in Human Cancer 8
2.5 DNA Methylation in Clinical Practice 8
2.6 DNA Methylation in Tumor Detection 9
2.7 The Use of DNA Methylation as a Prognostic Marker 10
2.8 DNA Methylation as a Predictive Factor 11
2.9 DNA Demethylating Agents 12
2.10 Conclusions 13
References 13

2.1 INTRODUCTION

Cancer can be defined in many different ways, depending on the area in which the disease is studied, and it can be understood to encompass a group of about 100 different and distinctive diseases. These diseases are characterized by an abnormal growth of cells that generally lead to an uncontrolled proliferation that, in some cases, can metastasize to other organs and tissues. In recent decades researchers have concentrated their efforts on identifying a wide variety of genomic changes, such as amplifications, translocations, deletions, and point mutations, that are involved in this uncontrolled proliferation, and thus in the development of cancer.

In the past, analysis of these genomic alterations has led to the identification of oncogenes and tumor-suppressor genes involved in tumor development. However, the occurrence of cancer is due not only to the genetic changes described above, but also to epigenetic changes. While genetics is concerned with the information transmitted on the basis of gene sequence, epigenetics deals with the inheritance of information based on gene expression levels. The main epigenetic modifications in mammals are DNA methylation and histone modification. The most widely studied epigenetic modification in humans to date has been the cytosine methylation of DNA. This consists of the covalent addition of a methyl group from the methyl donor S-adenosylmethionine to the carbon-5 position of cytosine within the CpG dinucleotide. This enzymatic reaction occurs after DNA synthesis and is performed by a family of enzymes called DNA methyltransferases (DNMTs). The proportion of CpG dinucleotides in the human genome is lower (1.2%) than expected (4%) from the abundance of cytosine and guanine (42% of the DNA bases). This lack of CpGs in our genome can be explained by a phenomenon known as CpG suppression, in which methylated CpG dinucleotides are progressively depleted due to the spontaneous deamination of methylated cytosines to thymidines during evolution.[1] The distribution of CpGs in vertebrate genomes is not uniform; they are concentrated in short stretches or clusters (500–2000 bp) called CpG islands, and are located mainly in the promoter region of approximately half of all human genes. However, the bulk of CpGs are found at low density within the intergenic and intronic regions of DNA, particularly within repeat sequences and transposable elements.

To date, most studies have focused on the role of DNA methylation in gene expression regulation under normal and pathological conditions.[2] Several examples illustrate the involvement of DNA methylation in disease. Rett syndrome is characterized by mutations in the methyl-binding protein MeCP.[2] With lupus, patients suffer severe degrees of DNA hypomethylation. It features in neurological diseases, for example, where the methylation of the fragile X mental retardation-1 (FMR) gene is the catalyst of the disorder of the same name. Aberrant patterns of DNA methylation can also be found in atherosclerosis, where protective cardiovascular genes are aberrantly hypermethylated. Patients with ICF (immunodeficiency, centromere instability, and facial anomalies) have mutations in a major DNA methyltransferase (DNMT3b). Finally, DNA methylation is also an important player in cancer development.

In this chapter we will focus on the role of DNA methylation in cancer and on the use of this epigenetic modification in clinical practice. To understand the role of DNA methylation in cancer development, first let us begin with a short introduction to the DNA methylation pattern in healthy cells.

2.2 DNA METHYLATION IN HEALTHY CELLS

In healthy cells, while repetitive genomic sequences are heavily methylated, most of the CpG islands are unmethylated, which allows genes to be expressed in the presence of the necessary transcriptional activators. However, in specific instances, gene-promoter regions are methylated in normal cells as part of normal developmental processes: imprinted genes, X chromosome genes in women, and germline- and tissue-specific genes.[3] Genomic or parental imprinting is a process involving acquisition of a closed chromatin state and DNA hypermethylation in one allele of a gene (for example, a growth-suppressor gene) early in the male and female germline, which leads to monoallelic expression. A similar phenomenon of gene-dosage reduction can also be invoked with regard to the methylation of CpG islands in one X chromosome in women, where only one of two copies is active. Methylation of regulatory regions is involved in repression of expression of the silent loci.[4] Finally, although DNA methylation is not widely used for regulating "normal" gene expression, and we certainly have more complex and specialized molecular networks to achieve this aim, sometimes DNA methylation can fulfill this purpose. There is the case, for example, of those genes whose expression is restricted to the male or female germline and which are not subsequently expressed in any adult tissue, such as the MAGE and LAGE gene families.[5] A more controversial case may be cited for the classical tissue-specific genes; some of them contain CpG islands, while others contain only a few CpG dinucleotides scattered throughout in their 5′ regulatory region. Methylation has been postulated as a mechanism for silencing these tissue-specific genes in those cell types where they should not be expressed. A well-characterized example of this type of regulation is the methionine adenosyl transferases 1A and 2A of rodents.[6]

As will be discussed in other chapters of this book, DNA methylation regulates gene transcription in conjunction with other epigenetic modifications. Methylated DNA is recognized by methyl-CpG binding proteins. These proteins and DNMTs recruit multiprotein complexes containing chromatin remodeling enzymes such as histone deacetylases and histone methyltransferases, which are key regulators of histone modifications.[7] However, in this chapter we will focus on the role of DNA methylation in cancer progression and its use in clinical practice.

2.3 DNA METHYLATION PATTERNS CHANGE IN CANCER CELLS

The pattern of DNA methylation changes substantially when cells became cancerous, as a result of two major phenomena. First, the tumoral genome becomes globally hypomethylated, unlike in normal cells, due mainly to the generalized demethylation in the CpGs scattered throughout the body of the genes. Second, local and discrete regions situated at the promoter region of tumor-suppressor genes undergo intense hypermethylation (Figure 2.1).

2.3.1 GLOBAL DNA HYPOMETHYLATION IN HUMAN CANCER

One of the first reports linking aberrant DNA methylation to cancer came from Lapeyre and Becker, who used HPLC to determine the 5-methylcytosine content in normal rat liver and hepatocellular carcinomas induced in rats by acetyl aminofluorene or diethylnitrosamine.[8] The carcinogen-induced cancers displayed a decrease in overall genomic methylation of about 20–40% relative to normal liver. We now know that the genome of a cancer cell loses 20–60% of its 5-methylcytosine content in comparison to the normal tissue. The loss of methyl groups is accomplished mainly by hypomethylation of the coding regions and introns of the genes and demethylation of repetitive sequences that account for 20–30% of the human genome.[9] Moreover, genome hypomethylation is an early event in cancer development and accumulates throughout all tumorigenic steps, from benign proliferation to invasive cancer.[10] In this study, the authors described a decrease in the 5-methylcytosine content associated with the degree of tumor aggressiveness using a multistage skin cancer progression model.

Although gene-specific demethylation occurs in the context of global DNA hypomethylation, many of the effects are thought to arise through the activation of the transposable elements and endogenous retroviruses present in the human genome,

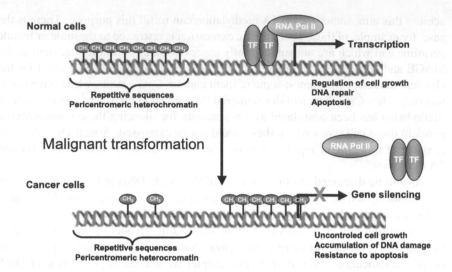

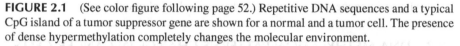

FIGURE 2.1 (See color figure following page 52.) Repetitive DNA sequences and a typical CpG island of a tumor suppressor gene are shown for a normal and a tumor cell. The presence of dense hypermethylation completely changes the molecular environment.

and through loss of imprinting. Potentially, the reactivation of the strong promoters associated with transposable elements can globally modify the expression levels of transcription factors and/or the gene expression levels of the growth regulatory genes in which these factors reside.[11] It has also been suggested that unmethylated transposable elements permit genomic mutations and anomalous chromosomal recombinations. Additionally, hypomethylation of centromeric sequences is common in human tumors and may play a role in producing aneuploidy. For example, patients with ICF syndrome, which is characterized by germline mutations in DNA methyltransferase 3b (DNMT3B), display numerous chromosome aberrations.[12] Finally, global DNA demethylation induces the expression of the silent allele of imprinted genes. The best-studied case affects the H19/IGF-2 locus in chromosome 11p15,[13] where the disturbance of methylation may cause overexpression of an anti-apoptotic growth factor (IGF-2) and loss of a transformation-suppressing RNA (H19) in certain childhood tumors.

2.3.2 ABERRANT GENE HYPERMETHYLATION IN HUMAN CANCER

The best-studied epigenetic event related to cancer development is probably the hypermethylation of CpG islands associated with the promoter region of tumor-suppressor genes. These CpG islands are unmethylated in normal tissues, but often become hypermethylated in cancer cells, leading to gene silencing. Growing evidence suggests that de novo methylation of CpG islands induces the silencing of associated tumor-suppressor genes and may, in fact, be a critical step towards tumor formation. The presence of CpG island promoter hypermethylation affects genes regulating almost all cellular functions, such as cell cycle (p16INK4a, p15INK4b, Rb, p14ARF), DNA repair (BRCA1, hMLH1, MGMT, WRN), cell adherence and invasion (CDH1,

CDH13, EXT1, SLIT2, EMP3), apoptosis (DAPK, TMS1, SFRP1), carcinogen metabolism (GSTP1), hormonal response (RARB2, ER, PRL, TSH receptors), Ras signaling (RASSF1A, NOREIA), and microRNAs, among others. Table 2.1 shows the most important hypermethylated genes in human cancer reported so far. The deregulation of most of these genes has been implicated in cancer development. For this reason it is very important to identify the role of any of the newly methylated genes in the biology of the tumor. Several functional assays restore gene function and help us to identify the role of genes that undergo CpG island hypermethylation in human cancer. For example, it is possible to test the tumor-suppressor features and to determine whether the reintroduction of the gene reduces colony formation and/or xenograft growth in nude mice. It is also possible to rescue the functionality of the genes by using DNA methylating agents, as has been shown for p14ARF, hMLH1, DAPK, EXT1, and WRN.[14,15] These types of approaches have been used to demonstrate, for example, that the methylation of O6-methylguanine-DNA methyltransferase (MGMT) is associated with the appearance of transition mutations and the chemosensitivity of alkylating agents.[16]

The particular genes that are hypermethylated in tumor cells are strongly specific to the tissue of origin.[17,18] For example, BRCA1 hypermethylation is characteristic of breast and ovarian tumors[19] but does not occur in other tumor types. hMLH1 methylation-mediated silencing is typical of colorectal, gastric, and endometrial neoplasm but is almost unmethylated in other solid tumors. EXT1 is almost exclusively methylated in acute promyelocytic leukemias and to a lesser extent in acute myelocytic and lymphoblastic leukemias, but it is unmethylated in other hematological malignancies and solid tumors,[14] suggesting that EXT1 has a key role in the origin of these types of leukemias. Furthermore, we know not only that this carefully respected pattern of epigenetic inactivation is a property of the sporadic tumors, but also that neoplasms appearing in inherited cancer syndromes display a pattern of CpG island hypermethylation that is specific to the tumor type.[20] By this we mean that the same tumor type features a common group of hypermethylated tumor-suppressor genes.

A number of reports have provided a picture of the hypermethylated genes of particular tumor types, and from these studies it is evident that there are also tumor types whose known CpG islands are more methylated than others. For example, the most hypermethylated tumor types originate in the gastrointestinal tract (esophagus, stomach, colon), while significantly less hypermethylation has been reported in ovarian tumors and sarcomas.[21] There is a clear gradient in the distribution of tumors with different degrees of CpG island methylation, from tumors with few hypermethylated CpG islands to neoplasms with a very large number of hypermethylated islands. We do not fully understand the reason for these differences, but it is possible that tumors with more hypermethylated tumor-suppressor genes are more exposed to external carcinogenic agents. Another possible explanation is that in tumors with few hypermethylated CpG islands the hypermethylated genes have not yet been found.

Other questions that remain to be answered are why some genes become hypermethylated in certain tumor types while others with similar characteristics remain unmethylated, and why genes regulating crucial steps in cell cycle regulation are methylated only in certain tumor types. As has been done previously with genetic changes, here we can hypothesize that the specific profile of hypermethylated genes confers a selective advantage. Other authors have proposed that the hypermethylation is directly targeted, for example, with fusion proteins such as PML-RAR, which can contribute to gene hypermethylation by recruiting the epigenetic machinery (DNMTs and HDACs) to promoter genes.[22] However, this does not seem to be a general mechanism, at least in leukemias.[23] Finally, it is possible that other epigenetic genes, such as Polycomb proteins, are essential players that mark the CpG islands that have to be methylated.[24,25]

2.4   miRNA SILENCING IN HUMAN CANCER

In recent years great effort has been expended on the study of the involvement of miRNA in cancer. These small RNAs are short noncoding RNA molecules that function as transcriptional regulators of gene expression. miRNAs regulate genes, including those that themselves regulate important cellular functions, such as cell proliferation, differentiation, and apoptosis.[26,27] Thus, the deregulation of miRNA expression could be an important event in cancer development. The available data indicate that miRNA expression profiles differ between normal tissues and derived tumors and between different tumor types. Some of these studies show that down-regulation of subsets of miRNAs is a common event in cancer, suggesting that some of them may act as putative tumor-suppressor genes.[28,29] The loss of expression of some of these miRNAs could be explained in different ways, one of which is CpG island hypermethylation of their regulatory region. Two recent studies have addressed this possibility. The first showed that treatment with demethylating agents induces the expression of miR-127 and the downregulation of the proto-oncogene BCL-6, a putative target of miR-127.[30] In the second study, we showed that the epigenetic silencing of miR-124a induces the activation of cyclin D kinase 6 (CDK6), a bona fide oncogenic factor, and the phosphorylation of the retinoblastoma (Rb) tumor-suppressor gene.[31]

2.5   DNA METHYLATION IN CLINICAL PRACTICE

Knowledge of the molecular alterations associated with cancer development, in particular with respect to DNA changes, will lead to the development of new strategies for assessing cancer risk status, detecting tumors as early as possible, and monitoring prognosis. The detection of hypermethylated promoter region CpG islands may be one of the most promising approaches towards achieving these goals. DNA methylation changes are known to occur early on in carcinogenesis and therefore are potentially good indicators of existing disease[32] and even of the risk of developing the disease in the future.

The use of epigenetic changes and in particular DNA methylation in clinical practice requires the use of quick, easy, nonradioactive and sensitive ways of detecting hypermethylation in the CpG islands of tumor-suppressor genes. Until a few years ago, DNA methylation detection was based almost entirely on the use of enzymes that distinguished methylated and unmethylated sequences. This approach has several drawbacks, such as the incomplete restriction cutting of the region of study and

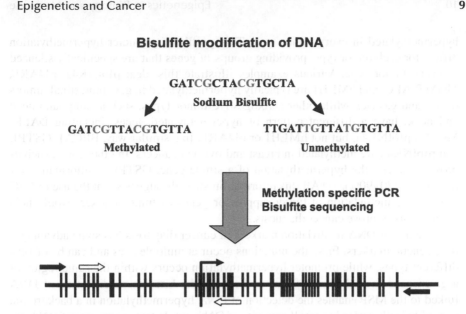

Bisulfite modification of DNA

GATCGCTACGTGCTA

Sodium Bisulfite

GATCGTTACGTGTTA   TTGATTGTTATGTGTTA

Methylated   Unmethylated

Methylation specific PCR
Bisulfite sequencing

FIGURE 2.2 (See color figure following page 52.) A schematic representation of the bisulfite modification of DNA and the detection of DNA methylation by methylation-specific PCR and bisulfite sequencing. CpG dinucleotides are represented as short vertical lines. Locations of primers for bisulfite genomic sequencing PCR are indicated by black arrows and locations of methylation-specific PCR primers by white arrows.

the use of Southern-blot technologies, which require substantial amounts of DNA of high molecular weight, and 32P-labeled DNA. The emergence of new technology based on the bisulfite modification of DNA, which converts unmethylated C to T but maintains methylated C as C, associated with amplification by methylation-specific PCR (MSP)[33] (using specific primers for methylated and unmethylated alleles) (Figure 2.2), TaqMan, restriction analysis, and genomic sequencing[34] has made it possible for most laboratories and hospitals to study DNA methylation, even using archived pathological material. Furthermore, using these methods to detect DNA methylation in small amounts of DNA allows gene hypermethylation to be detected in biological fluids. These methods can also be coupled with global genomic approaches, such as restriction landmark genomic scanning (RLGS), amplification of intermethylated sites (AIMS), CpG island microarrays, methyl DNA immunoprecipitation (MeDip), to establish the molecular signatures of tumors based on DNA methylation markers. Once the gene-hypermethylation profile specific to every tumor type has been defined, DNA methylation markers can be used in the clinical field for neoplasm detection, identification of tumor behavior, prediction of response to treatment, and therapies that target methylated tumor-suppressor genes.

2.6 DNA METHYLATION IN TUMOR DETECTION

As previously discussed, human tumors show the specific profile of hypermethylated tumor-suppressor genes. This means that usually one or more genes are hypermethylated in every tumor type. This profile of promoter hypermethylation differs for each tumor type, providing groups of genes that are specifically silenced in every tumor type. Various examples illustrate this idea: p16INK4a, p14ARF, MGMT, APC, and hMLH1 are typically hypermethylated in gastrointestinal tumors (colon and gastric), while other aerodigestive tumor types, such as lung, and head and neck, have a different pattern of hypermethylated genes, including DAPK, MGMT, p16INK4a, but not hMLH1 or p14ARF. In a similar way, BRCA1, GSTP1, and p16INK4a are methylated in breast and ovarian cancers. In other cases, such as prostate cancer, the hypermethylation of a single gene, GSTP1, is informative for most cases (70–90%).[35,36] All this information strongly suggests that the use of CpG island hypermethylation of tumor-suppressor genes as tumor markers would help pathologists improve cancer diagnosis.

The use of DNA methylation markers in cancer diagnosis has some advantages over genetic markers. First, the mutations occur at multiple sites and can be of very different types, while promoter hypermethylation occurs within the same region of a given gene in each form of cancer. Second, the bisulfite modification of DNA linked to the MSP enables the detection of DNA hypermethylation in a background of healthy cells and using small amounts of DNA, while larger amounts of DNA are needed to detect mutations. Exploiting these features, CpG island hypermethylation has been used as a tool to detect cancer cells in all types of biopsies and biological fluids, including bronchoalveolar lavage,[37] lymph nodes,[38] sputum,[39] urine,[40] semen,[41] ductal lavage,[42] saliva,[43] blood,[44] and stool.[45,46] An exciting new line of research was initiated in 1999 when our group showed for the first time that it was possible to screen for hypermethylated promoter loci in serum DNA from lung cancer patients.[47] This prompted many studies that have corroborated the feasibility of detecting CpG island hypermethylation of multiple genes in the serum DNA of a broad spectrum of tumor types, some of them even using semiquantitative and automated methodologies. Likewise, it has recently been demonstrated that septin 9, a single methylation marker for colorectal cancer, could be detected with high specificity (95%) in plasma with sensitivity of 50–65%.[48]

Another interesting aspect is that the gene silencing of tumor-suppressor genes associated with aberrant DNA CpG island hypermethylation is often found in the early stages of tumor progression and is present in premalignant lesions. Examples include the p16INK4a, p14ARF, APC, and MGFMt hypermethylation in colorectal carcinomas and the aberrant hypermethylation of hMLH1 in atypical endometrial hyperplasia.[49] Thus, DNA methylation is an epigenetic event present in premalignant or precursor lesions that could be very useful for the early detection of cancer.

2.7   THE USE OF DNA METHYLATION AS A PROGNOSTIC MARKER

One can define a prognostic marker of a specific tumor type as a molecule that provides information about the clinical outcome of cancer patients. This information is useful for physicians in their choice of clinical management of such patients.

Classical examples of DNA methylation markers of poor prognosis include death-associated protein kinase (DAPK) and p16INK4a hypermethylation, which have been linked to more aggressive tumor behavior in lung and colorectal cancer patients.[50,51] Several studies have described a number of DNA methylation markers associated with poor prognosis of ovarian cancer, such as BRCA1, insulin-like growth factor binding protein 3, and secreted frizzled-related protein1.[52] Finally, E-cadherin (CDH1), H-cadherin (CDH13), and thrombospondin-1 (THBS-1) are candidates associated with enhanced metastatic potential or angiogenic activity in primary tumors.

2.8 DNA METHYLATION AS A PREDICTIVE FACTOR

The anticancer properties of most of the therapies against cancer reside in their effects on DNA integrity, which interferes with DNA synthesis and replication, ultimately inducing cell death. Thus, treatment with alkylating agents that induce DNA adducts will be more effective in cancer cells than in slowly growing cells or growth-arrested cells (normal cells). The response to these treatments is associated with the expression of DNA repair enzymes, some of which are regulated by DNA methylation in cancer cells, thus affecting the response of these cancer cells to treatment with alkylating agents. One of the best-characterized examples is the methylation-associated silencing of MGMT in human cancer. The MGMT protein is directly responsible for repairing the addition of alkyl groups to the guanine (G) base of the DNA. This base is the preferred point of attack in the DNA of several alkylating chemotherapeutic drugs, such as BCNU (1,3-bis(2-chloroethyl)-1-nitrosourea), ACNU (1-(4-amino-2-methyl-5-pyrimidinyl)methyl-3-(2-chloroethyl)-3-nitrosourea), procarbazine, streptozotocin, and temozolomide. Thus the idea is that tumors that have lost MGMT due to hypermethylation[53] would be more sensitive to the action of these chemotherapeutic agents since their DNA lesions could not be repaired in the cancer cell, leading to cell death. In fact, we found that in gliomas inactivation of MGMT by promoter hypermethylation predicts a good response to chemotherapy, greater overall survival, and a longer time to progression in patients treated with the alkylating agent carmustina.[54] In further studies, the role of MGMT hypermethylation as a predictive factor was confirmed for two other alkylating agents, temozolomide and procarbazine.[55] However, it is important to note that MGMT hypermethylation without treatment with alkylating agents is a poor prognostic factor, probably because patients with inactivated-MGMT accumulate more mutations, as has been demonstrated for p53 and K-ras in colorectal, brain, and lung tumors. A recent study has elucidated the role of MGMT in mediating drug sensitivity. Its authors showed that MGMT repairs the O6-methylguanine induced after treatment with temozolomide, inhibiting the apoptosis signal.[56]

The case of hMLH1 provides another example of DNA repair and detoxifier genes that undergo aberrant DNA methylation. In colon cancer, the combined treatment with 5-fluorouracil (5-FU) and oxaliplatin or irinotecan is a standard chemotherapy for advanced disease. Several reports have described the role of hMLH1 hypermethylation in the response of different chemotherapeutic agents. In particular, hMLH1 aberrant methylation induces resistance to 5-FU in colorectal cancer cell lines and primary tumors.[57]

Similar cases to those for MGMT and hMLH1 can be described for other genes. For example, the response to adriamycine may be related to the methylation status of

GSTP1,[58] and the response to certain DNA-damaging drugs could be a function of the state of BRCA1 hypermethylation.[59]

Finally, gene inactivation by promoter hypermethylation may be the key to understanding the loss of hormone-responsive breast cancer. Endocrine therapy is well established for steroid hormone receptor-positive patients. Since several genes that code for steroid hormones are regulated by DNA methylation, the inefficacy of anti-steroid-related compounds, such as tamoxifen for the treatment of breast cancer, may be a direct consequence of the methylation-mediated silencing of their respective cellular receptors. Something similar could be suggested to explain the loss of response to endocrine therapies of endometrial, ovarian, and prostate cancers that display aberrant hypermethylation of the receptors that control the response to this treatment.

2.9   DNA DEMETHYLATING AGENTS

One interesting aspect of epigenetic alterations, in particular DNA methylation, in the initiation and progression of human cancer is their use as novel therapeutic targets. Unlike genetic alterations, which are almost impossible to revert, DNA methylation is a reversible event. Thus, reactivation of hypermethylated tumor-suppressor genes can be considered as a possible therapeutic target; indeed, this has resulted in the development of pharmacological inhibitors of DNA methylation. In fact, for several years we have been able to reactivate hypermethylated genes in vitro using demethylating agents such as 5-azacytidine or 5-aza-2-deoxycytidine (Decitabine). These demethylating agents are cytosine analogs that are incorporated into DNA in place of the natural cytosine during DNA replication. Once incorporated into the DNA, these analogs trap the DNA methyltransferases and target them for degradation, thus inhibiting the restoration of the DNA methylation pattern in daughter cells after several cell divisions.

If we consider that only tumor-suppressor genes are hypermethylated, the use of DNA methyltransferase inhibitors is good news for cancer treatment because the restoration of expression of tumor-suppressor genes could restore the protective effect of these genes on tumor progression. In other cases the restoration of expression of silenced receptors will sensitize cancer cells to treatment with their corresponding ligands. Under these circumstances this therapy could be very specific. However, we do not know if we have disrupted some essential methylation at certain sites, and global hypomethylation may be associated with even greater chromosomal instability. Thus, one of the obstacles to the transfer of this technique to human primary cancers is the lack of specificity of the drugs used.[60] These drugs inhibit the DNMTs and cause global hypomethylation, because one cannot reactivate only the particular gene of interest.

5-Azacytidine (Vidaza) and 5-aza-2′-deoxycytidine (Decitabine) are two DNA methylation inhibitors that are effective hypomethylating agents that inhibit cell proliferation. These two drugs represent the two most prominent DNMT inhibitors being used in clinical practice.[61,62] In fact, two of them have been approved by the U.S. Food and Drug Administration for the treatment of myelodysplastic syndrome.[63]

Hypermethylation of the CpG island is not a solitary epigenetic event. DNA methylation and regulation of histone deacetylase activity work together to silence gene expression inappropriately in cancer. The simultaneous inhibition of both processes would be the most efficacious approach to reactivating key genes for therapeutic purpose. The synergy between these two families of compounds might allow the reduction of individual doses, which would minimize toxic effects and optimize the therapeutic response of these combinations. In fact, some clinical studies have examined the effects of demethylating agents in combination with HDAC inhibitors on patients with hematological malignancies that have achieved complete or partial responses.[64,65]

2.10 CONCLUSIONS

A picture has emerged in recent years of cancer as not only a polygenetic disease, but also a polyepigenetic disease, in which genes involved in multiple pathways, from cell cycle to apoptosis, and from cellular adhesion to hormonal response, are inactivated by promoter hypermethylation. The specific pattern of gene hypermethylation in every tumor type combined with the use of bisulfite-related methods for DNA methylation detection enables DNA hypermethylation to be used for cancer detection and prognosis. Finally, the reexpression of silenced tumor-suppressor genes in cancer by the use of demethylating agents opens up new and promising possibilities for the management of cancer patients.

REFERENCES

1. Bird AP. CpG-rich islands and the function of DNA methylation. *Nature* 1986; 321: 209 213.
2. Robertson KD, Wolffe AP. DNA methylation in health and disease. *Nat. Rev. Genet.* 2000; 1: 11–19.
3. Baylin SB, Herman JG, Graff JR, Vertino PM, Issa JP. Alterations in DNA methylation: a fundamental aspect of neoplasia. *Adv. Cancer Res.* 1998; 72: 141–196.
4. Avner P, Heard E. X-chromosome inactivation: counting, choice and initiation. *Nat. Rev. Genet.* 2001; 2 (1): 59–67.
5. De Smet C, Lurquin C, Lethe B, Martelange V, Boon T. DNA methylation is the primary silencing mechanism for a set of germ line- and tumor-specific genes with a CpG-rich promoter. *Mol. Cell. Biol.* 1999; 19: 7327–7335.
6. Torre L, Lopez-Rodas G, Latasa MU, Carretero MV, Boukaba A, Rodriguez JL. DNA methylation and histone acetylation of rat methionine adenosyltransferase 1A and 2A genes is tissue-specific. *Int. J. Biochem. Cell. Biol.* 2000; 32: 397–404.
7. Esteller M. Cancer epigenomics: DNA methylomes and histone-modification maps. *Nat. Rev. Genet.* 2007; 8(4): 286–298.
8. Lapeyre JN, Becker FF. 5-methylcytosine content of nuclear DNA during chemical hepatocarcinogenesis and in carcinomas which result. *Biochem. Biophys. Res. Comm.* 1979; 87: 698–705.
9. Ehrlich M. DNA hypomethylation and cancer. In: *DNA alterations in cancer: genetic and epigenetic changes.* Melanie Ehrlich, Ed., Eaton, Natick, 2000. Pp. 273–291.
10. Fraga MF, Herranz M, Espada J, Ballestar E, Paz MF, Ropero S et al. A mouse skin multistage carcinogenesis model reflects the aberrant DNA methylation patterns of human tumors. *Cancer Res.* 2004; 64: 5527–5534.
11. Whitelaw E, Martin DI. Retrotransposons as epigenetic mediators of phenotypic variation in mammals. *Nat. Genet.* 2001; 27: 361–365.

12. Xu GL, Bestor TH, Bourc'his D, Hsieh CL, Tommerup N, Bugge M, Hulten M, Qu X, Russo JJ, Viegas-Péquignot E. Chromosome instability and immunodeficiency syndrome caused by mutations in a DNA methyltransferase gene. *Nature* 1999; 402(6758): 187–191.

13. Feinberg AP, Vogelstein B. Hypomethylation distinguishes genes of some human cancers from their normal counterparts. *Nature* 1999; 301: 89–92.

14. Ropero S, Setien F, Espada J, Fraga MF, Herranz M, Asp J et al. Epigenetic loss of the familial tumor-suppressor gene exostosin-1 (EXT1) disrupts heparan sulfate synthesis in cancer cells. *Hum. Mol. Genet.* 2004; 13: 2753–2765.

15. Agrelo R, Cheng WH, Setien F, Ropero S, Espada J, Fraga MF et al. Epigenetic inactivation of the premature aging Werner syndrome gene in human cancer. *Proc. Natl. Acad. Sci. USA* 2006; 103: 8822–8827.

16. Esteller M, Herman JG. Generating mutations but providing chemosensitivity: the role of O6-methylguanine DNA methyltransferase in human cancer. *Oncogene* 2004; 23: 1–8.

17. Esteller M., Corn PG, Baylin SB, Herman JG. A gene hypermethylation profile of human cancer. *Cancer Res.* 2001; 61: 3225–3229.

18. Costello JF, Fruhwald MC, Smiraglia DJ, Rush LJ, Robertson GP, Gao X et al. Aberrant CpG-island methylation has non-random and tumour-type-specific patterns. *Nat. Genet.* 2000; 24: 132–138.

19. Esteller M, Fraga MF, Guo M, Garcia-Foncillas J, Hedelfank I, Godwin AK et al. DNA methylation patterns in hereditary human cancers mimic sporadic tumorigenesis. *Hum. Mol. Genet.* 2001; 10: 3001–3007.

20. Esteller M, Fraga MF, Guo M, Garcia-Foncillas J, Hedelfank I, Godwin AK et al. *Hum. Mol. Genet.* 2001b; 10: 3001–3007.

21. Esteller M. Epigenetic gene silencing in cancer: the DNA hypermethylome. *Hum. Mol. Genet.* 2007; 16 (1): R50–R59.

22. Di Croce L, Raker VA, Corsaro M, Fazi F, Fanelli M et al. Methyltransferase recruitment and DNA hypermethylation of target promoters by an oncogenic transcription factor. *Science* 2002; 295:1079–1082.

23. Esteller M, Fraga MF, Paz MF, Campo E, Colomer D, et al. Cancer epigenetics and methylation. *Science* 2002; 297: 1807–1808.

24. Vire E, Brenner C, Deplus R, Blanchon L, Fraga M, Didelot C et al. The Polycomb group protein EZH2 directly controls DNA methylation. *Nature* 2006; 439: 871–874.

25. Schlesinger Y, Straussman R, Keshet I, Farkash S, Hecht M, Zimmerman J et al. Polycomb-mediated methylation on Lys27 of histone H3 pre-marks genes for de novo methylation in cancer. *Nat. Genet.* 2006; 39: 232–236.

26. He L, Hannon GJ. MicroRNAs: small RNAs with a big role in gene regulation. *Nat. Rev. Genet.* 2004; 5: 522–531.

27. Miska EA. How microRNAs control cell division, differentiation and death. *Curr. Opin. Genet.* 2005; 5: 563–568.

28. Lu J, Getz G, Miska EA, Alvarez-Saavedra E, Lamb J, Peck D et al. MicroRNA expression profiles classify human cancers. *Nature* 2005; 435: 834–838.

29. Calin GA, Croce CM. MicroRNA signatures in human cancers. *Nat. Rev. Cancer* 2006; 6: 857–866.

30. Saito Y, Liang G, Egger G, Friedman JM, Chuang JC, Coetzee GA, Jones PA. Specific activation of microRNA-127 with downregulation of the proto-oncogene BCL6 by chromatin-modifying drugs in human cancer cells. *Cancer Cell* 2006; 9: 435–443.

31. Lujambio A, Ropero S, Ballestar E, Fraga MF, Cerrato C, Setien F et al. Genetic unmasking of an epigenetically silenced microRNA in human cancer cells. *Cancer Res.* 2007; 67: 1424–1429.

32. Laird PW. Oncogenic mechanism mediated by DNA methylation. *Mol. Med. Today* 1997; 3: 223–229.

33. Herman JG, Graff JR, Myohanen S, Nelkin BD, Baylin SB. Methylation-specific PCR: a novel PCR assay for methylation status of CpG islands. *Proc. Natl. Acad. Sci. USA* 1996; 93: 9821–9826.

34. Fraga MF, Esteller M. DNA methylation: a profile of methods and applications. *Biotechniques* 2002; 632: 636–649.

35. Jerónimo C, Usadel H, Henrique R, Oliveira J, Lopes C, Nelson WG, Sidransky D. Quantitation of GSTP1 methylation in non-neoplastic prostatic tissue and organ-confined prostate adenocarcinoma. *J. Nat. Cancer Inst.* 2001; 93(22): 1747–1752.

36. Cairns P. Gene methylation and early detection of genitourinary cancer. *Nat. Rev. Cancer* 2007; 7: 531–543.

37. Ahrendt SA, Chow JT, Xu L, Yang SC, Eisenberger CF, Esteller M et al. Molecular detection of tumor cells in bronchoalveolar lavage fluid from patients with early stage lung cancer. *J. Nat. Cancer Inst.* 1999; 91: 332–339.

38. Sanchez-Cespedes M, Esteller M, Hibi K, Cope FO, Westra WH, Piantadosi S, et al. Molecular detection of neoplastic cells in lymph nodes of metastatic colorectal cancer patients predicts recurrence. *Clin. Cancer Res.* 1999; 5: 2450–2454.

39. Belinsky SA. Gene-promoter hypermethylation as a biomarker in lung cancer. *Nat. Rev. Cancer* 2004; 4 (9): 707–717.

40. Cairns P, Esteller M, Herman JG, Schoenberg M, Jeronimo C, Sanchez-Cespedes M et al. Molecular detection of prostate cancer in urine by GSTP1 hypermethylation. *Clin. Cancer Res.* 2001; 7: 2727–2730.

41. Goessl C, Krause H, Muller M, Heicappell R, Schrader M, Sachsinger J, Miller K. Fluorescent methylation-specific polymerase chain reaction for DNA-based detection of prostate cancer in bodily fluids. *Cancer Res.* 2000; 60: 5941–5945.

42. Evron E, Dooley WC, Umbricht CB, Rosenthal D, Sacchi N, Gabrielson E et al. Detection of breast cancer cells in ductal lavage fluid by methylation-specific PCR. *Lancet* 2001; 357: 1335–1336.

43. Rosas SL, Koch W, da Costa Carvalho MG, Wu L, Califano J, Westra W et al. Promoter hypermethylation patterns of p16, O6-methylguanine-DNA-methyltransferase, and death-associated protein kinase in tumors and saliva of head and neck cancer patients. *Cancer Res.* 2001; 61: 939–942.

44. Bremnes RM, Sirera R, Camps C. Circulating tumour-derived DNA and RNA markers in blood: A tool for early detection, diagnostics, and follow-up? *Lung Cancer* 2005; 49 (1): 1–12.

45. Muller HM, Oberwalder M, Fiegl H, Morandell M, Goebel G, Zitt M. Methylation changes in faecal DNA: A marker for colorectal cancer screening? *Lancet* 2004; 363(9417): 1283–1285.

46. Jubb AM, Quirke P, Oates AJ. DNA methylation, a biomarker for colorectal cancer: Implications for screening and pathological utility. *Ann. N.Y. Acad. Sci.* 2003; 983: 251–267.

47. Esteller M, Sanchez-Cespedes M, Rosell R, Sidransky D, Baylin SB, Herman JG. Detection of aberrant promoter methylation of tumor suppressor genes in serum DNA from non-small cell lung cancer patients. *Cancer Res.* 1999; 59: 67–70.

48. Lofton-Day C, Model F, Tetzner R, DeVos T, Schuster M, Lesche R, Sledziewski A, Day RW. A real-time PCR test for septin 9 gene methylation identifies early stage colorectal cancer in plasma. *Proc. Am. Assoc. Cancer Res.* 2006; 47: LB–224.

49. Esteller M. Aberrant DNA methylation as a cancer-inducing mechanism. *Annu. Rev. Pharmacol. Toxicol.* 2005; 45: 629–656.

50. Tang X, Khuri FR, Lee JJ, Kemp BL, Liu D, Hong WK, Mao L. Hypermethylation of the death-associated protein (DAP) kinase promoter and aggressiveness in stage I non-small-cell lung cancer. *J. Nat. Cancer. Inst.* 2000; 92: 1511–1516.

51. Esteller M, Gonzalez S, Risques RA, Marcuello E, Mangues R, Germa JR, Herman JG, Capella G, Peinado MA. K-ras and p16 alterations confer poor prognosis in human colorectal cancer. *J. Clin. Oncol.* 2001c; 19: 299–304.

52. Lesche R, Eckhardt F. DNA methylation markers: a versatile diagnostic tool for routine clinical use. *Curr. Opin. Mol. Ther.* 2006; 9: 222–230.

53. Esteller M, Hamilton SR, Burger PC, Baylin SB, Herman JG. Inactivation of the DNA repair gene O6-methylguanine-DNA methyltransferase by promoter hypermethylation is a common event in primary human neoplasia. *Cancer Res.* 1999b; 59: 793–797.

54. Esteller M, Garcia-Foncillas J, Andion E, Goodman SN, Hidalgo OF, Vanaclocha V et al. Activity of the DNA repair gene MGMT and the clinical response of gliomas to alkylating agents. *New Engl. J. Med.* 2000a; 343: 1350–1354.

55. Paz MF, Yaya-Tur R, Rojas-Marcos I, Reynes G, Pollan M, Aguirre-Cruz L et al. CpG island hypermethylation of the DNA repair enzyme methyltransferase predicts response to temozolomide in primarygliomas. *Clin. Cancer Res.* 2004; 10(15): 4933–4938.

56. Roos WP, Batista LF, Naumann SC, Wick W, Weller M, Menck CF, Kaina B. Apoptosis in malignant glioma cells triggered by the temozolomide-induced DNA lesion O6-methylguanine. *Oncogene* 2006; 26(2): 186–197.

57. Arnold CN, Goel A, Boland CR. Role of hMLH1 promoter hypermethylation in drug resistance to 5 fluorouracil in colorectal cancer cell lines. *Int. J. Cancer* 2003; 106 (1): 66–73.

58. Tew KD, Ronai Z. GST function in drug and stress response. *Drug Resist. Updat.* 1999; 2: 143–147.

59. Lafarge S, Sylvain V, Ferrara M, Bignon YJ. Inhibition of BRCA1 leads to increased chemoresistance to microtubule-interfering agents, an effect that involves the JNK pathway. *Oncogene* 2001; 20: 6597–6606.

60. Villar-Garea A, Esteller M. DNA demethylating agents and chromatin-remodelling drugs: which, how and why. *Curr. Drug. Metabol.* 2003; 4: 11–31.

61. Egger G, Liang G, Aparicio A, Jones PA. Epigenetics in human disease and prospects for epigenetic therapy. *Nature* 2004; 429: 457–463.

62. Fenaux P. Inhibitors of DNA methylation: beyond myelodysplastic syndromes, *Nat. Clin. Pract. Oncol.* 2005; 2 (Suppl. 1): S36–S44.

63. Kaminskas E, Farrell A, Abraham S, Baird A, Hsieh LS, Lee SL et al. Approval summary: azacitidine for treatment of myelodysplastic syndrome subtypes, *Clin. Cancer Res.* 2005; 11: 3604–3608.

64. Rudek MA, Zhao M, He P, Hartke C, Gilbert J, Gore SD et al. Pharmacokinetics of 5-azacitidine administered with phenylbutyrate in patients with refractory solid tumors or hematologic malignancies. *J. Clin. Oncol.* 2005; 23: 3906–3911.

65. Gore SD. Changes in promoter methylation and gene expression in patients with MDS and MDS-AML treated with 5-azacitidine and sodium phenylbutyrate [abstract]. *Blood* 2004; 104: 469.

3 Epigenetics and Cancer
Histone Modifications

Mario F. Fraga and Manel Esteller

CONTENTS

3.1 Introduction 17
3.2 Histone Posttranslational Modification 17
3.3 Global Alterations of Histone Modifications in Cancer 18
3.4 Site-Specific Alterations of Histone Modification in Cancer 19
3.5 Histone-Modifying Enzymes and Cancer 19
3.6 Histone Modification and Proliferation 22
References 22

3.1 INTRODUCTION

Anomalous epigenetic signaling plays a critical role in cancer [1]. The best-known epigenetic modifications are DNA methylation and histone posttranscriptional mod ifications, including methylation, acetylation, ubiquitination, and phosphorylation. In this chapter, we discuss how epigenetic alterations in the histone modification pattern of a normal cell contribute to the process of its transformation.

3.2 HISTONE POSTTRANSLATIONAL MODIFICATION

Histone tails can have various fates, including acetylation, methylation, phosphory-lation, poly-ADP ribosylation, ubiquitination, and glycosylation (reviewed in [2]). The combination of these modifications determines the histone–DNA interaction and the interaction of nonhistone proteins with chromatin through what we know as the histone code [3]. The chromatin structure and, thus, gene expression can be modified by means of the histone code.

One of the best-studied histone modifications is acetylation. This is catalyzed by histone acetyltransferases (HATs) such as MORF, MOZ, MOF, TIP60, and HBO1[4], using acetyl-coenzyme A as a donor group. Histone acetylation occurs primarily at lysine residues of the histones H4 and H3 and has two main biological consequences: alteration of histone–DNA binding, since the lysine loses a positive charge in the process, and alterations of the binding codes of chromatin-interacting transcription factors [2]. The histone acetylation level depends on the precise balance between the action of HATs and histone deacetylases (HDACs). There are four families of

HDACs: class I (HDAC1, 2, 3, and 8), class II (HDAC4, 5, 6, 7, 9, and 10), class III (sirtuins 1–7), and class IV (HDAC11). Those of classes I, II, and IV have a similar sequence and structure, but the sirtuins have a different structural homology and use a different catalytic mechanism that is dependent on NAD+ (nicotinamide adenine dinucleotide) [5].

Histone methylation is another well-studied histone modification. It is linked to both transcriptional activation and repression [2]. Histone tails can become methylated at several lysine and arginine residues. Among the lysine residues, the best studied are K4, -9, -27, -36, and -79 for H3 and K20 for H4. Lysine can be mono-, di-, and trimethylated, whereas arginine can only be monomethylated. Histone methylation is catalyzed by a family of enzymes called histone methyltransferases (HMTs) and the methyl group can be removed by the recently identified group of proteins called histone demethylases (HDMs) [2]. The numerous possible points of modification and enzymes that can catalyze the reaction in both directions indicate the huge complexity of the system and the very many possible levels of regulation.

3.3   GLOBAL ALTERATIONS OF HISTONE MODIFICATIONS IN CANCER

The histone code (i.e., the combination of histone modifications at a certain region of the chromatin) determines its structure and function [3]. Thus, the histone code will differ depending on the region of the chromatin, the cell type, the tissue type, and the external conditions of a cell. Thus, before determining the altered patterns of histone modification in cancer, it is essential to identify the normal patterns in the corresponding region. Once these have been established, the altered patterns of histone modification in cancer can be sought by various means. One of the best-studied alterations is the acetylation of histone H4, which is hypoacetylated in esophageal squamous cell carcinoma [6,7], gastric cancer [8–10], testicular cancer [11], and acute promyelocytic leukemia (APL) [12]. Furthermore, monoacetylated K16-H4 is commonly reduced in various types of tumors [13], and lower levels of AcK12-H4 are an indicator of recurrence in prostate cancer [14]. Interestingly, exposure to the carcinogen Ni^{2+} induces a huge decrease in histone acetylation [15–18].

Another histone H4 modification, the trimethylation of K20-H4, which is enriched in differentiated cells [19], increases with age [20,21], is commonly reduced in cancer cells [13,21–24] and, interestingly, is also reduced after treatment with the hepatocarcinogen tamoxifen [25].

Global alterations of histone H3 modifications in cancer have been less thoroughly investigated. One study found that low levels of acetylation at lysines 9 and 18 of histone H3 are associated with high recurrence of prostate cancer [14]. In two recent studies, H3 acetylation has been found to be reduced in human colon primary tumors [26] and in several human colon cancer cell lines [27].

3.4 SITE-SPECIFIC ALTERATIONS OF HISTONE MODIFICATION IN CANCER

One of the most important epigenetic routes to carcinogenesis involves the aberrant pattern of histone modifications at gene promoters. Histone posttranslational modifications (acetylation, methylation, phosphorylation and ubiquitination, among others) are read by different proteins and complexes involved in chromatin remodeling and transcriptional activation or repression [2,28] within the context of what we previously introduced as the "histone code" [3,28]. One of the best-studied histone modifications at the promoter level is the acetylation of lysine residues that is controlled by HATs and HDACs. In general, the presence of acetylated lysines within the histone tails is associated with less-condensed chromatin and a transcriptionally active gene status, whereas the deacetylated residues are associated with heterochromatin and transcriptional gene silencing [2,28]. Cancer cells have aberrant gene expression, which has multiple causes: genetic (gene mutations, homozygous deletions, loss of heterozygosity, etc.), cytogenetic (monosomies, trisomies, homogenous staining regions, double minutes, etc.), and, of course, epigenetics [5]. Apart from aberrant promoter DNA hypermethylation, epigenetic gene silencing can also occur by aberrant targeting of HDACs to the gene promoter, which causes histone hypoacetylation.

The typical gene silenced in this manner in human cancer is the cyclin-dependent kinase inhibitor p21WAF1 [5]. It has many features that qualify it as a bona fide tumor-suppressor gene: p21WAF1 knockout mice develop tumors, p21WAF1 expression is lost in a broad spectrum of tumor types, and its overexpression in deficient cancer cells can cause growth arrest [29]. However, the lack of any gene mutation or CpG island promoter methylation may cause some concern to classical researchers: the epigenetic inactivation of p21WAF1 by promoter hypoacetylation-mediated silencing offers an attractive alternative. Other genes altered in this manner in cancer cells continue to be identified, but more work is needed in this area. HDAC inhibitors may achieve some of their antitumoral effects through reactivation of these new types of dormant tumor-suppressor gene.

3.5 HISTONE-MODIFYING ENZYMES AND CANCER

The analysis of expression patterns of histone-modifying enzymes enables the discrimination of tumor samples from their normal counterparts and the clustering of the tumor samples according to cell type [30], which implies that these proteins have tumor-specific roles in cancer development. In the case of one of the histone-modification changes in human neoplasia previously mentioned—the reduction of monoacetylated H4K16—some clues may be found within the sirtuin family of histone deacetylases as to how these cancer-specific alterations arise. This family of epigenetic enzymes has HDAC activity, which plays a key role in transcriptional regulation and in the control of chromatin structure. Sirtuins comprise the class III family of HDACs. They are NAD+-dependent and are involved in multiple cellular events, including transcriptional silencing, chromatin remodeling, mitosis, and lifespan duration [31]. The founding member of the family, Sir2, was initially described in yeast [32]. Sir2-like enzymes catalyze a reaction in which the cleavage of NAD(+)

and histone and/or protein deacetylation is coupled to the formation of O-acetyl-ADP-ribose. The dependence of the reaction on both NAD(+) and the generation of this potential second messenger suggests new avenues to explore in our attempt to understand the function and regulation of Sir2-like enzymes. In yeast, deletion of SIR2 shortened the lifespan, whereas an extra copy of this gene increased it, indicating the importance of the Sir2 family in aging [33]. The NAD+-dependent HDAC activity of this family of proteins is highly conserved and may be relevant to their role in chromatin silencing [34]. NAD is one of the critical molecules involved in many metabolic pathways, and sirtuins can control the activity of many other proteins involved in cell growth, thus implying that proteins of the Sir2 family are involved in the elongation of lifespan mediated by caloric restriction. It has been suggested that carbon flow in the glycolysis and TCA (trichloroacetic acid) cycles is much reduced under caloric restriction, and so less NAD is available for Sir2. In this way, the Sir2 proteins may link metabolic rate and aging through NAD-dependent gene regulation and chromatin remodeling [31]. The extension of lifespan by caloric restriction requires Sir2 and is accompanied by an increase in respiration, which, in turn, increases Sir2 activity [35]. These findings have generated considerable interest in the mammalian orthologs of Sir2, known as sirtuins. Mammalian cells contain seven homologs of the yeast Sir2 protein (SIRT1 to SIRT7), and these have roles in the regulation of gene expression, apoptosis, stress responses, DNA repair, cell cycle, genomic stability, insulin regulation, and so on. Taken together, these observations make it clear that the sirtuins are the central proteins for the control of critical metabolic pathways and for the regulation of cell growth and cancer. Here, we pay special attention to two members of the family, Sirt1 and Sirt2, for two reasons: first, they are able to deacetylate lysine 16 of histone H4 (monoacetylated lysine 16 of histone H4 is commonly decreased in cancer cells [13]), and, second, they are known to change in cancer [30,36–40].

Mammalian sirtuin 1 (SIRT1) has been shown to possess a similar NAD+-dependent deacetylase activity to that of yeast Sir2 [39]. In mammals, substrates for SIRT1 include not only histones (mainly, K16-H4 and K9-H3 positions [41,42]), but also key transcription factors, such as p53 [39], forkhead transcriptional factors [43], p300 histone acetyltransferase [44], p73 [45], E2F1 [46], the DNA repair factor Ku70 [47], NF-κB [37], and the androgen receptor [48] (Figure 3.2). SIRT1 is expressed in most mammalian somatic and germ tissues [49]. Sirt1 has been found to be upregulated in mouse lung carcinomas, lymphomas, soft-tissue sarcomas [36], human lung cancer [37–39], prostate cancer [40], and leukemia [50], and sometimes downregulated in colon tumors [30]. The histone targets of Sirt1 (K16-H4 and K9-H3) [41] have sometimes been found to be altered in different types of tumors. Cancer cells have a lower level of monoacetylated K16-H4 [13], and underacetylation of K9-H3 is associated with a higher risk of recurrence in prostate cancer [14]. Since Sirt1 can specifically deacetylate these positions, its upregulation in tumorigenesis could contribute to the establishment of the cancer-specific histone-modification profile [13,14]. In the case of monoacetylated K16-H4, Sirt1 acts mainly on the promoter genes [42]. Thus, the impact of Sirt1 alterations over global acetyl K16-H4 must be moderate. Actually, a substantial portion of K16-monoacetylated histones H4 lost in cancer cells might result from the acetylation of other lysines of the histone tail and thus the increase of polyacetylated H4 isoforms [51]. The upregulation of Sirt1 in some tumor types and its relationships with proliferation and lifespan suggest that this NAD+-dependent deacetylase might be directly involved in tumorigenesis. The oncogenic potential of Sirt1 is indicated by its role in controlling many different molecular pathways in the cell. Probably the most direct association is its ability to deacetylate and inactivate the tumor-suppressor genes p53 and p73. Sirt1 can also inactivate a series of tumor-suppressor genes by deacetylation of K16-H4 at histones located within the promoters of these genes [42]. Sirt1 can also induce gene inactivation through the deacetylation of the transcriptional coactivators p300 and E2F1. The interaction with E2F1 can also contribute to tumorigenesis through the abolition of E2F1-dependent apoptosis [46]. SIRT1 deacetylates the DNA repair factor Ku70, thus inhibiting stress-induced apoptotic cell death, and promotes the long-term survival of irreplaceable cells [47], as occurs in cancer. Finally, Sirt1 may contribute to immortalization by inhibiting the ability of FOXOs to induce cell death [52]. In contrast, in some cases, like the hormonal control of androgen receptor [48] and the regulation of NF-κB [37], Sirt1 clearly shows antitumoral properties.

Sirtuin 2 (SIRT2) was identified for the first time using in silico and PCR cloning techniques [53]. Initially, it was reported to be cytoplasmic tubulin deacetylase [54]. Subsequently, its role in the control of cell-cycle progression [55] and, more recently, its involvement in the control of cell cycle progression through the deacetylation of K16-H4 at a global level [56] have been demonstrated. Changes of AcK16-H4 during the cell cycle have been widely reported, represented by a notable drop in the G2/M phase [13,56,57]. Before mitosis, chromatin needs to be deacetylated at K16-H4, most probably in order to produce the compact chromatin fiber of metaphase chromosomes [58]. Sirt2 could be responsible for most of this global H4-K16 deacetylation [56]. To recover the standard chromatin status and complete the cell cycle, H4-K16 must be re-acetylated, and the candidate for achieving this is the K16-H4-specific histone acetyltransferase hMOF [59]. There is functional evidence to support these statements: overexpression of Sirt2 leads to a decrease of acetyl K16-H4 and accumulation of cells in G2/M [55]; iRNA and knockouts of Sirt2 provoke an increase of acetyl K16-H4 and G0/G1 cell cycle arrest [56]; and iRNA of hMOF induces loss of acetyl K16-H4 and G2/M cell cycle arrest [60,61]. The loss of AcK16-H4 in transformed and proliferating cells [13] may be associated with the deregulation of Sirt2, but the alteration of this deacetylase during aging and cancer remains to be demonstrated. In fact, it has been reported recently that Sirt2 is frequently downregulated in human gliomas [62]. Subsequently, tubulin-dependent Sirt2 activity has been proposed as being a novel mitotic checkpoint that can prevent chromosomal instability [63]. Thus, Sirt2 may have a dual role in cancer, depending on the stress context and its primary molecular target.

For the trimethyl-H4K20 marker, the observed loss in cancer cells and the demonstration that knockout mice for the histone methyltransferase SUV39H are prone to developing cancer imply that HMTs for H4K20 could function as tumor-suppressor genes [4]. The loss of trimethylated K20-H4 in cancer can be caused by the loss of expression of the K20-H4-specific methyltransferase Suv4-20h [23,24], loss of the tumor suppressor RB [64,65], or deregulation of other histone-modifying enzymes. It is not known whether these histone-modifying enzymes are altered in cancer. The increase of trimethylated K20-H4 in aged-like cells has been associated with defects in the nuclear lamina [66], but there is little information available about the molecular mechanism linking the nuclear lamins and the histone-modifying machinery. However, there is an association between alterations of the nuclear lamina and altered nuclear morphology [67]. In this regard, a recent study in mice reported that disruption of FACE1, a metalloprotease involved in prelamin A proteolytic maturation, is associated with premature aging and with the disruption of the integrity of the nuclear envelope [68]. It is not known whether lamin-dependent nuclear alterations are associated with epigenetic alterations in these transgenic mice. From the epigenetic standpoint, an altered nuclear morphology is also observed in DNA methyltransferase-deficient human cells [69].

A recent study has identified HDAC2 as another component of the epigenetic machinery that is targeted for mutational inactivation in human cancer [70]. This enzyme deacetylates various histone-tail lysines, including those with an altered acetylation profile in cancer. Cancers associated with microsatellite instability were chosen for study, with the aim of screening those tumors with the highest probability of carrying mutations in any gene. When the exonic repeats of numerous genes involved in DNA methylation and histone modification were analyzed, only HDAC2 featured an inactivating frameshift mutation [70]. In light of the increasing interest in a human cancer genome project, it would be informative to include in the sequencing effort all of the genes known to encode components of the epigenetic machinery [4].

3.6   HISTONE MODIFICATION AND PROLIFERATION

The previous sections have described how several sirtuin-dependent histone modifications can be involved in the control of cell growth. The potential role of sirtuins as proteins promoting cell proliferation is also supported by the use of sirtuin inhibitors. These are thought to possess anticancer activity [71] and to induce senescence-like growth arrest [72]. The first known sirtuin inhibitors were general inhibitors of NAD reactions—like the noncompetitive inhibitor nicotinamide [73,74]—and the more specific sirtuin inhibitors like splitomicin [75]. More recently, a considerable number of sirtuin inhibitors have been identified, including sirtinol [72], cambinol [71], dihydrocoumarin [76], and indoles [77]. Their common characteristic is that they have strong antiproliferative properties that depend on the increase of acetylation of nonhistone targets and the primary histone target K16-H4 [42]. Thus, the acetylation status of H4 seems to play an important role in cell growth [51]. Transformed cells present altered patterns of acetylated H4 [13] and, occasionally, of sirtuins [30,36–40], a family of H4-specific histone deacetylases [41]. Finally, nickel compounds, which are known to be carcinogenic, induce substantial histone H4 deacetylation [18,78]. Thus, a general rule is that inhibiting any type of HDAC activity increases global acetylation levels and has antitumoral activity.

REFERENCES

1. Esteller M. CpG island hypermethylation and tumor suppressor genes: a booming present, a brighter future. *Oncogene.* 2002;21:5427–5440.

2. Kouzarides T. Chromatin modifications and their function. *Cell.* 2007;128:693–705.
3. Turner BM. Cellular memory and the histone code. *Cell.* 2002;111:285–291.
4. Esteller M. Cancer epigenomics: DNA methylomes and histone-modification maps. *Nat. Rev. Genet.* 2007;8:286–298.
5. Villar-Garea A, Esteller M. Histone deacetylase inhibitors: understanding a new wave of anticancer agents. *Int. J. Cancer.* 2004;112:171–178.
6. Toh Y, Yamamoto M, Endo K, Ikeda Y, Baba H, et al. Histone H4 acetylation and histone deacetylase 1 expression in esophageal squamous cell carcinoma. *Oncol. Rep.* 2003;10:333–338.
7. Toh Y, Ohga T, Endo K, Adachi E, Kusumoto H, et al. Expression of the metastasis-associated MTA1 protein and its relationship to deacetylation of the histone H4 in esophageal squamous cell carcinomas. *Int. J. Cancer.* 2004;110:362–367.
8. Yasui W, Oue N, Aung PP, Matsumura S, Shutoh M, et al. Molecular-pathological prognostic factors of gastric cancer: a review. *Gastric Cancer.* 2005;8:86–94.
9. Ono S, Oue N, Kuniyasu H, Suzuki T, Ito R, et al. Acetylated histone H4 is reduced in human gastric adenomas and carcinomas. *J. Exp. Clin. Cancer Res.* 2002;21:377–382.
10. Yasui W, Oue N, Ono S, Mitani Y, Ito R, et al. Histone acetylation and gastrointestinal carcinogenesis. *Ann. N.Y. Acad. Sci.* 2003;983:220–231.
11. Faure AK, Pivot-Pajot C, Kerjean A, Hazzouri M, Pelletier R, et al. Misregulation of histone acetylation in Sertoli cell-only syndrome and testicular cancer. *Mol. Hum. Reprod.* 2003;9:757–763.
12. Nouzova M, Holtan N, Oshiro MM, Isett RB, Munoz-Rodriguez JL, et al. Epigenomic changes during leukemia cell differentiation: analysis of histone acetylation and cytosine methylation using CpG island microarrays. *J. Pharmacol. Exp. Ther.* 2004;311:968–981.
13. Fraga MF, Ballestar E, Villar-Garea A, Boix-Chornet M, Espada J, et al. Loss of acetylation at Lys16 and trimethylation at Lys20 of histone H4 is a common hallmark of human cancer. *Nat. Genet.* 2005;37:391–400.
14. Seligson DB, Horvath S, Shi T, Yu H, Tze S, et al. Global histone modification patterns predict risk of prostate cancer recurrence. *Nature.* 2005;435:1262–1266.
15. Kang J, Zhang D, Chen J, Liu Q, Lin C. Antioxidants and trichostatin A synergistically protect against in vitro cytotoxicity of Ni2+ in human hepatoma cells. *Toxicol. In Vitro.* 2005;19:173–182.
16. Kang J, Zhang D, Chen J, Lin C, Liu Q. Involvement of histone hypoacetylation in Ni2+-induced bcl-2 down-regulation and human hepatoma cell apoptosis. *J. Biol. Inorg. Chem.* 2004;9:713–723.
17. Kang J, Zhang Y, Chen J, Chen H, Lin C, et al. Nickel-induced histone hypoacetylation: the role of reactive oxygen species. *Toxicol. Sci.* 2003;74:279–286.
18. Broday L, Peng W, Kuo MH, Salnikow K, Zoroddu M, et al. Nickel compounds are novel inhibitors of histone H4 acetylation. *Cancer Res.* 2000;60:238–241.
19. Biron VL, McManus KJ, Hu N, Hendzel MJ, Underhill DA. Distinct dynamics and distribution of histone methyl-lysine derivatives in mouse development. *Dev. Biol.* 2004;276:337–351.
20. Prokocimer M, Margalit A, Gruenbaum Y. The nuclear lamina and its proposed roles in tumorigenesis: projection on the hematologic malignancies and future targeted therapy. *J. Struct. Biol.* 2006;155:351–360.
21. Sarg B, Koutzamani E, Helliger W, Rundquist I, Lindner HH. Postsynthetic trimethylation of histone H4 at lysine 20 in mammalian tissues is associated with aging. *J. Biol. Chem.* 2002;277:39195–39201.
22. Olins DE, Olins AL. Granulocyte heterochromatin: defining the epigenome. *BMC Cell Biol.* 2005;6:39.

23. Pogribny IP, Ross SA, Tryndyak VP, Pogribna M, Poirier LA, et al. Histone H3 lysine 9 and H4 lysine 20 trimethylation and the expression of Suv4-20h2 and Suv-39h1 histone methyltransferases in hepatocarcinogenesis induced by methyl deficiency in rats. *Carcinogenesis*. 2006;27:1180–1186.
24. Tryndyak VP, Kovalchuk O, Pogribny IP. Loss of DNA methylation and histone H4 lysine 20 trimethylation in human breast cancer cells is associated with aberrant expression of DNA methyltransferase 1, Suv4-20h2 histone methyltransferase and methyl-binding proteins. *Cancer Biol. Ther.* 2006;5:65–70.
25. Tryndyak VP, Muskhelishvili L, Kovalchuk O, Rodriguez-Juarez R, Montgomery B, et al. Effect of long-term tamoxifen exposure on genotoxic and epigenetic changes in rat liver: implications for tamoxifen-induced hepatocarcinogenesis. *Carcinogenesis*. 2006;27:1713–1720.
26. Chen YX, Fang JY, Lu R, Qiu DK. Expression of p21(WAF1) is related to acetylation of histone H3 in total chromatin in human colorectal cancer. *World J. Gastroenterol.* 2007;13:2209–2213.
27. Chen YX, Fang JY, Zhu HY, Lu R, Cheng ZH, et al. Histone acetylation regulates p21WAF1 expression in human colon cancer cell lines. *World J. Gastroenterol.* 2004;10:2643–2646.
28. Jenuwein T, Allis CD. Translating the histone code. *Science*. 2001;293:1074–1080.
29. Herranz M, Esteller M. DNA methylation and histone modifications in patients with cancer: potential prognostic and therapeutic targets. *Methods Mol. Biol.* 2007;361:25–62.
30. Ozdag H, Teschendorff AE, Ahmed AA, Hyland SJ, Blenkiron C, et al. Differential expression of selected histone modifier genes in human solid cancers. *BMC Genomics*. 2006;7:90.
31. Guarente L. Sir2 links chromatin silencing, metabolism, and aging. *Genes Dev.* 2000;14:1021–1026.
32. Shore D, Squire M, Nasmyth KA. Characterization of two genes required for the position-effect control of yeast mating-type genes. *EMBO J.* 1984;3:2817–2823.
33. Kaeberlein M, McVey M, Guarente L. The SIR2/3/4 complex and SIR2 alone promote longevity in Saccharomyces cerevisiae by two different mechanisms. *Genes Dev.* 1999;13:2570–2580.
34. Imai S, Armstrong CM, Kaeberlein M, Guarente L. Transcriptional silencing and longevity protein Sir2 is an NAD-dependent histone deacetylase. *Nature*. 2000;403:795–800.
35. Lin SJ, Kaeberlein M, Andalis AA, Sturtz LA, Defossez PA, et al. Calorie restriction extends Saccharomyces cerevisiae lifespan by increasing respiration. *Nature*. 2002;418:344–348.
36. Chen WY, Wang DH, Yen RC, Luo J, Gu W, et al. Tumor suppressor HIC1 directly regulates SIRT1 to modulate p53-dependent DNA-damage responses. *Cell*. 2005;123:437–448.
37. Yeung F, Hoberg JE, Ramsey CS, Keller MD, Jones DR, et al. Modulation of NF-kappaB-dependent transcription and cell survival by the SIRT1 deacetylase. *EMBO J.* 2004;23:2369–2380.
38. Luo J, Nikolaev AY, Imai S, Chen D, Su F, et al. Negative control of p53 by Sir2alpha promotes cell survival under stress. *Cell*. 2001;107:137–148.
39. Vaziri H, Dessain SK, Ng Eaton E, Imai SI, Frye RA, et al. hSIR2(SIRT1) functions as an NAD-dependent p53 deacetylase. *Cell*. 2001;107:149–159.
40. Kuzmichev A, Margueron R, Vaquero A, Preissner TS, Scher M, et al. Composition and histone substrates of polycomb repressive group complexes change during cellular differentiation. *Proc. Natl. Acad. Sci. U.S.A.* 2005;102:1859–1864.

41. Vaquero A, Scher M, Lee D, Erdjument-Bromage H, Tempst P, et al. Human SirT1 interacts with histone H1 and promotes formation of facultative heterochromatin. *Mol. Cell.* 2004;16:93–105.
42. Pruitt K, Zinn RL, Ohm JE, McGarvey KM, Kang SH, et al. Inhibition of SIRT1 reactivates silenced cancer genes without loss of promoter DNA hypermethylation. *PLoS Genet.* 2006;2:e40.
43. Motta MC, Divecha N, Lemieux M, Kamel C, Chen D, et al. Mammalian SIRT1 represses forkhead transcription factors. *Cell.* 2004;116:551–563.
44. Bouras T, Fu M, Sauve AA, Wang F, Quong AA, et al. SIRT1 deacetylation and repression of p300 involves lysine residues 1020/1024 within the cell cycle regulatory domain 1. *J. Biol. Chem.* 2005;280:10264–10276.
45. Dai JM, Wang ZY, Sun DC, Lin RX, Wang SQ. SIRT1 interacts with p73 and suppresses p73-dependent transcriptional activity. *J. Cell Physiol.* 2006;210:161–166.
46. Wang C, Chen L, Hou X, Li Z, Kabra N, et al. Interactions between E2F1 and SirT1 regulate apoptotic response to DNA damage. *Nat. Cell Biol.* 2006;8:1025–1031.
47. Cohen HY, Miller C, Bitterman KJ, Wall NR, Hekking B, et al. Calorie restriction promotes mammalian cell survival by inducing the SIRT1 deacetylase. *Science.* 2004;305:390–392.
48. Fu M, Liu M, Sauve AA, Jiao X, Zhang X, et al. Hormonal control of androgen receptor function through SIRT1. *Mol. Cell Biol.* 2006;26:8122–8135.
49. Afshar G, Murnane JP. Characterization of a human gene with sequence homology to Saccharomyces cerevisiae SIR2. *Gene.* 1999;234:161–168.
50. Bradbury CA, Khanim FL, Hayden R, Bunce CM, White DA, et al. Histone deacetylases in acute myeloid leukemia show a distinctive pattern of expression that changes selectively in response to deacetylase inhibitors. *Leukemia.* 2005;19:1751–1759.
51. Fraga MF, Esteller M. Towards the human cancer epigenome: a first draft of histone modifications. *Cell Cycle.* 2005;4:1377–1381.
52. Brunet A, Sweeney LB, Sturgill JF, Chua KF, Greer PL, et al. Stress-dependent regulation of FOXO transcription factors by the SIRT1 deacetylase. *Science.* 2004;303:2011–2015.
53. Frye RA. Characterization of five human cDNAs with homology to the yeast SIR2 gene: Sir2-like proteins (sirtuins) metabolize NAD and may have protein ADP-ribosyltransferase activity. *Biochem. Biophys. Res. Commun.* 1999;260:273–279.
54. North BJ, Marshall BL, Borra MT, Denu JM, Verdin E. The human Sir2 ortholog, SIRT2, is an NAD+-dependent tubulin deacetylase. *Mol. Cell.* 2003;11:437–444.
55. Dryden SC, Nahhas FA, Nowak JE, Goustin AS, Tainsky MA. Role for human SIRT2 NAD-dependent deacetylase activity in control of mitotic exit in the cell cycle. *Mol. Cell Biol.* 2003;23:3173–3185.
56. Vaquero A, Scher MB, Lee DH, Sutton A, Cheng HL, et al. SirT2 is a histone deacetylase with preference for histone H4 Lys 16 during mitosis. *Genes Dev.* 2006;20:1256–1261.
57. Turner BM, Fellows G. Specific antibodies reveal ordered and cell-cycle-related use of histone-H4 acetylation sites in mammalian cells. *Eur. J. Biochem.* 1989;179:131–139.
58. Shogren-Knaak M, Ishii H, Sun JM, Pazin MJ, Davie JR, et al. Histone H4-K16 acetylation controls chromatin structure and protein interactions. *Science.* 2006;311:844–847.
59. Gupta A, Sharma GG, Young CS, Agarwal M, Smith ER, et al. Involvement of human MOF in ATM function. *Mol. Cell Biol.* 2005;25:5292–5305.
60. Taipale M, Rea S, Richter K, Vilar A, Lichter P, et al. hMOF histone acetyltransferase is required for histone H4 lysine 16 acetylation in mammalian cells. *Mol. Cell Biol.* 2005;25:6798–6810.
61. Smith ER, Cayrou C, Huang R, Lane WS, Cote J, et al. A human protein complex homologous to the Drosophila MSL complex is responsible for the majority of histone H4 acetylation at lysine 16. *Mol. Cell Biol.* 2005;25:9175–9188.

62. Hiratsuka M, Inoue T, Toda T, Kimura N, Shirayoshi Y, et al. Proteomics-based identification of differentially expressed genes in human gliomas: down-regulation of SIRT2 gene. *Biochem. Biophys. Res. Commun.* 2003;309:558–566.

63. Inoue T, Hiratsuka M, Osaki M, Yamada H, Kishimoto I, et al. SIRT2, a tubulin deacetylase, acts to block the entry to chromosome condensation in response to mitotic stress. *Oncogene.* 2006.

64. Isaac CE, Francis SM, Martens AL, Julian LM, Seifried LA, et al. The retinoblastoma protein regulates pericentric heterochromatin. *Mol. Cell. Biol.* 2006;26:3659–3671.

65. Siddiqui H, Fox SR, Gunawardena RW, Knudsen ES. Loss of RB compromises specific heterochromatin modifications and modulates HP1alpha dynamics. *J. Cell Physiol.* 2007;211:131–137.

66. Shumaker DK, Dechat T, Kohlmaier A, Adam SA, Bozovsky MR, et al. Mutant nuclear lamin A leads to progressive alterations of epigenetic control in premature aging. *Proc. Natl. Acad. Sci. U.S.A.* 2006;103:8703–8708.

67. Broers JL, Kuijpers HJ, Ostlund C, Worman HJ, Endert J, et al. Both lamin A and lamin C mutations cause lamina instability as well as loss of internal nuclear lamin organization. *Exp. Cell Res.* 2005;304:582–592.

68. Cadinanos J, Varela I, Lopez-Otin C, Freije JM. From immature lamin to premature aging: molecular pathways and therapeutic opportunities. *Cell Cycle.* 2005;4:1732–1735.

69. Espada J, Ballestar E, Fraga MF, Villar-Garea A, Juarranz A, et al. Human DNA methyltransferase 1 is required for maintenance of the histone H3 modification pattern. *J. Biol. Chem.* 2004;279:37175–37184.

70. Ropero S, Fraga MF, Ballestar E, Hamelin R, Yamamoto H, et al. A truncating mutation of HDAC2 in human cancers confers resistance to histone deacetylase inhibition. *Nat. Genet.* 2006;38:566–569.

71. Heltweg B, Gatbonton T, Schuler AD, Posakony J, Li H, et al. Antitumor activity of a small-molecule inhibitor of human silent information regulator 2 enzymes. *Cancer Res.* 2006;66:4368–4377.

72. Ota H, Tokunaga E, Chang K, Hikasa M, Iijima K, et al. Sirt1 inhibitor, Sirtinol, induces senescence-like growth arrest with attenuated Ras-MAPK signaling in human cancer cells. *Oncogene.* 2006;25:176–185.

73. Bitterman KJ, Anderson RM, Cohen HY, Latorre-Esteves M, Sinclair DA. Inhibition of silencing and accelerated aging by nicotinamide, a putative negative regulator of yeast sir2 and human SIRT1. *J. Biol. Chem.* 2002;277:45099–45107.

74. Avalos JL, Bever KM, Wolberger C. Mechanism of sirtuin inhibition by nicotinamide: altering the NAD(+) cosubstrate specificity of a Sir2 enzyme. *Mol. Cell.* 2005;17:855–868.

75. Bedalov A, Gatbonton T, Irvine WP, Gottschling DE, Simon JA. Identification of a small molecule inhibitor of Sir2p. *Proc. Natl. Acad. Sci. U.S.A.* 2001;98:15113–15118.

76. Olaharski AJ, Rine J, Marshall BL, Babiarz J, Zhang L, et al. The flavoring agent dihydrocoumarin reverses epigenetic silencing and inhibits sirtuin deacetylases. *PLoS Genet.* 2005;1:e77.

77. Napper AD, Hixon J, McDonagh T, Keavey K, Pons JF, et al. Discovery of indoles as potent and selective inhibitors of the deacetylase SIRT1. *J. Med. Chem.* 2005;48:8045–8054.

78. Golebiowski F, Kasprzak KS. Inhibition of core histones acetylation by carcinogenic nickel(II). *Mol. Cell Biochem.* 2005;279:133–139.

4 Epigenetic Drugs
DNA Demethylating Agents

Ehab Atallah and Guillermo Garcia-Manero

CONTENTS

4.1   Introduction 27
4.2   Nucleoside Analogs 28
       4.2.1 5-Azacytidine 28
             4.2.1.1 Chemistry and Pharmacokinetics 28
             4.2.1.2 Clinical Experience with 5-Azacytidine in
                      MDS and Leukemia 28
       4.2.2 5-Aza-2′-Deoxycytidine 30
             4.2.2.1 Chemistry and Pharmacokinetics 30
             4.2.2.2 Clinical Experience with 5-Aza-2′-Deoxycytidine in
                      MDS and Leukemia 31
       4.2.3 Other Nucleoside Analogs 33
4.3   Hypomethylating Agents in Solid Tumors 33
4.4   Hypomethylating Agents as Immunomodulators 34
4.5   Hypomethylating Agents in Nonmalignant Hematological Disorders 34
4.6   Combination of DNA Methylation Inhibitors with Histone
       Deacetylase Inhibitors 35
4.7   In Vitro Effects of Nucleoside Analog Hypomethylating Agents 37
4.8   In Vivo Effects of Nucleoside Analog Hypomethylating Agents 37
4.9   Non-Nucleoside Analogs 38
4.10   Conclusion 41
References 41

4.1 INTRODUCTION

DNA methylation plays a critical role in the control of gene expression regulation. Aberrant DNA methylation of promoter-associated CpG islands, and its associated aberrant epigenetic gene silencing, is currently considered a phenomenon functionally equivalent to physical genetic inactivation via mutations or deletions, and therefore it has a major role in oncogenesis. Reversal of aberrant DNA methylation, with agents with hypomethylating activity, results not only in gene expression reactivation but also in clinical anticancer activity, in particular in leukemias and myelodysplastic syndromes (MDSs). This is now being extensively studied in multiple human clinical trials.[1–4] Hypomethylating agents are either nucleoside analogs (e.g., 5-azacytidine,[2]

5-aza-2'-deoxycytidine,[2] 5-flouro-2'-deoxycytidine,[2] 5,6-dihydro-5-azacytidine,[2] 1-β-d-arabinofuranosyl-5-azacytosine [fazarabine],[5] and 1-(β-d-ribofuranosyl)-1,2-dihydropyrimidin-2-one [zebularine][6]) or non-nucleoside analogs (e.g., MG98,[7] RG108,[8] tea polyphenol (-)-epigallocatechin-3-gallate,[9] genestein,[10] arsenic,[11] hydralazine,[12] procainamide,[13] and psammaplin[14]). Nucleoside analogs require DNA and/ or RNA incorporation to exert their hypomethylating properties. However, this is not a prerequisite for non-nucleoside analogs. Currently, of all the above-mentioned hypomethylating agents, only two nucleoside analogs, 5-azacytidine and 5-aza-2'-deoxycytidine, are FDA approved for the therapy of hematological disorders, namely MDS. Both of those hypomethylating agents were initially developed as cytarabine derivatives with disappointing results. Later, with lower dose schedules, both drugs showed efficacy in MDS. Our aim in this chapter is to review the current knowledge on the therapeutic role of hypomethylating agents.

4.2 NUCLEOSIDE ANALOGS

4.2.1 5-AZACYTIDINE

4.2.1.1 Chemistry and Pharmacokinetics

5-Azacytidine[15] was synthesized in the 1960s. Its anti-leukemic effect was first noted in 1968.[16,17] The chemical name for 5-azacytidine is 4-amino-1-β-d-ribofuranosyl-1,3,5-triazin-2(1H)-one. 5-Azacytidine is uptaken into the cell by a nucleotide-specific transport system and activated by sequential phosphorylation to cytidine triphosphate (CTP).[18] CTP is incorporated into RNA and DNA, forming irreversible covalent adducts with DNA methyltransferase (DNMT). The active compound CTP is degraded by deaminase.[19] The pharmacokinetic characteristics of intravenous (IV) vs. subcutaneous (SQ) 5-azacytidine at a dose of 75 mg/m^2 were evaluated in six patients. The bioavailability of the SQ route was 89% of the IV route. The median half-life was 0.36 ± 0.02 and 0.69 ± 0.14 hours for the SQ and IV routes, respectively. Interestingly, 5-azacytidine clearance exceeded the glomelular filtration rate and total renal blood flow, which suggests other nonrenal elimination pathways.[20] In a study using radioactive 5-azacytidine, 73–98% was excreted in urine, with <1% excretion in feces.[21]

4.2.1.2 Clinical Experience with 5-Azacytidine in MDS and Leukemia

The Southwest Oncology Group (SWOG) conducted several early clinical trials in patients with relapsed/refractory acute leukemia. 5-Azacytidine was administered at doses ranging from 750 mg/m^2 to 1500 mg/m^2, given either as a bolus or by continuous intravenous infusion (CIV). Complete remission (CR) rates ranged from 0% to 24%, with myelosuppression and somnolence being the dose-limiting toxicities.[22–25] Later, when the hypomethylating properties of low-dose 5-azacytidine were demonstrated, responses were seen in MDS patients with lower doses of azacytidine[26] but not in acute leukemia.[27] This was followed by several trials by the Cancer and Leukemia Group B (CALGB) in patients with MDS. In two phase II studies (protocols 8421 and 8921) and a randomized phase III trial (protocol 9221), 75 mg/m^2 5-azacytidine was administered for 7 days every 28 days by CIV in protocol 8421 and

TABLE 4.1
Phase II and III Studies of 5-Azacytidine in MDS

Author	n	Dose	CRn (%)	PRn (%)	HIn (%)	ORn (%)	MR (m)	MS (m)
Silverman (protocol 8421)[29]	48	75 mg/m²/d IV × 7d q 28 d	7 (15)	1 (2)	13 (27)	21 (44)	NA	NA
Silverman (protocol 8921)[29]	70	75 mg/m²/d SQ × 7d q 28 d	12 (17)	0	16 (23)	28 (40)	NA	NA
Silverman (protocol 9221)[29]	99	75 mg/m2/d SQ × 7d q 28 d	10 (10)	1 (1)	36 (36)	47 (47)	15	20
Lyons[37]	106	5-2-2 : 75 mg/m²/d SQ × 7d q 28 d 5-2-5 : 75 mg/m²/d SQ × 10d q 28 d 5 : 75 mg/m²/d SQ × 5d q 28 d	0	0	46 (58)	46 (58)	NA	NA

n: number of patients; CR: complete remission; PR: partial remission; HI: hematological improvement; OR: overall response; MR: median response; MS: median survival; m: months; NA: not available

SQ in protocols 8921 and 9221. The classification of MDS has changed since those trials were published, so the authors have recently reanalyzed the data using the new World Health Organization (WHO) classification and the International Working Group (IWG) criteria[28] for response (Table 4.1). A total of 118 patients were enrolled in the phase II studies (48 with IV and 70 with SQ).[29] In the IV study, all patients had either refractory anemia with excess blasts (RAEB) or RAEB in transformation, whereas in the SQ study, 16% of patients had either refractory anemia (RA) or RA with ringed sideroblasts (RARS). Response rates were comparable in both studies. In protocol 8421 with IV 5-azacytidine, the CR, PR (partial remission), and HI (hematological improvement) rates were 15%, 2%, and 27%, respectively. Similarly, in protocol 8921 with SQ 5-azacytidine, the CR, PR, and HI rates were 17%, 0%, and 23%, respectively.

After the encouraging results seen in the phase II studies, the CALGB initiated a randomized phase III trial: 191 patients were randomized to 5-azacytidine (75 mg/m² SQ daily × 7 days every 28 days) vs. best supportive care in a crossover design.[30] The median time to acute myeloid leukemia (AML) transformation or death for patients receiving supportive care vs. 5-azacytidine was 12 vs. 21 months, respectively (p = .007). When the data was reanalyzed using the IWG criteria, the overall response rate was 47% (CR 10%, PR 1%, and HI 36%) vs. 17 % (HI 17%) in patients who received 5-azacytidine or supportive care only (with no crossover to 5-azacytidine), respectively. The median number of cycles to response was 3 cycles and the median duration of response was 5 cycles. Of the 65 packed red blood cell (PRBC) transfusion-dependent patients, 29 (45%) became transfusion independent with a median duration of 9 months. When results from both phase II and III studies with SQ 5-aza-cytidine were combined, the overall response rate was 44% (CR 13%, PR 1%, and HI 31%) in 169 patients. The most common nonhematological toxicity was nausea and vomiting, occurring in 4% of patients, with <1% treatment-related death.[29]

Because the study had a crossover design, it was difficult to evaluate the effects on overall survival (OS). A landmark analysis at 6 months showed median survival of an additional 18 months for 5-azacytidine-treated patients and 11 months for supportive care (p = 0.03).[30,31] In addition, patients on the 5-azacytidine arm experienced statistically significantly greater improvement in fatigue, dyspnea, physical functioning, positive affect, and psychological distress than those in the supportive care arm.[32] 5-Azacytidine administration resulted in significant improvement of both time to AML transformation and OS in patients with high-risk MDS.[33,34] In addition, 5-azacytidine benefited patients with AML. When data from the three CALGB trials were reanalyzed using the WHO criteria, the percentage of patients with AML was 52%, 37%, and 27% in protocols 8421, 8921, and 9221 (azacytidine arm), respectively. The response rate (CR and PR) ranged from 7% to 16% according to the MDS IWG criteria.[28] This benefit was also observed in a retrospective review by Sudan et al. Of 20 patients with AML who had received 5-azacytidine, 9 (45%) achieved a CR (20%) or PR (25%) by IWG AML response criteria.[35] The median response duration was 8 months and the overall survival was significantly longer in responders (15+ vs. 2.5 months).[36] Although both studies were retrospective analyses, they do suggest that patients who are not candidates for intensive chemotherapy can respond to 5-azacytidine therapy.

One of the major logistical problems with the 5-azacytidine administration schedule has been the 7-day regimen, which requires weekend injections. To investigate the necessity of the weekend injections, Lyons et al. recently reported a randomized phase II study of three different dose schedules for 5-azacytidine: 106 patients were randomized to 5, 7, or 10 days of 5-azacytidine administration without weekend injections. Of those patients, 42% had RA and 30% had RAEB. Responses were equivalent in the three arms, with 71% of the 38 transfusion-dependent evaluable patients achieving transfusion independence. Hematological improvement was observed in 65%, 52%, and 55% in the 5, 7, and 10 day arms, respectively, suggesting that the 5-day schedule is as effective as the other two schedules.[37] Currently, we are assessing the pharmacokinetic characteristics of oral 5-azacytidine. An oral formulation would be easier to administer and would also avoid the local skin reactions associated with the SQ injections. An IV formulation of 5-azacytidine was recently approved in the United States.

4.2.2   5-Aza-2′-Deoxycytidine
4.2.2.1   Chemistry and Pharmacokinetics

5-Aza-2′-deoxycytidine was synthesized around the same time as 5-azacytidine. Its chemical name is 4-amino-1-(2-deoxy-β-d-erythro-pentofuranosyl)-1,3,5-triazin-2(1H). 5-Aza-2′-deoxycytidine is uptaken into the cell by a nucleotide-specific transport system and activated by sequential phosphorylation to dCTP.[18] dCTP is incorporated only into DNA, forming irreversible covalent adducts with DNMT. The active compound dCTP is degraded by deaminase.[19] Unlike 5-azacytidine, 5-aza-2'-deoxycytidine is not incorporated into RNA.[38] Because of lack of reproducible methodology until recently, there have been few pharmacokinetic studies of 5-aza-2'-deoxycytidine.[39–42] In a recent study using the 3-hour infusion of 5-aza-2'-deoxycytidine 15 mg/m^2 over 3-hour infusion every 8 hours for 3 days, the peak plasma concentration was 49.0 ± 22.2 ng/mL and 62.7 ± 45.2 in the first and second cycles, respectively, with no change in pharmacokinetics between the first and second cycle.[40] In another phase I study using one dose between 25 and 100 mg/m^2 infused over 1 hour separated by 7-hour intervals, there were no detectable serum levels in most patients receiving doses between 25 and 60 mg/m^2. However, between 75 and 100 mg/m^2, the peak plasma concentrations were 0.93 and 2.01 µM, respectively. The half-life was 7 min in the initial phase and 35 min in the second phase.[42] 5-Aza-2'-deoxycytidine is rapidly metabolized in the liver, with <1% urinary excretion,[43] and effectively crosses the blood–brain barrier, with a cerebrospinal fluid concentration of 58% of the plateau plasma level in dogs.[44]

4.2.2.2   Clinical Experience with 5-Aza-2'-Deoxycytidine in MDS and Leukemia

The initial phase I studies defined the maximum tolerated dose (MTD) of the drug as between 1500 and 2250 mg/m^2.[41,42,45,46] Considering the hypomethylating effects of 5-aza-2'-deoxycytidine and the possibility that a lower dose may be more efficacious in exploiting that effect, Zagonel et al.[47] evaluated two low-dose regimens in MDS patients: 45 mg/m^2 infused over 4 hours daily × 3 days and 50 mg/m^2 CIV daily × 3 days. Ten patients with advanced MDS were treated. The overall response rate was 50%, with four patients achieving CR. The median duration of CR was 11 months (10 to 14+ months). Both regimens were well tolerated, with 50% of patients experiencing transient marrow hypoplasia. Another phase I study by Issa et al.[48] attempted to identify the lowest effective biological dose (Table 4.2). 5-Aza-2'-deoxycytidine was administered in doses ranging from 5 to 20 mg/m^2 over 1 hour. The duration of therapy ranged from 10 to 20 days, with total dose per course ranging from 50 to 300 mg/m^2. Fifty patients were enrolled: 35 with AML, 7 with MDS, 5 with chronic myeloid leukemia (CML), and 1 with acute lymphocytic leukemia (ALL). The overall response rate was 32%. Responses were observed in 11 of the 17 (67%) patients enrolled on the 15 mg/m^2 × 10 days dose. Of interest, response rates dropped at dose levels below or above 15 mg/m^2 × 10 doses indicating a narrow dose range of activity. In the initial phase II study by Wijermans et al.,[49] 5-aza-2'-deoxycytidine was administered at a dose of 50 mg/m^2 every 24 hrs × 3 days every 6 weeks in 29 elderly patients with high-risk MDS. The response rate was 54% (CR 28% and PR 26%), with a median response duration of 31 weeks. This led to a multi-institutional phase II trial. In that study 66 patients with MDS were treated with 5-aza-2'-deoxycytidine at a dose of 15 mg/m^2 over 4 hours IV every 8 hours daily × 3 days every 6 weeks. The overall response rate was 49% (CR 20%, PR 4%, and 24% HI) (Table 4.2). The induction mortality was 8%.[50] Major cytogenetic responses were observed in 31% of patients with abnormal cytogenetics at initial presentation. Patients who achieved a complete cytogenetic remission had a statistically

TABLE 4.2

Phase II and III Studies of 5-Aza-2'-Deoxycytidine in MDS

Study	n	Dose	CRn (%)	PRn (%)	HIn (%)	ORn (%)	MR (m)	MS (m)
Wijermans[49] (Phase II)	29	50 mg/m²/day for 72 hrs q 6 wks	8 (27)	5 (17)	2 (6)	15 (54)	8.2	10
Wijermans[50] (Phase II)	66	15 mg/m² over 4 hrs q 8 hrs × 3d q 6 wks	13 (20)	3 (4)	16 (24)	32 (49)	8	15
Kantarjian[53] (Phase III)	89	15 mg/m² over 3hrs q 8 hrs × 3d q 6 wks	8 (9)	7 (8)	12 (13)	27 (30)	10.3	14
Kantarjian[54] (Phase II)	115	20 mg/m² IV × 5 d q 4 wks 20 mg/m² SQ × 5 d q 4 wks 10 mg/m² IV × 10 d q 4 wks	40 (35)	2 (2)	26 (22)	80 (70)	20	22

n: number of patients; CR: complete remission; PR: partial remission; HI: hematological improvement; OR: overall response; MR: median response; MS: median survival; m: months; NA: not available

significantly better survival compared to those who did not (24 vs. 11 months, p = .02).[51] Platelet responses were reported in the three consecutive phase II trials separately. Of the 126 thrombocytopenic patients, 58% (47% major HI and 11% minor HI) achieved a response after one cycle of therapy. The median survival for patients with either stable or rising platelet counts (13 and 25 months, respectively) was better than for patients with decreasing counts (4 months).[52]

The encouraging phase II results described above led to a multi-institutional randomized phase III trial in the United States. 5-Aza-2'-deoxycytidine was administered at a dose of 15 mg/m² IV over 4 hours every 8 hours daily × 3 days every 6 weeks (Table 4.2). One hundred seventy patients were randomized to 5-aza-2'-deoxycytidine with best supportive care vs. supportive care only. Of the patients on the 5-aza-2'-deoxycytidine arm, 61 (69%) were intermediate-2/high risk by IPSS and 74% were transfusion dependent. The median number of cycles administered was three (range: 0–9). The overall response rate was 30% (9% CR, 8% PR, and 13% HI). The median time to AML or death was not statistically different in both groups (12.1 vs. 7.8 months). However, subgroup analysis revealed a longer median time to AML transformation or death in patients with de novo MDS (12.6 vs. 9.4 months, p = .04), patients with high-risk MDS (9.3 vs. 2.8 months, p = .01), and patients who were treatment naive (12.3 vs. 7.3 months, p = .08).[53] In addition, responding patients had a longer median time to AML progression or death (17.5 vs. 9.8 months, p = .01). The incidence of death was lower on the 5-aza-2'-deoxycytidine arm compared to the supportive care arm (14 vs. 22%). Therapy was very well tolerated, with myelosuppression being the most common side effect. The incidences of grade III or IV hematological adverse events were as follows: neutropenia (87%), thrombocytopenia (85%), febrile neutropenia (23%), and leukopenia (22%), with a decreasing incidence over the first four cycles. Other grade III or IV nonhematological toxicities included pneumonia (15%), hyperbilirubinemia (6%), and constipation (2%).[53] This study led to the approval of 5-aza-2'-deoxycytidine in the United States.

Following the initial phase I studies of low-dose 5-aza-2'-deoxycytidine in advanced leukemia, researchers at the M.D. Anderson Cancer Center[54] conducted a phase II study of different low-dose schedules of 5-aza-2'-deoxycytidine. They used a 5-aza-2'-deoxycytidine low dose of 100 mg/m² per course. Eligible patients were randomized following a Bayesian adaptive design to one of three arms: (1) 20 mg/m² IV over 1 hour daily × 5 days; (2) 20 mg/m² daily × 5, given in two SQ doses daily × 5 ; or (3) 10 mg/m² IV over 1 hour daily × 10. Regimen #1 had the best response rate, and 84 of the 115 patients were enrolled on that arm. Cycles were repeated every 4 weeks regardless of counts as long as there was evidence of residual marrow disease and no life-threatening complications. With a median follow-up time of >14 months, the median number of cycles was 7 (range: 1–23). The overall response rate by the modified IWG criteria was 80%, with 40 patients (35%) achieving CR. Median survival was 22 months, with an estimated 2-year survival of 47%. Complete cytogenetic response occurred in 24 of 69 patients with pretreatment cytogenetic abnormalities. Clinically insignificant transient elevations of alanine aminotransferase (ALT) or aspartate aminotransferase (AST) occurred in 4% of patients. Hospitalization for myelosuppression-related symptoms was necessary in 14% of courses; however, 34% of patients were never hospitalized. Responses have also been observed on retreatment with 5-aza-2'-deoxycytidine in patients who had previously responded to 5-aza-2'-deoxycytidine.[55] 5-Aza-2'-deoxycytidine has shown clinical activity in CML as well. A recently published phase II study enrolled 35 patients with imatinib-resistant CML: 12 in chronic phase, 17 in accelerated phase, and 6 in blastic phase. The administration schedule was 15 mg/m² IV over 1 hour daily, 5 days a week for 2 weeks. The overall hematologic response rate was 54% (CHR 34% and PHR 20%), and 6 patients achieved a complete cytogenetic response. Median response duration was 3.5 months (range: 2 to 13+ months).[56]

4.2.3 OTHER NUCLEOSIDE ANALOGS

Dihydroazacytidine was developed to overcome the instability of 5-azacytidine in solution. It was extensively evaluated in several phase I[57] and II[58–62] studies with little efficacy and its development was abandoned. Similarly, fazarabine, another nucleoside analog that had undergone extensive phase I[63–67] and II[68–76] evaluation, showed no proven clinical efficacy and its development was abandoned. Of note, both drugs were evaluated at their MTD and neither was evaluated at low doses.

4.3 HYPOMETHYLATING AGENTS IN SOLID TUMORS

Results of hypomethylating agents in the therapy of solid tumors have been disappointing, with several trials reporting little to no activity. With 5-aza-2'-deoxycytidine, one study enrolled 101 patients with various tumor types (colorectal cancer, melanoma, squamous cell carcinoma of the head and neck, and renal cell carcinoma), and only 1 patient with malignant melanoma had a PR.[77] 5-Aza-2'-deoxycytidine has been tested alone,[45] in combination with cisplatin,[46] and as a low-dose continuous infusion[39,78] in patients with non-small cell lung cancer (NSCLC) with no significant responses. Interestingly, in a follow-up analysis of the study by Momparler et al.,[45] they reported on six patients who had received 2–5 cycles of 5-aza-2′-deoxycytidine. Although none of the patients had an objective response, three patients survived more than 15 months, suggesting some benefit in the form of disease stabilization.

Similarly, 5-azacytidine was extensively evaluated in several phase II studies using different schedules and doses. Three large phase II studies enrolled patients with various refractory tumors.[79-81] In total, 551 patients were enrolled and 459 were evaluable. 5-Azacytidine was administered as a bolus in all three studies, with toxicity being severe nausea, vomiting, and myelosuppression. In one study,[80] when the investigators noted the severe nausea and vomiting, 5-azacytidine was administered by CIV to 29 patients. They noted an increase in myelosuppression in those patients with almost no gastrointestinal toxicity. In all three studies 16 patients had short-lived responses. Other phase II studies evaluated the efficacy of 5-azacytidine in specific tumor types such as gastrointestinal tumors,[82] testicular tumors,[83] bone sarcoma,[84] and malignant melanoma,[85] with little to no activity. Based on that, it can be concluded that 5-azacytidine is not an active agent in solid tumors with the dose and schedule used in those studies. Dihydroazacytidine, the 5-azacytidine derivative, had an unusual dose-limiting toxicity (DLT) of pleuritic chest pain in the phase I study. It was hypothesized that it would be effective in patients with malignant mesothelioma. And in fact, it did show some activity in patients with malignant mesothelioma as a single agent when administered at or close to the MTD,[62] but no increase in efficacy was observed when combined with cisplatin.[61] It has also been evaluated in patients with NSCLC[60] and malignant melanoma[58] with little to no activity. Fazarabine has also been evaluated in several solid tumors, with no activity.[68,71-76] As previously mentioned, neither drug is being actively developed.

4.4   HYPOMETHYLATING AGENTS AS IMMUNOMODULATORS

In malignant melanoma and renal cell carcinoma, interleukin-2 activates cell apoptosis through stimulation of lymphocytes. It has been previously shown that malignant cells develop resistance either by downregulating HLA or by decreased expression of apoptotic proteins. A phase I study was conducted in which 5-aza-2′-deoxycytidine was administered prior to IL-2, with the aim of inducing protein reexpression and hence enhancing cytotoxicity of IL-2. The investigators showed that the combination was safe; there was an increase in Hb F and a decrease in DNA methylation. Major responses were seen in 23% of patients, which is comparable to the response rates observed with IL-2 alone.[86]

4.5   HYPOMETHYLATING AGENTS IN NONMALIGNANT
HEMATOLOGICAL DISORDERS

5-Azacytidine was able to induce gamma globin synthesis both in vitro and in vivo in anemic baboons.[87] Building on that research, several investigators conducted small studies evaluating the abililty of hypomethylating agents to induce gamma globin synthesis and consequently Hb F production in patients with thalassemia and sickle cell anemia. In the first reported case,[88] 5-azacytidine was administered to a patient with thalassemia. The patient experienced an improvement in his hemoglobin as well as hypomethylation of bone marrow DNA near both the gamma globin and epsilon globin gene. A similar effect was seen in a patient with sickle cell disease.[89] This was followed by several small studies in patients with sickle cell disease (SSD), and the findings were confirmed.[90-92] Similarly, 5-aza-2'-deoxycytidine was evaluated in a phase I/II study in which eight patients with SSD received 5-aza-2'-deoxycytidine 0.2 mg/kg SQ 1–3 times per week in two cycles of 6 weeks' duration.[93] The hemoglobin increased from a mean of 7.6 g/dl to 9.6 g/dl (p < .001). More importantly, the hemoglobin F increased from 6.5 to 20.4% (p < .0001). Again these findings were confirmed in other small studies.[94] Due to safety concerns about the long-term toxicity of these agents,[95] they have not been employed in the therapy of patients with hemoglobinopathies. Further studies are needed to evaluate the long-term toxicity of such therapy in these patients.

4.6 COMBINATION OF DNA METHYLATION INHIBITORS WITH HISTONE DEACETYLASE INHIBITORS

Histone acetylation leads to an open chromosome configuration and consequently to gene transcription and cell differentiation. Several enzymatic activities control this process of histone acetylation/deacytelation.[96] Currently, several compounds with histone deacytelation (HDAC) inhibitor properties are undergoing clinical evaluation in the treatment of MDS.[97-101] A synergistic effect of demethylation and histone acetylation in reexpression of genes was first reported by Cameron et al. using trichostatin and 5-aza-2'-deoxycytidine.[102] A similar in vitro synergistic effect with 5-aza-2'-deoxycytidine and valproic acid (VPA) was observed.[103] A phase I/II study to evaluate the MTD and efficacy of 5-aza-2'-deoxycytidine and VPA in patients with AML or MDS has been conducted.[104] Patients participating in phase I of the study received a fixed dose of 5-aza-2'-deoxycytidine, 15 mg/m^2, over 1-hour IV infusion daily × 10 days with escalating doses of VPA (25, 35, and 50 mg/kg daily for 10 days) concomitantly with 5-aza-2'-deoxycytidine. The MTD of valproic acid was 50 mg/kg for 10 days. The median age was 60, and most patients had AML (89%). Of 53 evaluable patients, 12 responded, with 10 achieving CR and 2 achieving CRp (CR with incomplete platelet recovery). Of the previously untreated patients, 50% responded (5 of 10 patients). The median remission duration was 7.2 months (range: 1.3 to 12.6 months) (Table 4.3). The combination of 5-azacytidine, VPA, and all-trans retinoic acid (ATRA) in high-risk MDS (=10% blasts), relapsed/refractory AML, and patients >60 years with untreated AML has also been evaluated.[105] A fixed dose of 5-azacytidine was used: 75 mg/m^2 SQ daily × 7. The ATRA dose was 45 mg/m^2 PO daily × 5 starting on day 3 of 5-azacytidine, with dose escalation of the VPA. The MTD was 50 mg/kg daily × 7. Of the 31 evaluable patients, 9 achieved CR (ANC 10^9/L, platelets 100 × 10^9/L, and marrow blasts =5%) and 3 achieved CRp, with an overall response (OR) of 39% (Table 4.4). Of the 18 patients treated at the MTD, 9 responded (50%). To assess the hypomethylating effect of 5-azacytidine, the LINE test was used. Histone acetylation and transient global hypomethylation were observed, but no correlation was seen between degree of hypomethylation and

TABLE 4.3

Clinical Activity of the Combination of 5-Aza-2'-Deoxycytidine and Valproic Acid (VPA) by VPA Dose

Group	N	CR	CRp	OR (%)
VPA dose				
20 mg/kg	3	1	0	33
35 mg/kg	9	1	0	11
50 mg/kg	41	8	2	23
Total	53	10	2	22
Untreated AML/MDS	10	4	1	50

N: number of patients; CR: complete remission; CRp: complete remission without complete platelet recovery; OR: overall response; AML: acute myeloid leukemia; MDS: myelodysplastic syndrome

response. Higher levels of VPA were found in responders. Another combination therapy of hypomethylating agents and HDAC has been recently reported: 5-azacytidine and phenylbutarate. Of the 32 patients enrolled, 13 patients had MDS, 1 CMML, 15 AML-TLD (MDS-AML), and 3 relapsed AML. Of 29 evaluable patients, 11 responded (4 CR, 1 PR, and 4 major HI). Using this combination, all responding patients showed evidence of p15 demethylation, whereas none of the nonresponders had any demethylation. In addition, 5-azacytidine induced histone deacytelation, which has not been shown with 5-aza-2'-deoxycytidine.[106] The combination of 5-azacytidine and VPA has also been evaluated in a phase I study in patients with solid tumors.[107] In that study, 4 of 22 patients experienced stable disease, albeit for a short duration. MGCD0,[103] another HDACi, is currently being evaluated at our institution in combination with 5-azacytidine.[108] In the phase I/II trial, the combination appears to be safe and effective, with 2 of 12 patients responding to therapy.

TABLE 4.4

Clinical Activity of the Combination of 5-Azacytidine, Valproic Acid (VPA) and All-Trans Retinoic Acid (ATRA) by VPA Dose

VPA (mg/kg)	N	CR	CRp	OR%
50	18	7	2	50
62.5	7	1	0	14
75	6	1	1	33
Total	31	9	3	38
Untreated AML/MDS >60 years	16	7	2	56

N: number of patients; CR: complete remission; CRp: complete remission without complete platelet recovery; OR: overall response

4.7   IN VITRO EFFECTS OF NUCLEOSIDE ANALOG HYPOMETHYLATING AGENTS

The hypomethylating properties of cytosine derivatives were first noted by Jones and Taylor.[2] In that study, cytidine analogs containing a modification at the 5 position induced myocyte differentiation in mouse embryo cells after a 24-hour incubation period. However, this effect was seen neither with cytarabine nor with 6-azacytidine. These nucleoside analogs are incorporated into DNA, RNA, or both. This leads to the formation of irreversible DNA–DNMT adducts. It is postulated that hypomethylating agents have two mechanisms of action. At high doses the DNA–DNMT adducts trigger apoptosis and cell death. However, at lower doses DNA hypomethylation is induced, a process that is dependent on cell division. Two important observations were noted: first, hypomethylation occurred at narrow concentrations between 2 and 5 μm, with decreased differentiation at levels both lower and higher than that; second, the differentiation effect occurred after 8 to 11 cell divisions. Interestingly, this recapitulates clinical observations, as best responses with either 5-azacytidine or 5-aza-2′-deoxycytidine are often observed after 4 to 6 cycles of therapy. As a consequence in global[109] and gene-specific DNA hypomethylation, reexpression of cell differentiation, antiangiogenesis,[110] antiproliferative,[1,3,4,111] hormone receptor,[112,113] and proapoptotic[114] genes can occur. In addition, hypomethylating agents reexpress tumor-associated antigens, leading to improved immune recognition,[115] a mechanism that has been exploited together with immunotherapy in the treatment of cancer.[86] A new nucleoside analog, zebularine, was originally synthesized as a cytidine deaminase inhibitor.[116–118] Its hypomethylating properties were subsequently recognized and tested in vitro and in animal models.[6,119,120] It is an attractive new DNMT inhibitor because it is stable and can be administered orally.[121] In mice, zebularine administered by oral gavage induced DNA hypomethylation with shrinkage of the bladder tumors in those animals.[6] Human trials with this compound have not been conducted yet.

4.8   IN VIVO EFFECTS OF NUCLEOSIDE ANALOG HYPOMETHYLATING AGENTS

Several investigators have set out to study the hypomethylating effects of these compounds in vivo. Both global and gene-specific hypomethylation have been evaluated in several studies. Global hypomethylation was evaluated in 35 CML patients who received 5-aza-2′-deoxycytidine 15 mg/m^2 over 1 hour 5 days a week for 2 weeks. The methylation status of long interspersed nucleotide elements (LINE) quantified by bisulfite pyrosequencing was used as a surrogate for global DNA methylation prior to, on day 5, and on day 12 of therapy. This was measured in peripheral blood monocytes. Surprisingly, hypomethylation was less pronounced in responders than nonresponders.[56] The authors explained this paradoxical finding by suggesting increased cell death by hypomethylation in responders, leading to the cell death of hypomethylated leukemic cells and consequently leading to overall less pronounced hypomethylation in those patients. Global and gene-specific hypomethylation were evaluated in a study of 10 patients with solid tumors in the peripheral whole blood.

5-Aza-2′-deoxycytidine was administered at a dose of 2 mg/m²/d for 168 hours. Both MAGE-1 and global DNA hypomethylation were induced, and recovery occurred by day 28–35. There was no correlation between response and hypomethylation because none of the patients responded.[78]

In another study by Daskalakis et al.,[1] using a methylation-sensitive primer extension assay, p15 hypermethylation and expression in bone marrow samples was reversed by 5-aza-2′-deoxycytidine. Of 19 evaluable patients, 12 exhibited hypermethylation of p15 prior to therapy. Of those, 9 had a decrease of p15 promotor methylation and all 9 responded to therapy. However, 4 of the 7 patients with no evidence of p15 promoter hypermethylation prior to therapy also responded. Interestingly, p15 expression (measured by immunohistochemistry) was reversed in 4 of 8 patients with no expression prior to therapy. This finding was not reproduced in other studies.[48] In a study by Issa et al., no change in peripheral blood cell p15 methylation levels before and after 5-aza-2′-deoxycytidine therapy in MDS patients was seen. In addition, no correlation between p15 hypomethylation and response was observed. The difference between the two studies could be secondary to the less sensitive technique[122] used and/or the evaluation of hypomethylation status in the peripheral blood as compared to the bone marrow. Using a more sensitive technique, the same investigators demonstrated a decrease in p15 methylation status, but again with no correlation to response. However, pretreatment p15 methylation levels correlated with response to therapy, with no patient with a p15 methylation level >25% responding to therapy, whereas 11 of 13 responding patients had a p15 methylation level <10% prior to therapy. Interestingly, the decrease in global DNA methylation was greater for responders than nonresponders in patients with CML receiving high-dose 5-aza-2′-deoxycytidine, while the opposite was true in acute leukemia patients receiving low-dose 5-aza-2′-deoxycytidine.[4]

As for 5-azacytidine, in one study, escalating doses of 5-azacytidine were given in combination with phenylbutarate.[123] In that study, the methylation status of the p15 and/or CDH-1 promoter region was examined using methylation-specific PCR. All six patients who showed reversal of the methylation status of one or both genes responded, whereas the six nonresponders did not show methylation reversal. In another study by Soriano et al.,[105] 5-azacytidine was given in combination with ATRA and VPA. Global hypomethylation decreased by day 7 and returned to baseline by day 0 of the next cycle (using the LINE test). However, there was no correlation with response (Table 4.5).

4.9 NON-NUCLEOSIDE ANALOGS

Following the realization of the hypomethylating properties and activities of the nucleoside analogs, several other non-nucleoside compounds were found to have DNMT inhibitor activity, including MG987, RG1088, tea polyphenol (-)-epigallo-catechin-3-gallate,[9] genistein,[10] hydralazine,[12] procainamide,[13] arsenic trioxide,[11] and psammaplin.[14] RG108 [2-(1,3-dioxo-1,3-dihydro-2H-isoindol-2-yl)-3-(1H-indol-3-yl) propanoic acid] is currently the only known direct inhibitor of DNMT. It was identified by screening virtual databases for small molecules utilizing a three-dimensional model of the human DNMT-1 catalytic domain.[8] Unlike nucleoside analogs, RG108

TABLE 4.5

In Vivo Studies of Hypomethylating Agents

Study	Disease	Drug	Sample		Global		Cytosine	Gene Specific
					LINE-1	Alu		p15
					Pyro	COBRA	LC	Pyro
Yang[4]		DAC	PB	Pre vs. post	↓*	↓*	↓*	→
High dose	CML/AML			R vs. NR	NR>R	NR>R	NR>R	No difference
Low dose	AML			R vs. NR	R>NR*	R>NR	R>NR	No difference
Issa[56]	CML	DAC	PB	Pre vs. post	↓*	—	—	—
Week 1				R vs. NR	No difference	—	—	—
Week 2				R vs. NR	NR>R*	—	—	—
Samlowski[78]	Solid tumors	DAC	PB	Pre vs. post	—	—	↓*	—
Issa[48]	Heme malig	DAC	PB	Pre vs. post	—	—	—	↔¥
				R vs. NR	—	—	—	No difference
Daskalakis[1]	MDS	DAC	BM	Pre vs. post	—	—	—	↓#$
				Resp	R>NR#	—	—	R>NR#
Gore[123]	MDS/AML	AZA/PBT	BM	Pre vs. post	—	—	—	—†
				R vs. NR	—	—	—	R>NR*
Garcia-Manero[104]	MDS/AML	DAC/VPA	PB	Pre vs. post	→	—	—	↔*
				R vs. NR	No difference	—	—	No difference
Gollob[86]	Renal cancer	DAC/IL-2	PB	Pre vs. post	↓*	—	—	—
				R vs. NR	No difference	—	—	—
Soriano[105]	MDS/AML	AZA/VPA/ATRA	PB	Pre vs. post	↓*	—	—	—
				R vs. NR	No difference	—	—	—

LINE: long interspersed element-1; DAC: 5-aza-2′-deoxycytidine; Pyro: pyrosequencing; PB: peripheral blood; PBT: phenylbutarate; BM: bone marrow; COBRA: combined bisulfite restriction analysis; LC: liquid chromatography; AZA: 5-azacytidine; VPA: valproic acid; IL-2: interleukin-2; Heme malig: hematological malignancies; MDS: myelodysplastic syndrome; AML: acute myeloid leukemia; ATRA: all-trans retinoic acid; R: responders; NR: nonresponders; *: statistically significant; #: no p value reported; $: p15 by Ms-SNuPE; ¥: p15 by COBRA; †: p15 by PCR does not require DNA incorporation, hence has no cytotoxic effect, and thus may have fewer side effects when compared to nucleoside analogs.[8,124]

Another method to inhibit DNMT activity is by antisense oligonucleotides. These are synthetic nucleic acids that hybridize with mRNA. This leads to either RNA destruction or transcription inhibition. MG98 is an antisense oligonucleotide that specifically hybridizes with and inhibits DNMT1 mRNA. In preclinical studies, MG98 was shown to specifically inhibit DNMT-1 with no effect on DNMT3a or b in multiple cell lines. In addition, MG98 induced reexpression of p16ink4a, which was associated with growth inhibition of cells. This activity was confirmed in nude mice bearing the A549 human NSCL or Colo205 xenografts.[125] In addition, the combination of MG98 and 5-aza-2-deoxycytidine was synergistic. This compound has been evaluated in both phase I and phase II studies, as detailed later.

Two studies have directly compared hypomethylating agents in vitro. In the study by Chuang et al.,[126] only 5-aza-2'-deoxycytidine induced global and gene-specific DNA hypomethylation with reexpression of the p16 gene when compared to ECCG, procaine, and hydralazine. Similar results were seen by Stresemann et al.,[124] who evaluated the effect of DNMT inhibitors in four cell lines (TK6, Jurkat, KG-1, and HCT116). Six DNMT inhibitors were compared: 5-azacytidine, 5-aza-2'-deoxycytidine, EGCG, RG108, zebularine, and procaine. The researchers assessed the cytotoxic effect of the different compounds by their ability to induce apoptosis, micronuclei formation, and the fraction of hypodiploid cells. All DNMT inhibitors other than RG108 had a cytotoxic and genotoxic effect. Utilizing a micellar electrokinetic capillary electrophoresis technique to analyze genomic DNA methylation, they found that all DNMT compounds exhibited hypomethylating properties except procaine and ECCG. However, only 5-azacytidine and 5-aza-2'-deoxycytidine induced gene-specific hypomethylation of the TIMP-3 gene, resulting in gene reexpression. In addition, only RG108 completely inhibited DNMT in a cell-free in vitro assay, which is in agreement with the fact that 5-azacytidine, 5-aza-2'-deoxycytidine, and zebularine require DNA incorporation for their activity. It remains to be seen whether the other hypomethylating agents will be beneficial clinically.

Two phase I studies were conducted by the National Cancer Institute of Canada. In one, MG98 was administered as a continuous infusion for 21 days every 28 days. In that study, the MTD was 80 mg/m^2/day with DLT mainly in the form of drug-related fatigue and transaminitis. There was no change in the methylation patterns of peripheral blood monocytes (PBMC), no increase in Hb F levels, and more importantly, no clinical responses.[127] The second phase I trial evaluated a 2-hour infusion twice weekly schedule 3 out of 4 weeks. DLT, in the form of fever, rigors, and confusion, was reached at 360 mg/m^2/week. Again, there was no effect on methylation levels in PBMC, but there was one partial response in a patient with renal cell cancer.[128] Based on these results, a phase II study was conducted in patients with renal cell cancer with the twice weekly schedule at 360 mg/m^2/d.[129] With no responses seen in the first 17 patients, the trial was terminated early.

4.10 CONCLUSION

Hypomethylating agents are at the forefront of cancer research. Currently, two hypomethylating agents, 5-azacytidine and 5-aza-2′-deoxycytidine, are effective in the therapy of patients with MDS. However, the optimal dosing and schedule of these agents remains to be determined. In addition, the long-term toxicity remains unknown at this time, as several controversial reports have been published.[95,130,131] Finally, the dose, sequence, and role of epigenetic drug combinations in other tumors remain to be elucidated.

REFERENCES

1. Daskalakis M, Nguyen TT, Nguyen C et al. Demethylation of a hypermethylated P15/INK4B gene in patients with myelodysplastic syndrome by 5-aza-2′-deoxycytidine (Decitabine) treatment. *Blood.* 2002;100:2957–2964.
2. Jones PA, Taylor SM. Cellular differentiation, cytidine analogs and DNA methylation. *Cell.* 1980;20:85–93.
3. Uchida T, Kinoshita T, Nagai H et al. Hypermethylation of the p15INK4B gene in myelodysplastic syndromes. *Blood.* 1997;90:1403–1409.
4. Yang AS, Doshi KD, Choi SW et al. DNA methylation changes after 5-aza-2′-deoxycytidine therapy in patients with leukemia. *Cancer Res.* 2006;66:5495–5503.
5. Glazer RI, Knode MC. 1-Beta-D-arabinosyl-5-azacytosine. Cytocidal activity and effects on the synthesis and methylation of DNA in human colon carcinoma cells. *Mol Pharmacol.* 1984;26:381–387.
6. Cheng JC, Matsen CB, Gonzales FA et al. Inhibition of DNA methylation and reactivation of silenced genes by zebularine. *J Natl Cancer Inst.* 2003;95:399–409.
7. Beaulieu N, Fournel M, and MacLeod AR. Antitumor activity of MG98, an antisense oligodeoxynucleotide targeting DNA methyltransferase 1 (DNMT1) [abstract]. *Clin Cancer Res.* 2001;7:3800S.
8. Brueckner B, Garcia Boy R, Siedlecki P et al. Epigenetic reactivation of tumor suppressor genes by a novel small-molecule inhibitor of human DNA methyltransferases. *Cancer Res.* 2005;65:6305–6311.
9. Fang MZ, Wang Y, Ai N et al. Tea polyphenol (-)-epigallocatechin-3-gallate inhibits DNA methyltransferase and reactivates methylation-silenced genes in cancer cell lines. *Cancer Res.* 2003;63:7563–7570.
10. Fang MZ, Chen D, Sun Y et al. Reversal of hypermethylation and reactivation of p16INK4a, RAR{beta}, and MGMT genes by genistein and other isoflavones from soy. *Clin Cancer Res.* 2005;11:7033–7041.
11. Cui X, Wakai T, Shirai Y et al. Arsenic trioxide inhibits DNA methyltransferase and restores methylation-silenced genes in human liver cancer cells. *Hum Pathol.* 2006;37:298–311.
12. Segura-Pacheco B, Trejo-Becerril C, Perez-Cardenas E et al. Reactivation of tumor suppressor genes by the cardiovascular drugs Hydralazine and Procainamide and their potential use in cancer therapy. *Clin Cancer Res.* 2003;9:1596–1603.
13. Lee BH, Yegnasubramanian S, Lin X, Nelson WG. Procainamide is a specific inhibitor of DNA methyltransferase 1. *J Biol Chem.* 2005;280:40749–40756.
14. Pina IC, Gautschi JT, Wang GY et al. Psammaplins from the sponge Pseudoceratina purpurea: inhibition of both histone deacetylase and DNA methyltransferase. *J Org Chem.* 2003;68:3866–3873.

15. Sorm F, Piskala A, Cihak A, Vesely J. 5-Azacytidine, a new, highly effective cancer-ostatic. *Experientia.* 1964;20:202–203.
16. Sorm F, Vesely J. The activity of a new antimetabolite, 5-azacytidine, against lymphoid leukaemia in AK mice. *Neoplasma.* 1964;11:123–130.
17. Sorm F, Vesely J. Effect of 5-aza-2'-deoxycytidine against leukemic and hemopoietic tissues in AKR mice. *Neoplasma.* 1968;15:339–343.
18. Momparler RL. Pharmacology of 5-Aza-2'-deoxycytidine (Decitabine). *Semin Hematol.* 2005;42:S9–16.
19. Momparler RL, Rossi M, Bouchard J et al. Kinetic interaction of 5-AZA-2'-deoxycytidine-5'-monophosphate and its 5'-triphosphate with deoxycytidylate deaminase. *Mol Pharmacol.* 1984;25:436–440.
20. Marcucci G, Silverman L, Eller M, Lintz L, Beach CL. Bioavailability of azacitidine subcutaneous versus intravenous in patients with the myelodysplastic syndromes. *J Clin Pharmacol.* 2005;45:597–602.
21. Israili ZH, Vogler WR, Mingioli ES et al. The disposition and pharmacokinetics in humans of 5-azacytidine administered intravenously as a bolus or by continuous infusion. *Cancer Res.* 1976;36:1453–1461.
22. Saiki JH, McCredie KB, Vietti TJ et al. 5-azacytidine in acute leukemia. *Cancer.* 1978;42:2111–2114.
23. McCredie KB, Bodey GP, Burgess MA et al. Treatment of acute leukemia with 5-azacytidine (NSC-102816). *Cancer Chemother Rep.* 1973;57:319–323.
24. Levi JA, Wiernik PH. A comparative clinical trial of 5-azacytidine and guanazole in previously treated adults with acute nonlymphocytic leukemia. *Cancer.* 1976;38:36–41.
25. Saiki JH, Bodey GP, Hewlett JS et al. Effect of schedule on activity and toxicity of 5-azacytidine in acute leukemia: a Southwest Oncology group study. *Cancer.* 1981;47:1739–1742.
26. Chitambar CR, Libnoch JA, Matthaeus WG et al. Evaluation of continuous infusion low-dose 5-azacytidine in the treatment of myelodysplastic syndromes. *Am J Hematol.* 1991;37:100–104.
27. Lee EJ, Hogge DE, Gallagher R, Schiffer CA. Low dose 5-azacytidine is ineffective for remission induction in patients with acute myeloid leukemia. *Leukemia.* 1990;4:835–838.
28. Cheson BD, Bennett JM, Kantarjian H et al. Report of an international working group to standardize response criteria for myelodysplastic syndromes. *Blood.* 2000;96:3671–3674.
29. Silverman LR, McKenzie DR, Peterson BL et al. Further analysis of trials with azacitidine in patients with myelodysplastic syndrome: Studies 8421, 8921, and 9221 by the Cancer and Leukemia Group B. *J Clin Oncol.* 2006;24:3895–3903.
30. Silverman LR, Demakos EP, Peterson BL et al. Randomized controlled trial of azacitidine in patients with the myelodysplastic syndrome: a study of the cancer and leukemia group B. *J Clin Oncol.* 2002;20:2429–2440.
31. Silverman LR, McKenzie DR, Peterson BL et al. Response rates using international working group (IWG) criteria in patients with myelodysplastic syndromes (MDS) treated with azacytidine. *Blood.* 2005;106:2526.
32. Kornblith AB, Herndon JE, Silverman LR et al. Impact of azacytidine on the quality of life of patients with myelodysplastic syndrome treated in a randomized phase III trial: a Cancer and Leukemia Group B study. *J Clin Oncol.* 2002;20:2441–2452.
33. Silverman LR, McKenzie DR, Peterson BL et al. Analysis of survival, AML transformation, and transfusion independence in patients with high-risk myelodysplastic syndromes (MDS) receiving azacitidine determined using a prognostic model. *Blood.* 2005;106:2523.

34. Silverman LR, McKenzie DR, Peterson BL et al. Azacytidine prolongs survival and time to AML transformation in high-risk myelodysplastic syndrome (MDS) patients =65 years of age. *Blood*. 2005;106:2524.

35. Cheson BD, Bennett JM, Kopecky KJ et al. Revised recommendations of the International Working Group for Diagnosis, Standardization of Response Criteria, Treatment Outcomes, and Reporting Standards for Therapeutic Trials in Acute Myeloid Leukemia. *J Clin Oncol*. 2003;21:4642–4649.

36. Sudan N, Rossetti JM, Shadduck RK et al. Treatment of acute myelogenous leukemia with outpatient azacitidine. *Cancer*. 2006;107:1839–1843.

37. Lyons RM, Cosgriff T, Modi S et al. Hematologic improvement, transfusion independence, and safety assessed using three alternative dosing schedules of azacitidine in patients with myelodysplastic syndromes [abstract]. *ASH Annu Meet Abstr*. 2006;108:2662.

38. Santini V, Kantarjian HM, Issa JP. Changes in DNA methylation in neoplasia: pathophysiology and therapeutic implications. *Ann Intern Med*. 2001;134:573–586.

39. Aparicio A, Eads CA, Leong LA et al. Phase I trial of continuous infusion 5-aza-2'-deoxycytidine. *Cancer Chemother Pharmacol*. 2003;51:231–239.

40. Cashen A, Shah A, Helget A et al. A Phase I pharmacokinetic trial of Decitabine administered as a 3-hour infusion to patients with acute myelogenous leukemia (AML) or myelodysplastic syndrome (MDS). *ASH Annu Meet Abstr*. 2005;106:1854.

41. Rivard GE, Momparler RL, Demers J et al. Phase I study on 5-aza-2'-deoxycytidine in children with acute leukemia. *Leuk Res*. 1981;5:453–462.

42. van Groeningen CJ, Leyva A, O'Brien AM, Gall HE, Pinedo HM. Phase I and pharmacokinetic study of 5-aza-2'-deoxycytidine (NSC 127716) in cancer patients. *Cancer Res*. 1986;46:4831–4836.

43. Chabot GG, Bouchard J, Momparler RL. Kinetics of deamination of 5-aza-2'-deoxycytidine and cytosine arabinoside by human liver cytidine deaminase and its inhibition by 3-deazauridine, thymidine or uracil arabinoside. *Biochem Pharmacol*. 1983;32:1327–1328.

44. Chabot GG, Rivard GE, Momparler RL. Plasma and cerebrospinal fluid pharmacokinetics of 5-Aza-2'-deoxycytidine in rabbits and dogs. *Cancer Res*. 1983;43:592–597.

45. Momparler RL, Bouffard DY, Momparler LF et al. Pilot phase I–II study on 5-aza-2'-deoxycytidine (Decitabine) in patients with metastatic lung cancer. *Anticancer Drugs*. 1997;8:358–368.

46. Schwartsmann G, Schunemann H, Gorini CN et al. A phase I trial of cisplatin plus Decitabine, a new DNA-hypomethylating agent, in patients with advanced solid tumors and a follow-up early phase II evaluation in patients with inoperable non-small cell lung cancer. *Invest New Drugs*. 2000;18:83–91.

47. Zagonel V, Lo RG, Marotta G et al. 5-Aza-2'-deoxycytidine (Decitabine) induces trilineage response in unfavourable myelodysplastic syndromes. *Leukemia*. 1993;7 Suppl 1:30–35.

48. Issa JP, Garcia-Manero G, Giles FJ et al. Phase 1 study of low-dose prolonged exposure schedules of the hypomethylating agent 5-aza-2'-deoxycytidine (Decitabine) in hematopoietic malignancies. *Blood*. 2004;103:1635–1640.

49. Wijermans PW, Krulder JW, Huijgens PC, Neve P. Continuous infusion of low-dose 5-aza-2'-deoxycytidine in elderly patients with high-risk myelodysplastic syndrome. *Leukemia*. 1997;11 Suppl 1:S19–S23.

50. Wijermans P, Lubbert M, Verhoef G et al. Low-dose 5-aza-2'-deoxycytidine, a DNA hypomethylating agent, for the treatment of high-risk myelodysplastic syndrome: a multicenter phase II study in elderly patients. *J Clin Oncol*. 2000;18:956–962.

51. Lubbert M, Wijermans P, Kunzmann R et al. Cytogenetic responses in high-risk myelo-dysplastic syndrome following low-dose treatment with the DNA methylation inhibitor 5-aza-2'-deoxycytidine. *Br J Haematol.* 2001;114:349–357.
52. van den BJ, Lubbert M, Verhoef G, Wijermans PW. The effects of 5-aza-2'-deoxy-cytidine (Decitabine) on the platelet count in patients with intermediate and high-risk myelodysplastic syndromes. *Leuk Res.* 2004;28:785–790.
53. Kantarjian H, Issa JP, Rosenfeld CS et al. Decitabine improves patient outcomes in myelodysplastic syndromes: results of a phase III randomized study. *Cancer.* 2006;106:1794–1803.
54. Kantarjian HM, O'Brien S, Shan J et al. Update of the Decitabine experience in higher risk myelodysplastic syndrome and analysis of prognostic factors associated with out-come. *Cancer.* 2007;109:265–273.
55. Ruter B, Wijermans PW, Lubbert M. Superiority of prolonged low-dose azanucleoside administration? Results of 5-aza-2'-deoxycytidine retreatment in high-risk myelodys-plasia patients. *Cancer.* 2006;106:1744–1750.
56. Issa JP, Gharibyan V, Cortes J et al. Phase II study of low-dose Decitabine in patients with chronic myelogenous leukemia resistant to imatinib mesylate. *J Clin Oncol.* 2005;23:3948–3956.
57. Curt GA, Kelley JA, Fine RL et al. A phase I and pharmacokinetic study of dihydro-5-azacytidine (NSC 264880). *Cancer Res.* 1985;45:3359–3363.
58. Creagan ET, Schaid DJ, Hartmann LC, Loprinzi CL. A phase II study of 5,6-dihydro-5-azacytidine hydrochloride in disseminated malignant melanoma. *Am J Clin Oncol.* 1993;16:243–244.
59. Dhingra HM, Murphy WK, Winn RJ, Raber MN, Hong WK. Phase II trial of 5,6-dihydro-5-azacytidine in pleural malignant mesothelioma. *Invest New Drugs.* 1991;9:69–72.
60. Holoye PY, Dhingra HM, Umsawasdi T et al. Phase II study of 5,6-dihydro-5-aza-cytidine in extensive, untreated non-small cell lung cancer. *Cancer Treat Rep.* 1987;71:859–860.
61. Samuels BL, Herndon JE, Harmon DC et al. Dihydro-5-azacytidine and cisplatin in the treatment of malignant mesothelioma: a phase II study by the Cancer and Leukemia Group B. *Cancer.* 1998;82:1578–1584.
62. Yogelzang NJ, Herndon JE, Cirrincione C et al. Dihydro-5-azacytidine in malignant mesothelioma. A phase II trial demonstrating activity accompanied by cardiac toxicity. Cancer and Leukemia Group B. *Cancer.* 1997;79:2237–2242.
63. Amato R, Ho D, Schmidt S, Krakoff IH, Raber M. Phase I trial of a 72-h continuous-infusion schedule of fazarabine. *Cancer Chemother Pharmacol.* 1992;30:321–324.
64. Bailey H, Tutsch KD, Arzoomanian RZ et al. Phase I clinical trial of fazarabine as a twenty-four-hour continuous infusion. *Cancer Res.* 1991;51:1105–1108.
65. Heideman RL, Gillespie A, Ford H et al. Phase I trial and pharmacokinetic evaluation of fazarabine in children. *Cancer Res.* 1989;49:5213–5216.
66. Surbone A, Ford H, Jr., Kelley JA et al. Phase I and pharmacokinetic study of arabino-furanosyl-5-azacytosine (fazarabine, NSC 281272). *Cancer Res.* 1990;50:1220–1225.
67. Wilhelm M, O'Brien S, Rios MB et al. Phase I study of arabinosyl-5-azacytidine (faz-arabine) in adult acute leukemia and chronic myelogenous leukemia in blastic phase. *Leuk Lymphoma.* 1999;34:511–518.
68. Ben Baruch N, Denicoff AM, Goldspiel BR, O'Shaughnessy JA, Cowan KH. Phase II study of fazarabine (NSC 281272) in patients with metastatic colon cancer. *Invest New Drugs.* 1993;11:71–74.
69. Casper ES, Schwartz GK, Kelsen DP. Phase II trial of fazarabine (arabinofuranosyl-5-azacytidine) in patients with advanced pancreatic adenocarcinoma. *Invest New Drugs.* 1992;10:205–209.

70. Hubbard KP, Daugherty K, Ajani JA et al. Phase II trial of fazarabine in advanced colorectal carcinoma. *Invest New Drugs.* 1992;10:39–42.

71. Kuebler JP, Metch B, Schuller DE, Keppen M, Hynes HE. Phase II study of fazarabine in advanced head and neck cancer. A Southwest Oncology Group study. *Invest New Drugs.* 1991;9:373–374.

72. Manetta A, Blessing JA, Mann WJ, Smith DM. A phase II study of fazarabine (NSC 281272) in patients with advanced squamous cell carcinoma of the cervix. A Gynecologic Oncology Group study. *Am J Clin Oncol.* 1995;18:439–440.

73. Manetta A, Blessing JA, Look KY. A phase II study of fazarabine in patients with advanced ovarian cancer. A Gynecologic Oncology Group study. *Am J Clin Oncol.* 1995;18:156–157.

74. Selby GB, Upchurch C, Townsend J, Eyre HJ. A phase II evaluation of fazarabine in high-grade gliomas: a Southwest Oncology Group study. *Cancer Chemother Pharmacol.* 1994;34:179–180.

75. Walters RS, Theriault RL, Holmes FA, Hortobagyi GN, Esparza L. Phase II trial of fazarabine (ARA-AC, arabinosyl-5-azacytosine) in metastatic breast cancer. *Invest New Drugs.* 1992;10:43–44.

76. Williamson SK, Crowley JJ, Livingston RB, Panella TJ, Goodwin JW. Phase II trial and cost analysis of fazarabine in advanced non-small cell carcinoma of the lung: a Southwest Oncology Group study. *Invest New Drugs.* 1995;13:67–71.

77. Abele R, Clavel M, Dodion P et al. The EORTC Early Clinical Trials Cooperative Group experience with 5-aza-2'-deoxycytidine (NSC 127716) in patients with colorectal, head and neck, renal carcinomas and malignant melanomas. *Eur J Cancer Clin Oncol.* 1987;23:1921–1924.

78. Samlowski WE, Leachman SA, Wade M et al. Evaluation of a 7-day continuous intravenous infusion of Decitabine: inhibition of promoter-specific and global genomic DNA methylation. *J Clin Oncol.* 2005;23:3897–3905.

79. Velez-Garcia E, Vogler WR, Bartolucci AA, Arkun SN. Twice weekly 5-azacytidine infusion in dissmeinated metastatic cancer: a phase II study. *Cancer Treat Rep.* 1977;61:1675–1677.

80. Weiss AJ, Metter GE, Nealon TF et al. Phase II study of 5-azacytidine in solid tumors. *Cancer Treat Rep.* 1977;61:55–58.

81. Quagliana JM, O'Bryan RM, Baker L et al. Phase II study of 5-azacytidine in solid tumors. *Cancer Treat Rep.* 1977;61:51–54.

82. Mocrtcl CG, Schutt AJ, Rcitcmcicr RJ, Hahn RG. Phase II study of 5-azacytidine (NSC-102816) in the treatment of advanced gastrointestinal cancer. *Cancer Chemother Rep.* 1972;56:649–652.

83. Roth BJ, Elson P, Sledge GW, Jr., Einhorn LH, Trump DL. 5-Azacytidine (NSC 102816) in refractory germ cell tumors. A phase II trial of the Eastern Cooperative Oncology Group. *Invest New Drugs.* 1993;11:201–202.

84. Srinivasan U, Reaman GH, Poplack DG, Glaubiger DL, LeVine AS. Phase II study of 5-azacytidine in sarcomas of bone. *Am J Clin Oncol.* 1982;5:411–415.

85. Bellet RE, Catalano RB, Mastrangelo MJ, Berd D. Phase II study of subcutaneously administered 5-azacytidine (NSC-102816) in patients with metastatic malignant melanoma. *Med Pediatr Oncol.* 1978;4:11–15.

86. Gollob JA, Sciambi CJ, Peterson BL et al. Phase I trial of sequential low-dose 5-aza-2'-deoxycytidine plus high-dose intravenous bolus interleukin-2 in patients with melanoma or renal cell carcinoma. *Clin Cancer Res.* 2006;12:4619–4627.

87. DeSimone J, Heller P, Hall L, Zwiers D. 5-Azacytidine stimulates fetal hemoglobin synthesis in anemic baboons. *PNAS.* 1982;79:4428–4431.

88. Ley TJ, Nienhuis AW. Induction of hemoglobin F synthesis in patients with beta thalassemia. *Annu Rev Med.* 1985;36:485–498.

89. Charache S, Dover G, Smith K et al. Treatment of sickle cell anemia with 5-azacytidine results in increased fetal hemoglobin production and is associated with nonrandom hypomethylation of DNA around the gamma-delta-beta-globin gene complex. *Proc Natl Acad Sci USA.* 1983;80:4842–4846.

90. Dover GJ, Charache SH. The effects of variable doses of 5-azacytidine on fetal hemoglobin production in sickle cell anemia. *Prog Clin Biol Res.* 1984;165:73–83.

91. Dover GJ, Charache S, Boyer SH. Increasing fetal hemoglobin in sickle cell disease: comparisons of 5-azacytidine (subcutaneous or oral) with hydroxyurea. *Trans Assoc Am Physicians.* 1984;97:140–145.

92. Dover GJ, Charache S, Boyer SH, Vogelsang G, Moyer M. 5-Azacytidine increases HbF production and reduces anemia in sickle cell disease: dose-response analysis of subcutaneous and oral dosage regimens. *Blood.* 1985;66:527–532.

93. Koshy M, Dorn L, Bressler L et al. 2-deoxy 5-azacytidine and fetal hemoglobin induction in sickle cell anemia. *Blood.* 2000;96:2379–2384.

94. DeSimone J, Koshy M, Dorn L et al. Maintenance of elevated fetal hemoglobin levels by Decitabine during dose interval treatment of sickle cell anemia. *Blood.* 2002;99:3905–3908.

95. Gaudet F, Hodgson JG, Eden A et al. Induction of tumors in mice by genomic hypomethylation. *Science.* 2003;300:489–492.

96. Garcia-Manero G, Issa JP. Histone deacetylase inhibitors: a review of their clinical status as antineoplastic agents. *Cancer Invest.* 2005;23:635–642.

97. Marks PA, Richon VM, Rifkind RA. Histone deacetylase inhibitors: inducers of differentiation or apoptosis of transformed cells. *J Natl Cancer Inst.* 2000;92:1210–1216.

98. Gore SD, Weng LJ, Figg WD et al. Impact of prolonged infusions of the putative differentiating agent sodium phenylbutyrate on myelodysplastic syndromes and acute myeloid leukemia. *Clin Cancer Res.* 2002;8:963–970.

99. Kuendgen A, Strupp C, Aivado M et al. Treatment of myelodysplastic syndromes with valproic acid alone or in combination with all-trans retinoic acid. *Blood.* 2004;104:1266–1269.

100. Pilatrino C, Cilloni D, Messa E et al. Increase in platelet count in older, poor-risk patients with acute myeloid leukemia or myelodysplastic syndrome treated with valproic acid and all-trans retinoic acid. *Cancer.* 2005;104:101–109.

101. Kuendgen A, Schmid M, Knipp S et al. Valproic acid (VPA) achieves high response rates in patients with low-risk myelodysplastic syndromes. *ASH Annu Meet Abstr.* 2005;106:789.

102. Cameron EE, Bachman KE, Myohanen S, Herman JG, Baylin SB. Synergy of demethylation and histone deacetylase inhibition in the re-expression of genes silenced in cancer. *Nat Genet.* 1999;21:103–107.

103. Yang H, Hoshino K, Sanchez-Gonzalez B, Kantarjian H, Garcia-Manero G. Antileukemia activity of the combination of 5-aza-2′-deoxycytidine with valproic acid. *Leuk Res.* 2005;29:739–748.

104. Garcia-Manero G, Kantarjian HM, Sanchez-Gonzalez B et al. Phase 1/2 study of the combination of 5-aza-2′-deoxycytidine with valproic acid in patients with leukemia. *Blood.* 2006;108:3271–3279.

105. Soriano AO, Yang H, Tong W et al. Significant clinical activity of the combination of 5-azacytidine, valproic acid and all-trans retinoic (ATRA) acid in leukemia: results of a Phase I/II study. *Blood.* 2006;108:160.

106. Gore SD, Baylin S, Sugar E et al. Combined DNA methyltransferase and histone deacetylase inhibition in the treatment of myeloid neoplasms. *Cancer Res.* 2006;66:6361–6369.

107. Braiteh F, Soriano A, Luis CH et al. Phase I study of low-dose hypomethylating agent azacitidine (5-AC) combined with the histone deacetylase inhibitor valproic acid (VPA) in patients with advanced cancers. *J Clin Oncol* (Meeting Abstracts). 2006;24:3060.

108. Garcia-Manero G, Yang AS, Giles F et al. Phase I/II study of the oral isotype-selective histone deacetylase (HDAC) inhibitor MGCD0103 in combination with azacitidine in patients (pts) with high-risk myelodysplastic syndrome (MDS) or acute myelogenous leukemia (AML). *ASH Annu Meet Abstr.* 2006;108:1954.

109. Mund C, Hackanson B, Stresemann C, Lubbert M, Lyko F. Characterization of DNA demethylation effects induced by 5-Aza-2'-deoxycytidine in patients with myelodysplastic syndrome. *Cancer Res.* 2005;65:7086–7090.

110. Hellebrekers DM, Jair KW, Vire E et al. Angiostatic activity of DNA methyltransferase inhibitors. *Mol Cancer Ther.* 2006;5:467–475.

111. Garcia-Manero G, Kantarjian HM, Sanchez-Gonzalez B et al. Phase I/II study of the combination of 5-aza-2'-deoxycytidine with valproic acid in patients with leukemia. *Blood.* 2006;blood-2006.

112. Izbicka E, MacDonald JR, Davidson K et al. 5,6 Dihydro-5'-azacytidine (DHAC) restores androgen responsiveness in androgen-insensitive prostate cancer cells. *Anticancer Res.* 1999;19:1285–1291.

113. Izbicka E, Davidson KK, Lawrence RA, MacDonald JR, Von Hoff DD. 5,6-Dihydro-5'-azacytidine (DHAC) affects estrogen sensitivity in estrogen-refractory human breast carcinoma cell lines. *Anticancer Res.* 1999;19:1293–1298.

114. Schmelz K, Wagner M, Dorken B, Tamm I. 5-Aza-2'-deoxycytidine induces p21WAF expression by demethylation of p73 leading to p53-independent apoptosis in myeloid leukemia. *Int J Cancer.* 2005;114:683–695.

115. Maio M, Coral S, Fratta E, Altomonte M, Sigalotti L. Epigenetic targets for immune intervention in human malignancies. *Oncogene.* 2003;22:6484–6488.

116. Driscoll JS, Marquez VE, Plowman J et al. Antitumor properties of 2(1H)-pyrimidinone riboside (zebularine) and its fluorinated analogues. *J Med Chem.* 1991;34:3280–3284.

117. Kim CH, Marquez VE, Mao DT, Haines DR, McCormack JJ. Synthesis of pyrimidin-2-one nucleosides as acid-stable inhibitors of cytidine deaminase. *J Med Chem.* 1986;29:1374–1380.

118. Barchi JJ, Jr., Cooney DA, Hao Z et al. Improved synthesis of zebularine [1-(beta-D-ribofuranosyl)-dihydropyrimidin-2-one] nucleotides as inhibitors of human deoxycytidylate deaminase. *J Enzyme Inhib.* 1995;9:147–162.

119. Scott SA, Lakshimikuttysamma A, Sheridan DP et al. Zebularine inhibits human acute myeloid leukemia cell growth in vitro in association with p15INK4B demethylation and reexpression. *Exp Hematol.* 2007;35:263–273.

120. Zhou L, Cheng X, Connolly BA et al. Zebularine: a novel DNA methylation inhibitor that forms a covalent complex with DNA methyltransferases. *J Mol Biol.* 2002;321:591–599.

121. Holleran JL, Parise RA, Joseph E et al. Plasma pharmacokinetics, oral bioavailability, and interspecies scaling of the DNA methyltransferase inhibitor, zebularine. *Clin Cancer Res.* 2005;11:3862–3868.

122. Oki Y, Aoki E, Issa JP. Decitabine–bedside to bench. *Crit Rev Oncol Hematol.* 2007;61:140–152.

123. Gore SD, Baylin S, Sugar E et al. Combined DNA methyltransferase and histone deacetylase inhibition in the treatment of myeloid neoplasms. *Cancer Res.* 2006;66:6361–6369.

124. Stresemann C, Brueckner B, Musch T, Stopper H, Lyko F. Functional diversity of DNA methyltransferase inhibitors in human cancer cell lines. *Cancer Res.* 2006;66:2794–2800.

125. Beaulieu N, Fournel M, Siders B, and MacLeod AR. Synergistic antitumor activity of MG98 (antisense oligodeoxynucleotide targeting DNMT1) in combination with 5-aza-deoxycytosine [abstract]. *Clin Cancer Res.* 2001;7:3800S.
126. Chuang JC, Yoo CB, Kwan JM et al. Comparison of biological effects of non-nucleoside DNA methylation inhibitors versus 5-aza-2'-deoxycytidine. *Mol Cancer Ther.* 2005;4:1515–1520.
127. Davis AJ, Gelmon KA, Siu LL et al. Phase I and pharmacologic study of the human DNA methyltransferase antisense oligodeoxynucleotide MG98 given as a 21-day continuous infusion every 4 weeks. *Invest New Drugs.* 2003;21:85–97.
128. Stewart DJ, Donehower RC, Eisenhauer EA et al. A phase I pharmacokinetic and pharmacodynamic study of the DNA methyltransferase 1 inhibitor MG98 administered twice weekly. *Ann Oncol.* 2003;14:766–774.
129. Winquist E, Knox J, Ayoub JP et al. Phase II trial of DNA methyltransferase 1 inhibition with the antisense oligonucleotide MG98 in patients with metastatic renal carcinoma: a National Cancer Institute of Canada Clinical Trials Group investigational new drug study. *Invest New Drugs.* 2006;24:159–167.
130. Yamada Y, Jackson-Grusby L, Linhart H et al. Opposing effects of DNA hypomethylation on intestinal and liver carcinogenesis. *PNAS.* 2005;102:13580–13585.
131. Yang AS, Estecio MRH, Garcia-Manero G, Kantarjian HM, Issa JP. Comment on "Chromosomal Instability and Tumors Promoted by DNA Hypomethylation" and "Induction of Tumors in Mice by Genomic Hypomethylation." *Science.* 2003;302:1153b.

5 Epigenetic Drugs
Histone Deacetylase Inhibitors

Melissa Peart and Ricky W. Johnstone

CONTENTS

5.1 Introduction 49
5.2 Histone Deacetylase Inhibitors (HDACi) 50
    5.2.1 Structural Basis of the Enzyme Inhibitory Activity of HDACi 50
    5.2.2 Characteristics of HDACi Classes 52
        5.2.2.1 Hydroxamic Acids 52
        5.2.2.2 Cyclic Peptides 52
        5.2.2.3 Short-Chain Fatty Acids 56
        5.2.2.4 Benzamides 56
        5.2.2.5 Ketones 56
5.3 Biological Effects of HDACi 56
    5.3.1 Apoptosis 56
        5.3.1.1 Role of the Death Receptor (Extrinsic) Pathway in
                HDACi-Induced Apoptosis 57
        5.3.1.2 Role of the Mitochondrial (Intrinsic) Pathway in
                HDACi-Induced Apoptosis 58
        5.3.1.3 Cell Cycle Arrest and Differentiation 59
    5.3.2 Angiogenesis and Immune Responses 60
    5.3.3 Transcription-Independent Effects of HDACi 61
        5.3.3.1 Cell Cycle Checkpoints 61
        5.3.3.2 DNA Damage and Repair 61
        5.3.3.3 Hyperacetylation of Nonhistone Proteins 61
    5.3.4 HDACi as Clinical Anticancer Agents 63
5.4 Conclusions 63
References 64

5.1 INTRODUCTION

Histone deacetylase inhibitors (HDACi) represent a novel class of anticancer compounds that inhibit tumor cell growth and/or survival by inducing cell cycle arrest at the G_1 or G_2/M phases, differentiation, or apoptosis, inhibiting angiogenesis and tumor cell metastasis, and activating the host immune response (1). Although several

HDACi are already in early phase clinical trials as single agents or in combination with other chemotherapeutic agents, their mechanism(s) of action remain largely undefined. Reports describing encouraging preliminary clinical results where HDACi exhibited potent antitumor efficacy with few or no side effects have prompted intense investigations into the molecular events that contribute to their antitumor activities. Since histone acetylation plays a fundamental role in chromatin remodeling and transcription, it was initially proposed that HDACi mediate their biological effects through the regulation of gene expression. More recently, it has been shown that acetylation regulates the activity of diverse nonhistone proteins and is important for other cellular processes that involve the DNA template, including mitosis, DNA replication, and DNA repair. These findings imply that HDACi may have a broader effect on cellular processes than originally understood, and establishing a molecular basis for the antitumor activities of HDACi will require an understanding of all of these pathways. Herein we describe the rationale behind the development of HDACi as novel anticancer compounds, the biological outcomes of HDAC inhibition, and the molecular mechanisms of HDACi-mediated antitumor activities.

5.2   HISTONE DEACETYLASE INHIBITORS (HDACi)

A large number of structurally diverse HDACi exist that have been purified from natural sources and synthetically developed. Interestingly, HDACi were initially identified based on their ability to induce differentiation of Friend erythroleukemia cells (2–4), and were only subsequently discovered to inhibit HDACs (5–7). HDACi can be divided into six classes based on their chemical structure (Figure 5.1):

1. Hydroxamic acid-derived compounds: trichostatin A (TSA), suberoylanilide hydroxamic acid (SAHA), m-carboxycinnamic acid bis-hydroxamide (CBHA), azelaic bishydroxamic acid (ABHA), LAQ824, LBH589, Oxamflatin, PXD101, scriptaid, pyroxamide, SK-7041, SK-7068, tubacin
2. Cyclic peptides: romidepsin (depsipeptide, FK228/FR901228), apicidin, CHAPs, Trapoxin
3. Short-chain fatty acids: valproic acid (VPA), phenylbutyrate, phenylacetate, AN-9 (Pivanex)
4. Benzamides: MS-275, CI-994
5. Ketones: trifluoromethyl ketone
6. Miscellaneous: Depudecin, MGCD-0103, SB-429201, SB-379872

5.2.1   STRUCTURAL BASIS OF THE ENZYME INHIBITORY ACTIVITY OF HDACI

With the exception of the Aoe-containing cyclic tetrapeptides, all HDACi appear to inhibit HDACs in a reversible fashion (8). The x-ray crystallographic structure of an HDAC homolog from the hyperthermophilic bacterium Aquifex aeolicus alone and in complex with the hydroxamate inhibitors TSA and SAHA have revealed the structure of the HDAC catalytic core, as well as the mechanism of HDAC inhibition at least by hydroxamic acids (9). The active site is an approximate 390-amino-acid region that is conserved across the HDAC family and consists of a curved tubular

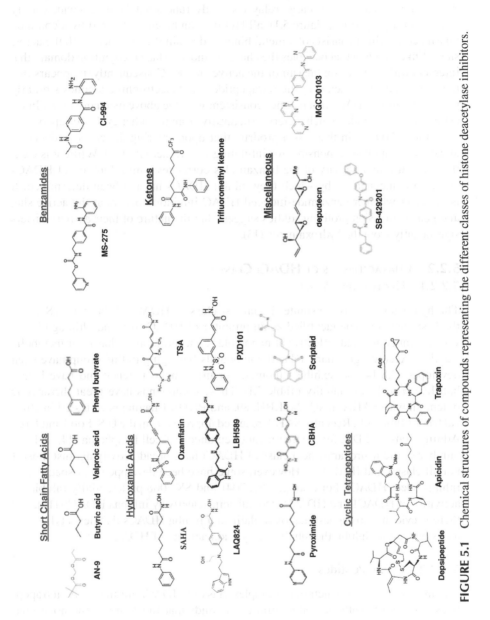

FIGURE 5.1 Chemical structures of compounds representing the different classes of histone deacetylase inhibitors.

pocket, two Asp-His charge-relay systems and a zinc binding site. The presence of a Zn^{2+} ion bound to the zinc binding site is an essential component of the charge-relay system. HDACi bind inside the pocket of the active site and function by displacing the zinc ion, rendering the charge-relay system dysfunctional. Despite a wide variety of structural properties (Figure 5.1), all HDACi can be characterized by a common pharmacore, which consists of a metal binding domain that interacts with the active site, a linker domain that occupies the channel, and a surface recognition domain that interacts with residues on the rim of the active site (8). Consequently, it appears that the short-chain fatty acids, cyclic tetrapeptides, and electrophilic ketones similarly bind and inhibit HDACs in a manner consistent with the above mechanism (8). Interestingly, depsipeptide, which differs structurally from the other cyclic tetrapeptides, is a unique HDACi in that it is a prodrug that upon entering the cell is reduced to an active compound responsible for inhibiting HDAC activity (10). While it is clear that the antitumor activity of the benzamides correlates with inhibition of HDACs (11,12), the mechanism by which they inhibit HDACs has not been determined. It is not clear whether benzamide-induced HDAC inhibition occurs at the active site, and gene expression profiling studies suggest that the nature of their activity differs significantly from the hydroxamates (13).

5.2.2 Characteristics of HDACi Classes
5.2.2.1 Hydroxamic Acids

The hydroxamic acids constitute the largest class of HDACi (Table 5.1). TSA was the first member to be identified as an inhibitor of HDACs (6), and although it is a very potent compound, effective at nanomolar concentrations, it has poor metabolic stability. Accordingly, many related compounds belonging to this group have been developed (1,8,14) that exhibit nanomolar or micromolar potency and have longer half-lives and bioavailability (Table 5.1). These compounds have great therapeutic potential, with SAHA, LAQ824, LBH589, and PXD101 having been tested in clinical trials (Table 5.1). Recently SAHA received approval from the U.S. Food and Drug Administration (FDA) for the treatment of cutaneous T cell lymphoma (14). A large number of the hydroxamic acid class of HDACi have broad activity against most if not all class I and II HDACs. However, some have been developed that specifically inhibit select HDACs. For example, SK-7041 and SK-7068 preferentially inhibit the activities of HDAC1 and HDAC2 (15), tubacin selectively inhibits HDAC6 (16), and there is evidence to suggest that oxamflatin can inhibit HDACs that deacetylate histones but cannot inhibit the tubulin acetylation activity of HDAC6 (17).

5.2.2.2 Cyclic Peptides

The most potent and structurally complex class of HDACi are the cyclic tetrapeptides—a mix of synthetic and natural compounds that function at nanomolar concentrations (Table 5.1). This class may be further divided into two subclasses: those that irreversibly bind to and inhibit HDAC enzymes (Aoe-containing inhibitors, such as trapoxin A/B, chlamydocin, and HC-toxins) and those that are reversible inhibitors (without Aoe moiety, such as depsipeptide, apicidin, and CHAPs). In terms of HDAC specificity, depsipeptide reportedly possesses stronger activity against the

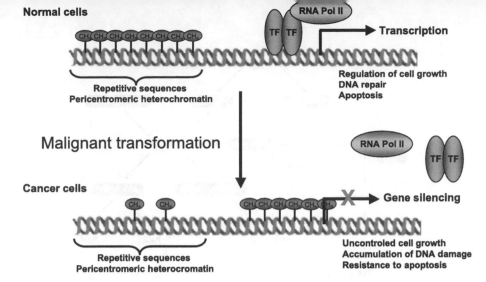

FIGURE 2.1 Repetitive DNA sequences and a typical CpG island of a tumor suppressor gene are shown for a normal and a tumor cell. The presence of dense hypermethylation completely changes the molecular environment.

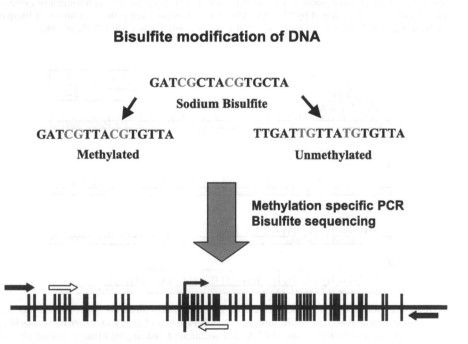

FIGURE 2.2 A schematic representation of the bisulfite modification of DNA and the detection of DNA methylation by methylation-specific PCR and bisulfite sequencing. CpG dinucleotides are represented as short vertical lines. Locations of primers for bisulfite genomic sequencing PCR are indicated by black arrows and locations of methylation-specific PCR primers by white arrows.

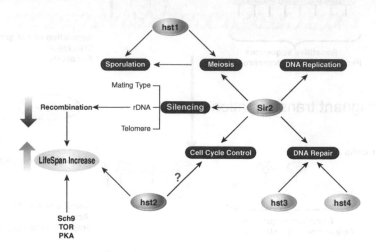

FIGURE 6.1 Sir2 family members in yeast are involved in many biological processes. Hst1 is mainly linked to meiosis and regulation of the sporulation program, whereas Hst3 and Hst4 are implicated in DNA repair signaling through deacetylation of H3K56Ac. Sir2 is involved in epigenetic silencing, DNA repair, DNA replication, cell cycle control, meiosis regulation, and lifespan control through its inhibitory effect on rDNA recombination. The cellular role of Hst2 is not clear, although it might be involved in cell cycle control (indicated by "?"). An additional Sir2-independent pathway promoter of lifespan increase has been described and involves the nutrient-dependent kinases Sch9, TOR, and PKA.

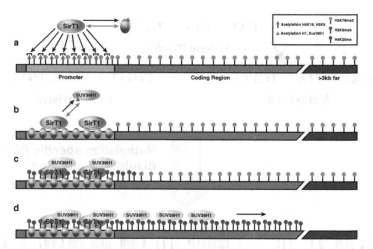

FIGURE 6.2 Model of SirT1's role in facultative heterochromatin formation. SirT1 promotes the formation of facultative heterochromatin through several steps: (*a*) SirT1 arrival to chromatin is associated with deacetylation of H4K16Ac and H3K9Ac and recruitment and deacetylation of histone H1 (*red arrows* indicate interaction; *black arrows*, deacetylation). These changes are restricted to the promoter regions. (*b*) SirT1 recruits the H3K9 methyltransferase Suv39h1 (*magenta*) and induces Suv39h1 enzymatic activity (*yellow* Suv39h1 in *c, d*) through a conformational change and deacetylation of the SET domain. (*c,d*) The spreading of Suv39h1 throughout the coding region is independent of SirT1. Similar to Suv39h1-dependent spreading of H3K9me3, H4K20me also spreads from the promoter throughout the coding region, but the role of SirT1 in the arrival and spreading of this heterochromatin mark is unknown. In parallel to this heterochromatin spreading, a loss of an active mark, H3K79me2, is detected, a phenomenon that spans at least 3 kb away from the coding region.

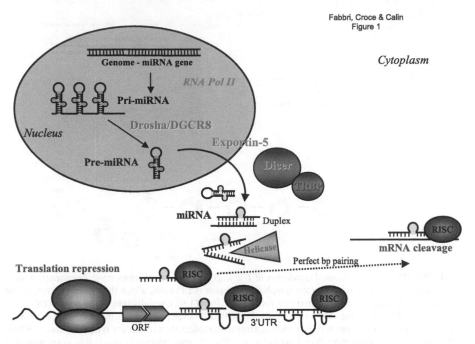

FIGURE 7.1 Mechanisms of biogenesis and action of microRNAs. For details see the text. (Modified with permission from Ref. 125.)

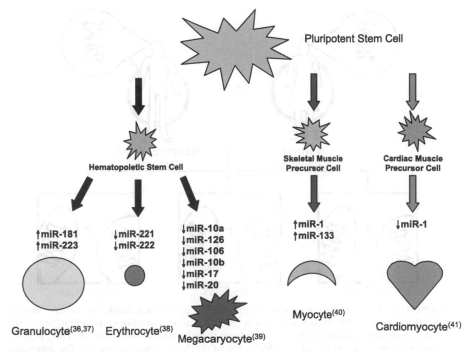

FIGURE 7.2 Examples of microRNAs involved in differentiation. For details see the text.

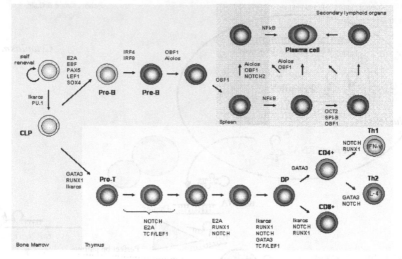

FIGURE 10.1 *Transcriptional regulation of T- and B-lymphocyte differentiation.* Bone marrow stem cells differentiate first into common lymphoid precursor (CLP) cells, which then either migrate to the thymus where they differentiate first into pro-T cells (Pro-T), then serially into CD4+ CD8+ (DP) cells, CD4+, or CD8+ cells, then Th1 or Th2 cells. Early B-cell development starts in the bone marrow with pro-B cells maturing to pre-B cells, which migrate to the spleen then secondary lymphoid tissues, where they differentiate into further B cells and plasma cells. The transcription factors involved are named at each step.

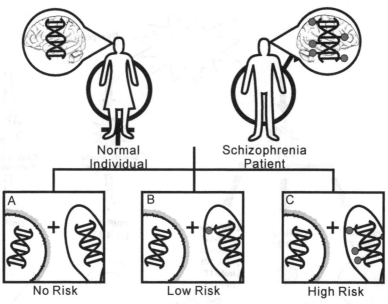

FIGURE 11.1 Partial meiotic epigenetic stability. In this hypothetical family, the father is affected with schizophrenia and has an epimutation on the gene predisposing to schizophrenia. (*A*) The epimutation is completely erased in the father's germline and the offspring has no disease. (*B*) There is partial erasure of epimutation, which results in a higher risk of developing schizophrenia. (*C*) The epimutation is meiotically stable; in which case, the offspring has a high chance of developing schizophrenia.

TABLE 5.1
Characteristics of Histone Deacetylase Inhibitors

HDACi	[range]	HDAC Specificity	Study Phase	Tumor Type	Number Studied	Responses Observed
Phenyl butyrate	mM	Class I, II (not HDAC6)				
			I	Solid	75	CR [1], SD [9] (120–122)
			I	AML/MDS	49	No CR or PR Haem. improvement [6] (123, 124)
Valproic acid	mM	Class I, II (not HDAC6)				
			I/II	AML/MDS	18	OR [8], PR [1] (125)
AN-9	μM	N/A				
			I	Solid	28	PR [1], SD [6] (126)
			II	NSCLC	47	PR [3], SD [14] (127)
SAHA (vorinostat)	μM	Classs I, II				
			I	Solid/Haem	37	Tumor regression [4] (128)
			I	Solid + Haem	73	OR [6], CR [1], PR[3] (129)
			I	AML/MDS	41	OR [9], CR [1], CRp [2] (130)
			I/II	Haem	35	CR [1], PR [4] (131)
			I	Mesothelioma	13	PR [2] (132)
			II	CTCL	74	OR [22] (116)
			II	CTCL	33	PR [8] (117)
LAQ824	nM	Classs I, II				
			I	Haem	21	CR [1], SD [6] (133)
			I	Solid	28	SD [3] (134)
LBH589 (panobinostat)	nM	Classs I, II				
			I	Solid	13	SD [6] (135)
			I	CTCL	11	CR [2], PR [3], SD [2] (119)

TABLE 5.1 (CONTINUED)
Characteristics of Histone Deacetylase Inhibitors

HDACi	HDAC [range]	HDAC Specificity	Study Phase	Tumor Type	Number Studied	Responses Observed
			I	AML/ALL/MDS	15	Transient reduction in blood blasts [8] (136)
PXD-101	µM	Class I, II				
			I	Haem	11	PR [1], SD [1] (137)
CBHA	µM	N/A	N/A			
Scriptaid	µM	N/A	N/A			
Pyroxamide	µM	Class I		Solid/Haem	32	N/A
ABHA/SBHA	µM	N/A	N/A			
SK-7041/SK-7068	nM	HDACs 1, 2	N/A			
CG-1521	µM	N/A	N/A			
Tubacin	µM	HDAC6	N/A			
Romsidepsin (depsipeptide)	nM	Class I				
			I	Solid	37	PR [1] (renal cell) (138)
			I	Solid	33	No CR or PR (139)
			I	CLL/AML	20	No CR or PR (140)
			II	CTCL	28	CR [2], PR [8], SD [16], pruritis improved in 45% (141)
			II	CTCL	27	CR [3; Sezary], PR [7] (118)
			II	PTCL	26	CR [3], PR [3] (118)
			II	RAI-resistant thyroid	14	Reversal of RAI resistance [1], SD [4] (142)
			II	Hormone refractory prostate	21	Radiological PR [1], fall in PSA [3], SD [6] (143)
			II	Advanced Colorectal	24	No CR or PR, SD [4] (144)
			II	Multiple myeloma	12	SD [11] (145)
			I	Paed. solid	24	No CR or PR, SD[3] (146)

TABLE 5.1 (CONTINUED)
Characteristics of Histone Deacetylase Inhibitors

HDACi	[range]	HDAC Specificity	Study Phase	Tumor Type	Number Studied	Responses Observed
Trapoxin	nM	Class I, II	N/A			
Apicidin	nM	HDAC1, 3, not 8	N/A			
CHAPs	nM	Class I	N/A			
MS-275	μM	HDACs 1, 2, 3, 8				
			I	Solid	17	PR [1, melanoma], SD (3) (147)
			I	Solid lymphoma	30	No CR or PR, SD [15] (25)
			I	AML	22	PR [2], SD [7] (148)
			II	Metastatic melanoma	28	No CR or PR, SD [7] (149)
			I	Advanced malignancies	16	SD [2] (150)
CI-994	μM	N/A				
			I	Solid	53	PR [1], SD [3] (151)
			II	NSCLC	29	PR [2], SD [8] (152)
			II	Renal cell	48	SD [26] (153)
			II	Pancreatic	15	No CR or PR, SD[2] (154)
MGCD0103	N/A	Class I				
			I	Leuk/MDS	20	Marrow responses [3] (155)
			I	Solid	28	SD [4] (156)
Depudecin	μM	Class I				

CR: Complete response, CRp: Complete response without complete platelet recovery; OR: overall response; PR: Partial response, Paed: paediatric; SD: Stable disease, RAI: radioactive iodine, AML: acute myeloid leukaemia; MDS: myelodysplastic syndrome; CTCL: cutaneous T cell lymphoma; ECG: electrocardiographic; CLL: chronic lymphocytic leukaemia; NSCLC: non-small cell lung cancer; ALL: acute lymphoblastic leukaemia; IV: intravenous; PO: oral; Haem: haematologic class I HDACs 1 and 2 than against the class II HDACs 4 and 6 (10), whereas apicidin can inhibit HDACs 1 and 3 but not HDAC8 (18). CHAP1 is reportedly less active against HDAC6 than HDAC1, and HDAC6 and HDAC10 are resistant to trapoxin, possibly due to the presence of their two catalytic domains (19,20).

5.2.2.3 Short-Chain Fatty Acids

The short-chain fatty acids are the least potent inhibitors, possess a short half-life, and may possess other targets in addition to HDACs (21,22). Nonetheless, this class of HDACi is being used in the clinic alone and in combination with other agents for the treatment of cancer (Table 5.1). The butyrates can inhibit class I and II HDACs, although these agents do not suppress the tubulin deacetylase activity of HDAC6 (17). VPA is a class I selective HDACi (23) and, in addition to inhibiting the enzymatic activity of HDACs, induces proteasomal degradation of HDAC2 (24).

5.2.2.4 Benzamides

The benzamides are less potent HDACi than the hydroxamate and cyclic tetrapeptide inhibitors but are more potent than the short-chain fatty acids (Table 5.1). Both MS-275 and CI-994 are active at micromolar concentrations, and MS-275 has a particularly long in vivo half-life ($t_{1/2}$ = 39–80 hrs) compared to all other HDACi (25). The HDAC specificity of CI-994 has not yet been defined, but MS-275 preferentially inhibits HDAC1, is less active against HDAC3, and only marginally inhibits HDAC8 (26).

5.2.2.5 Ketones

Electrophilic ketones function at submicromolar concentrations but possess inferior metabolic stability to most other HDACi classes (27). The enzyme specificity of electrophilic ketones has not yet been defined.

5.3 BIOLOGICAL EFFECTS OF HDACi

HDACi can induce diverse biological responses in tumor cells, including induction of apoptosis and suppression of cell proliferation by activation of cell cycle checkpoints at G_1/S or G_2/M (1,28,29). Moreover, the ability of HDACi to suppress angiogenesis and activate and enhance the host immune system may play important indirect roles in their therapeutic response (1,28,29). It is currently unclear which one or more of these biological effects are necessary for the anticancer responses attributed to these agents, and in many instances, the molecular events that lead to the biological effects of HDACi remain to be fully elucidated. This section will describe the biological outcomes of HDAC inhibition, with a focus on effects that occur downstream of dysregulated transcription following treatment with these agents.

5.3.1 APOPTOSIS

HDACi induce apoptosis in a variety of tumor cell types, including solid tumor cell lines and hematological transformed cell lines (1). The next three subsections will review data investigating the role of the death receptor pathway, the mitochondrial pathway, and key apoptotic regulators in HDACi-induced apoptosis.

5.3.1.1 Role of the Death Receptor (Extrinsic) Pathway in HDACi-Induced Apoptosis

The role of the death receptor pathway in drug-induced apoptosis is controversial (30), and the importance of an intact death receptor pathway for HDACi-induced apoptosis remains subject to debate due to conflicting reports in the literature. Studies by different groups using the cowpox virus serpin CrmA, c-FLIP, dominant negative FADD, or dominant negative caspase-8 that effectively block signaling through the death receptor pathway, RNA interference to knockdown specific death receptors or ligands, and neutralizing antibodies to inhibit receptor–ligand interaction have demonstrated that death receptor signaling is (31–35) or is not (36–41) required for HDACi-induced apoptosis. For example, three separate studies have tested a direct requirement for the death receptor pathway in HDACi-induced apoptosis in acute T-lymphoblastic leukemia (T-ALL) cells, namely by overexpressing the cowpox virus serpin CrmA (36–38). CrmA effectively blocks the death receptor pathway by binding with high affinity to the membrane proximal caspases 8 and 10. Inhibition of this pathway had no effect on the ability of butyrate, SAHA, oxamflatin, or depsipeptide to induce cell death, in contrast to Fas-induced apoptosis, demonstrating that these HDACi do not require an intact death receptor pathway to induce apoptosis. Similarly, MS-275-induced apoptosis proceeded independently of the death receptor pathway in AML cells using anti-FasL blocking antibodies and cells overexpressing CrmA or caspase-8 dominant negative (DN) (39). In contrast, Imai et al. showed that anti-FasL blocking antibody, or overexpression of FADD-DN or viral FLIP E8, significantly attenuated depsipeptide-induced apoptosis in osteosarcoma cells, indicating that depsipeptide induces apoptosis through activation of the Fas/FasL signaling pathway (32). In addition, inhibition of Fas signaling in APL cells using blocking antibodies antagonized apicidin-induced apoptosis (31), showing that Fas ligation is important. Glick et al. proposed that CBHA induces apoptosis through the regulation of Fas/Fas ligand, although it was not shown that inhibition of this pathway affected CBHA-induced apoptosis (42). Insinga and colleagues performed an in vivo study using PML-RAR transgenic mice that develop AML to try to address the importance of death receptor signaling in HDACi-mediated apoptosis (34). HDACi-induced expression of TRAIL and FAS in AML cells was suppressed in vivo using TRAIL- and FAS-selective siRNA delivered by hydrodynamic therapy resulting in a 50% reduction in apoptosis following treatment of mice with VPA. Whether loss of TRAIL- or FAS-mediated apoptosis suppressed the therapeutic effect of VPA in this model was not reported. Taken together, these data are conflicting with regard to the importance of the death receptor pathway in HDACi-induced apoptosis. Almost of all of these studies have been performed in vitro using human tumor cell lines. The current confusion in the field regarding the importance of the death receptor pathway in HDACi-induced tumor cell death may be alleviated through the use of sophisticated preclinical mouse models and through the analysis of samples obtained from HDACi clinical trials.

5.3.1.2   Role of the Mitochondrial (Intrinsic) Pathway
in HDACi-Induced Apoptosis

In contrast to the confusion regarding the importance of the death receptor pathway, a large number of independent studies strongly support a role for the mitochondrial apoptotic pathway in HDACi-mediated tumor cell death. For example, Bcl-2 overexpression protects T-ALL cells from butyrate- (36) SAHA- (37,38), oxamflatin- and depsipeptide-induced cell death (37), multiple myeloma (MM) cells from SAHA-induced apoptosis (43), melanoma cells from SBHA-induced apoptosis (44), and chronic lymphocytic leukemia (CLL) cells from MS-275-induced apoptosis (45). These data strongly suggest a key role for the intrinsic apoptotic pathway in HDACi-induced cell death. How HDACi trigger activation of the intrinsic apoptotic cascade is a major question that remains to be fully elucidated. One possibility is that HDACi induce global changes in gene expression that alter the balance of expression of pro- and anti-apoptotic genes in favor of a pro-apoptotic biological response. Indeed, gene expression profiling analyses and more refined studies analyzing the expression of specific apoptotic genes and proteins showing upregulation of a number of pro-apoptotic and downregulation of anti-apoptotic genes/proteins support this notion (41,44,46–50). However, it is equally possible that the triggering of apoptosis by HDACi is a more specific process, mediated by the activation of a defined protein or signaling pathway upstream of the mitochondria, as outlined below.

Given the central role of the Bcl-2 family of proteins as regulators of the intrinsic apoptotic pathway, a number of studies have explored the role of these proteins in HDACi-induced apoptosis. Several studies have demonstrated the cleavage and activation of the BH3-only pro-apoptotic protein Bid in response to HDACi. Bid is cleaved and activated in response to SAHA in T-ALL cells (37,38), MM cells (43), breast carcinoma cells, and fibroblast cells transformed with the E1A oncogene (51), oxamflatin and depsipeptide in T-ALL cells (37), and MS-275 in CLL cells (45). The events leading to Bid cleavage may depend on cell type, as one group showed that SAHA- and TSA-induced Bid cleavage was dependent on caspase-9, suggesting that Bid is cleaved via the amplification loop as a late event in HDACi-induced apoptosis (51), whereas other studies showed that HDACi-induced Bid cleavage occurred during the initial phase apoptosis upstream of the mitochondria (37,38,43).

Other BH3-only proteins that have been directly implicated in HDACi-induced apoptosis include Bim and Bmf. HDACi induce expression of Bim through enhanced binding of E2F-1 to the Bim promoter (52,53), while Bmf is transcriptionally activated by depsipeptide and CBHA and inhibition of Bmf expression by siRNA-suppressed HDACi-induced mitochondrial membrane damage and apoptosis (54).

Treatment with HDACi can result in elevated ROS levels and addition of free-radical scavengers can suppress the apoptotic activity of HDACi (38,55–59). Recently, it has been proposed that selective regulation of ROS production may confer the tumor-cell selective apoptotic effects of HDACi (29). In normal cells, HDACi transcriptionally activate the ROS scavenger Trx, while TBP-2, a negative regulator of Trx and enhancer of ROS production is selectively induced by HDACi in transformed cells (60,61). Collectively, this would result in ROS-mediated death of transformed cells only, a claim supported by experiments showing that siRNA-mediated knockdown of Trx increased the sensitivity of tumor cells to HDACi (61). Importantly, it has been shown that the increase in ROS levels precedes changes in mitochondrial membrane potential (55), consistent with a role for enzymes located outside of the mitochondria, such as in peroxisomes, endoplasmic reticulum, and cytoplasm, and on the plasma membrane being involved in ROS generation. Interestingly, ROS production can lead to the transcriptional induction and activation of Bim (62), raising an intriguing possible link between HDACi treatment, elevated ROS, BH3-only protein activation, and triggering of the intrinsic apoptosis pathway, which needs to be formally evaluated. How free-radical levels are initially elevated in response to HDACi remains unclear, and whether this is an active process of enhanced ROS production or an increase in levels due to altered expression of ROS-regulatory proteins such as TBP-2 and Trx remains to be determined.

5.3.1.3   Cell Cycle Arrest and Differentiation

HDACi inhibit cell cycle progression in a diverse array of cell lines, most often in the G_1 phase of the cell cycle, but also in the G_2/M phase (1). The cyclin-dependent kinase (CDK) inhibitor p21[WAF1Cip1] is an important regulator of cell fate in response to HDACi and may determine whether the cell undergoes cytostasis or apoptosis. CDKN1A encoding p21[WAF1Cip1] is upregulated in most cells treated with HDACi, resulting in a G_1 phase arrest (63–68). The cytostatic activity of HDACi is important for their ability to induce differentiation in a diverse array of leukemias and solid tumors (69). In some settings, p21[WAF1Cip1] induction and subsequent Rb dephosphorylation may promote terminal differentiation marked by the induction of the maturation markers CD11b and CD14 (4,68,70). However, HDACi can also induce differentiation independent of Rb (71), although the underlying mechanisms are unknown. In leukemias that are characterized by fusion proteins that recruit HDAC complexes to aberrantly silence gene expression and promote a maturation block, the ability of HDACi to induce differentiation may relate to the transcriptional reactivation of genes required for cell differentiation, rather than to their ability to regulate p21[WAF1Cip1]. For example, treatment of AML1-ETO-positive cells with depsipeptide induces histone acetylation and restores gene transcription, which may directly contribute to depsipeptide-induced morphological changes and upregulation of CD11b (72). Similarly, TSA induces differentiation of PML-RARα cells, as determined by an increase in CD14+ cells, and is accompanied by transcriptional activation of aberrantly silenced target genes (73).

HDACi can mediate G_2/M phase arrest by activating a G_2 checkpoint, although this is a much rarer event than HDACi-induced G_1 arrest (74,75). Loss of the G_2-phase checkpoint may be a major determinant of the apoptotic sensitivity to HDACi. Most tumor cells have a defective G_2-phase checkpoint, and these cells that are treated with HDACi initially accumulate in the G_2/M phase of the cell cycle with a 4N DNA content, and then move through this defective checkpoint and undergo apoptosis (37,38). Cell lines sensitive to HDACi-induced apoptosis often have a defective G_2 checkpoint, and reintroduction of a G_2 arrest can protect these cells from HDACi-induced apoptosis (74). Unlike HDACi-mediated G_1 arrest, where there is a documented role for the transcriptional activation of CDKN1A, the underlying mechanisms responsible for HDACi-mediated G_2 arrest are poorly understood. It has been proposed that the HDACi-associated G_2 checkpoint may be related to hyperacetylation of pericentric heterochromatin and loss of this checkpoint can result in HP1 spreading, abnormal chromosomal segregation, and nuclear fragmentation (76,77).

5.3.2 ANGIOGENESIS AND IMMUNE RESPONSES

HDACi possess additional antitumor properties that may affect tumor growth and/or survival. For example, HDACi exhibit antiangiogenic, anti-invasive, and immuno-modulatory activities in vitro and in vivo. The antiangiogenic properties of HDACi may stem from the differential expression of genes encoding proteins that directly regulate blood vessel development. TSA, VPA, butyric acid, MS-275, and depsipep-tide downregulate angiogenesis factors such as VEGF, bFGF, and/or eNOS to inhibit angiogenesis (78–82). TSA also suppresses HIF-1α activity (83), and depsipeptide induces the expression of angiogenic-inhibiting factors such as von Hippel Lindau and neurofibromin 2 (84). SAHA can suppress the expression of VEGF receptors 1 and 2 and neurophilin-1. Moreover, SAHA could induce expression of the VEGF competitor, semaphorin III to inhibit VEGF signaling and suppress endothelial cell growth and neoangiogenesis (85). In addition, HDACi have been shown to down-regulate expression of the chemokine receptor CRCX4 in both untransformed endo-thelial cells and tumor cell lines (86,87). CXCR4 is important for the homing of bone marrow progenitor and circulating endothelial cells to sites of angiogenesis (88). Finally, HDACi have been shown to suppress endothelial cell progenitor cell dif-ferentiation (87,89). Taken together, these studies provide evidence supporting a role for HDACi in suppressing neovascularization through alteration of genes directly involved in angiogenesis, which, in addition to affecting nutrient supply to the pri-mary tumor, may also inhibit metastatic spread of the tumor.

The importance of immune responses in mediating host antitumor responses and the selective pressure on tumor cells to downregulate expression of immune-stimulating molecules has been well documented (90). HDACi may acti-vate cancer immunosurveillance mechanisms by inducing expression of immune-stimulatory proteins such as MHC class I and II, co-stimulatory molecules CD40 and CD86, and the adhesion molecule ICAM-1 on the surface of tumor cells (91,92). HDACi also exhibit significant anti-inflammatory properties at lower con-centrations than required for their antitumor activities. For example, SAHA can reduce acute graft-versus-host disease (GVHD) following allogeneic bone mar-row transplantation by suppressing pro-inflammatory cytokines such as TNF-α, IL-1, and IFN-γ, which have a central role in the pathogenesis of GVHD (93). At the same time, SAHA maintains beneficial graft-versus-leukemia (GVL) effects, required to eradicate residual malignant cells, by preserving donor T cell prolifera-tive and cytotoxic responses to host antigens (93). The importance of an immune-regulatory function of HDACi in mediating antitumor effects has not yet been adequately addressed; however, the data that has already been obtained indicate that these drugs may possess counteracting immune-stimulatory and immunosup-pressive effects.

5.3.3 TRANSCRIPTION-INDEPENDENT EFFECTS OF HDACI
5.3.3.1 Cell Cycle Checkpoints

Cells treated with HDACi that have a defective G_2 checkpoint enter mitosis; however, the condensed chromosomes fail to align at the midline of the mitotic cells and form a proper metaphase plate (74). The improper alignment of chromosomes should then trigger a mitotic arrest through activation of the mitotic spindle checkpoint to allow spindle defects to be resolved before exit from mitosis. However, HDACi are able to overcome the mitotic spindle checkpoint (17), allowing cells to prematurely exit mitosis before proper chromosome alignment. This can result in mitotic catastrophe and apoptosis, or in the formation of polyploid and multinucleated cells. The mechanisms by which HDACi induce an aberrant mitosis and overcome the mitotic spindle checkpoint remain to be elucidated. It is known that histone acetylation status is tightly controlled during mitosis, because the chromatin of mitotic cells is hypoacetylated compared with interphase cells and is only transiently acetylated during S phase when DNA replication occurs (94). Treatment with TSA induces hyperacetylation of normally hypoacetylated centromeric heterochromatin during mitosis, which prevents proper chromosome condensation and segregation (95,96). This suggests that the antitumor effects of HDACi may be mediated through disruption of histone acetylation status during mitosis as a result of HDACi-induced hyperacetylation of heterochromatin. Alternatively, it is possible that HDACi might regulate an unknown gene(s) that may activate the G_2 checkpoint.

5.3.3.2 DNA Damage and Repair

Remodeling of chromatin can activate ATM, a serine/threonine protein kinase that is normally activated in response to double-stranded break damage to DNA (97), and HDACi may mimic the effects of genotoxic insults to induce a DNA damage response (98). HDACi can augment irradiation-induced apoptosis (99,100), and recent evidence suggests that HDACi may in fact induce DNA strand breaks (101), although the precise mechanisms underlying this effect are not known. Interestingly, HDACi appear to suppress DNA repair concomitant with decreased expression of DNA repair proteins Ku70, Ku86, and DNA-PKcs (102), and cells with knockout of DNA repair proteins Ku70, Ku80, or DNA Ligase IV were hypersensitive to HDACi-induced apoptosis (103,104). Many questions remain regarding the molecular mechanisms of HDACi-induced ATM activation and the importance of the DNA damage-like effect in HDACi-mediated antitumor effects. However, it is possible that the ability of HDACi to induce a DNA damage response while concomitantly suppressing DNA repair mechanisms may play an important primary role in HDACi-induced apoptosis.

5.3.3.3 Hyperacetylation of Nonhistone Proteins

Another transcription-independent mechanism that may contribute to the antitumor effects induced by HDACi, independent of their ability to inhibit histone deacetylation, is through the regulation of nonhistone protein activity. One such example may be the HDACi-induced hyperacetylation of Ku70, which regulates Bax-mediated

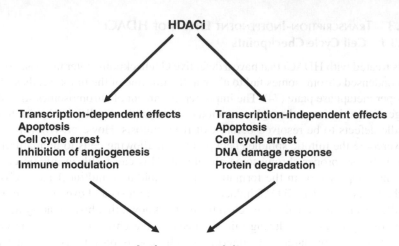

FIGURE 5.2 Antitumor effects of HDACi. HDACi can affect tumor cell growth and survival through multiple biological effects involving transcription-dependent and -independent processes. Gene expression profiling of tumor cells treated with HDACi demonstrates that pro-apoptotic genes involved in the death receptor pathway (e.g., TRAIL, DR5) and/or the intrinsic apoptotic pathway (e.g., Bax, Bak, Apaf-1) are generally upregulated, whereas pro-survival genes (e.g., Bcl-2, XIAP) are generally downregulated thereby lowering the apoptotic threshold within the treated cell. HDACi induce cell cycle arrest at the G_1/S boundary through upregulation of CDKN1A encoding p21WAF1 and/or through downregulation of cyclins. HDACi can suppress angiogenesis concomitant with decreased expression of pro-angiogenic factors VEGF, HIF-1α, and CXCR4. HDACi can also have immunomodulatory effects by enhancing tumor cell antigenicity (upregulation of MHC class I and II, MICA, CD40) and by altering the expression of key cytokines, including TNF-α, IL-1, and IFN-γ. In addition to regulating the acetylation state of histones, HDAC can bind to, deacetylate, and regulate the activity of a number of other proteins, including transcription factors (e.g., p53, E2F-1, NF-κB) and proteins with diverse biological functions (e.g., α-tubulin, Ku70, Hsp90). Hyperacetylation of transcription factors with HDACi can augment their gene regulatory activities and contribute to the changes in gene expression observed following direct HDACi-mediated histone hyperacetylation. Hyperacetylation of proteins such as Ku70 and Hsp90 or disruption of PP1–HDAC interactions by HDACi may have no direct effect or an indirect effect on gene expression but may be important for certain biological effects of HDACi, in particular, induction of apoptosis and cell cycle arrest.

apoptosis (105). Under normal growth conditions, Ku70 is maintained in an unacetylated state by HDACs and/or sirtuin deacetylases and sequesters the pro-apoptotic protein Bax from the mitochondria. Upon HDACi treatment or acetylation by CBP and/or PCAF, Ku70 releases Bax, allowing it to initiate apoptosis via the mitochondrial death pathway (105). HDACi can also disrupt the interactions between HDACs and other cellular proteins (106). HDACs interact with and form complexes containing protein serine/threonine phosphatases (107,108), which may allow for coordinated phosphorylation and acetylation of common target proteins to regulate cell growth and function. HDACi selectively disrupt cellular HDAC/phosphatase complexes, resulting in attenuated dephosphorylation or enhanced acetylation of target proteins (107,108). For example, HDACi disrupt a cytosolic complex formed by HDAC6, PP1, and α-tubulin that controls microtubule dynamics, resulting in enhanced tubulin acetylation that remains associated with PP1 (107).

Hsp90 is a cellular chaperone that binds a diverse array of client proteins including key oncogenic and anti-apoptotic proteins, preventing their ubiquitination and proteosomal degradation (109,110). Hsp90 is deacetylated by HDAC6 (111), and HDACi capable of inhibiting HDAC6 induce the hyperacetylation of Hsp90, resulting in proteosomal degradation of Her-2, ErbB1, ErbB2, Akt, c-Raf, Bcr-Abl, and Flt-3 (112–115). It is therefore possible that loss of expression of these important oncoproteins through hyperacetylation of Hsp90 plays an important role in the anti-cancer activities of HDACi. Further work is required to define the contribution of these effects to the biological activities of HDACi, as well as the role of other potential effector proteins that are regulated by acetylation and/or phosphorylation.

5.3.4 HDACi AS CLINICAL ANTICANCER AGENTS

A large number of HDACi have been tested as anticancer agents in clinical trials (Table 5.1) with encouraging results. Phase I clinical trials to date have delineated the more common toxicities as fatigue and dose-related transient cytopenias. When administered as single agents these drugs induce responses in T cell lymphoproliferative disorders, and this has been demonstrated in published studies with SAHA (vorinostat) (116,117), romidepsin (118), and LBH589 (119). This has culminated in the recent FDA approval of vorinostat for cutaneous T cell lymphomas. These trials have also demonstrated clear biological activity in myeloid malignancies with accumulating evidence of activity in some solid tumors (Table 5.1). To date there are sporadic reports of solid tumors responding to single-agent HDACi, and target solid tumors that are a focus of ongoing and future studies include pancreatic cancer, melanoma, NSCLC, prostate cancer, and melanoma.

Larger phase II single-agent studies have been completed or are close to completion for vorinostat (SAHA) and romidepsin, and a number of other HDACi are in the advanced stages of development. Combination studies with conventional chemotherapeutic agents, radiation therapies, and a range of targeted molecular therapeutic agents are underway (1), and these will ultimately need to be tested in large prospective randomized clinical trials.

5.4 CONCLUSIONS

HDACi are promising anticancer agents, and at least eleven different HDACi are currently in phase I or II clinical trials either as single agents or in combination with other chemotherapeutics (1). Two major features of HDACi provide the basis for their potential clinical use: (1) HDACi show selective cytotoxicity against tumor cells, and (2) HDACi can activate a number of molecular pathways to mediate a range of biological responses that impinge on tumor cell development, growth, and survival. The molecular events that underpin these important features of HDACi are slowly emerging, and it appears that their mechanisms of action may be far more complex than previously thought. In addition to their ability to regulate gene transcription, it now appears that HDACi may also mediate biological responses due to their effects on histone-dependent but transcription-independent molecular events such as mitosis and DNA repair, as well as histone-independent events such as the regulation of protein kinases, phosphatases, and pro-apoptotic molecules (Figure 5.2). If it would turn out that a significant proportion of the cytotoxic effects of HDACi were independent of histone acetylation, it may be that structurally diverse inhibitors kill cells by different mechanisms. Addressing this issue would require experiments comprehensively comparing a range of different HDACi for their molecular and biological effects.

REFERENCES

1. Bolden, J. E., Peart, M. J., and Johnstone, R. W. (2006) *Nat Rev Drug Discov* 5(9), 769–784.
2. Leder, A., Orkin, S., and Leder, P. (1975) *Science* 190(4217), 893–894.
3. Yoshida, M., Nomura, S., and Beppu, T. (1987) *Cancer Res* 47(14), 3688–3691.
4. Richon, V. M., Webb, Y., Merger, R., Sheppard, T., Jursic, B., Ngo, L., Civoli, F., Breslow, R., Rifkind, R. A., and Marks, P. A. (1996) *Proc Natl Acad Sci USA* 93(12), 5705–5708.
5. Riggs, M. G., Whittaker, R. G., Neumann, J. R., and Ingram, V. M. (1977) *Nature* 268(5619), 462–464.
6. Yoshida, M., Kijima, M., Akita, M., and Beppu, T. (1990) *J Biol Chem* 265(28), 17174–17179.
7. Richon, V. M., Emiliani, S., Verdin, E., Webb, Y., Breslow, R., Rifkind, R. A., and Marks, P. A. (1998) *Proc Natl Acad Sci USA* 95(6), 3003–3007.
8. Miller, T. A., Witter, D. J., and Belvedere, S. (2003) *J Med Chem* 46(24), 5097–5116.
9. Finnin, M. S., Donigian, J. R., Cohen, A., Richon, V. M., Rifkind, R. A., Marks, P. A., Breslow, R., and Pavletich, N. P. (1999) *Nature* 401(6749), 188–193.
10. Furumai, R., Matsuyama, A., Kobashi, N., Lee, K. H., Nishiyama, M., Nakajima, H., Tanaka, A., Komatsu, Y., Nishino, N., Yoshida, M., and Horinouchi, S. (2002) *Cancer Res* 62(17), 4916–4921.
11. Saito, A., Yamashita, T., Mariko, Y., Nosaka, Y., Tsuchiya, K., Ando, T., Suzuki, T., Tsuruo, T., and Nakanishi, O. (1999) *Proc Natl Acad Sci USA* 96(8), 4592–4597.
12. Kraker, A. J., Mizzen, C. A., Hartl, B. G., Miin, J., Allis, C. D., and Merriman, R. L. (2003) *Mol Cancer Ther* 2(4), 401–408.
13. Glaser, K. B., Staver, M. J., Waring, J. F., Stender, J., Ulrich, R. G., and Davidsen, S. K. (2003) *Mol Cancer Ther* 2(2), 151–163.
14. Marks, P. A., and Breslow, R. (2007) *Nat Biotechnol* 25(1), 84–90.
15. Park, J. H., Jung, Y., Kim, T. Y., Kim, S. G., Jong, H. S., Lee, J. W., Kim, D. K., Lee, J. S., Kim, N. K., and Bang, Y. J. (2004) *Clin Cancer Res* 10(15), 5271–5281.
16. Haggarty, S. J., Koeller, K. M., Wong, J. C., Grozinger, C. M., and Schreiber, S. L. (2003) *Proc Natl Acad Sci USA* 100(8), 4389–4394.
17. Warrener, R., Beamish, H., Burgess, A., Waterhouse, N. J., Giles, N., Fairlie, D., and Gabrielli, B. (2003) *Faseb J* 17(11), 1550–1552.
18. Vannini, A., Volpari, C., Filocamo, G., Casavola, E. C., Brunetti, M., Renzoni, D., Chakravarty, P., Paolini, C., De Francesco, R., Gallinari, P., Steinkuhler, C., and Di Marco, S. (2004) *Proc Natl Acad Sci USA* 101(42), 15064–15069.
19. Furumai, R., Komatsu, Y., Nishino, N., Khochbin, S., Yoshida, M., and Horinouchi, S. (2001) *Proc Natl Acad Sci USA* 98(1), 87–92.
20. Guardiola, A. R., and Yao, T. P. (2002) *J Biol Chem* 277(5), 3350–3356.
21. Boffa, L. C., Gruss, R. J., and Allfrey, V. G. (1981) *J Biol Chem* 256(18), 9612–9621.

22. de Haan, J. B., Gevers, W., and Parker, M. I. (1986) *Cancer Res* 46(2), 713–716.
23. Gottlicher, M., Minucci, S., Zhu, P., Kramer, O. H., Schimpf, A., Giavara, S., Slee-man, J. P., Lo Coco, F., Nervi, C., Pelicci, P. G., and Heinzel, T. (2001) *Embo J* 20(24), 6969–6978.
24. Kramer, O. H., Zhu, P., Ostendorff, H. P., Golebiewski, M., Tiefenbach, J., Peters, M. A., Brill, B., Groner, B., Bach, I., Heinzel, T., and Gottlicher, M. (2003) *Embo J* 22(13), 3411–3420.
25. Ryan, Q. C., Headlee, D., Acharya, M., Sparreboom, A., Trepel, J. B., Ye, J., Figg, W. D., Hwang, K., Chung, E. J., Murgo, A., Melillo, G., Elsayed, Y., Monga, M., Kalnits-kiy, M., Zwiebel, J., and Sausville, E. A. (2005) *J Clin Oncol* 23(17), 3912–3922.
26. Hu, E., Dul, E., Sung, C. M., Chen, Z., Kirkpatrick, R., Zhang, G. F., Johanson, K., Liu, R., Lago, A., Hofmann, G., Macarron, R., de los Frailes, M., Perez, P., Krawiec, J., Winkler, J., and Jaye, M. (2003) *J Pharmacol Exp Ther* 307(2), 720–728.
27. Frey, R. R., Wada, C. K., Garland, R. B., Curtin, M. L., Michaelides, M. R., Li, J., Pease, L. J., Glaser, K. B., Marcotte, P. A., Bouska, J. J., Murphy, S. S., and Davidsen, S. K. (2002) *Bioorg Med Chem Lett* 12(23), 3443–3447.
28. Bhalla, K. N. (2005) *J Clin Oncol* 23(17), 3971–3993.
29. Dokmanovic, M., and Marks, P. A. (2005) *J Cell Biochem* 96(2), 293–304.
30. Debatin, K. M., and Krammer, P. H. (2004) *Oncogene* 23(16), 2950–2966.
31. Kwon, S. H., Ahn, S. H., Kim, Y. K., Bae, G. U., Yoon, J. W., Hong, S., Lee, H. Y., Lee, Y. W., Lee, H. W., and Han, J. W. (2002) *J Biol Chem* 277(3), 2073–2080.
32. Imai, T., Adachi, S., Nishijo, K., Ohgushi, M., Okada, M., Yasumi, T., Watanabe, K., Nishikomori, R., Nakayama, T., Yonehara, S., Toguchida, J., and Nakahata, T. (2003) *Oncogene* 22(58), 9231–9242.
33. Sutheesophon, K., Nishimura, N., Kobayashi, Y., Furukawa, Y., Kawano, M., Itoh, K., Kano, Y., and Ishii, H. (2005) *J Cell Physiol* 203(2), 387–397.
34. Insinga, A., Monestiroli, S., Ronzoni, S., Gelmetti, V., Marchesi, F., Viale, A., Altucci, L., Nervi, C., Minucci, S., and Pelicci, P. G. (2005) *Nat Med* 11(1), 71–76.
35. Nebbioso, A., Clarke, N., Voltz, E., Germain, E., Ambrosino, C., Bontempo, P., Alva-rez, R., Schiavone, E. M., Ferrara, F., Bresciani, F., Weisz, A., de Lera, A. R., Grone-meyer, H., and Altucci, L. (2005) *Nat Med* 11(1), 77–84.
36. Bernhard, D., Ausserlechner, M. J., Tonko, M., Loffler, M., Hartmann, B. L., Csordas, A., and Kofler, R. (1999) *Faseb J* 13(14), 1991–2001.
37. Peart, M. J., Tainton, K. M., Ruefli, A. A., Dear, A. E., Sedelies, K. A., O'Reilly, L. A., Waterhouse, N. J., Trapani, J. A., and Johnstone, R. W. (2003) *Cancer Res* 63(15), 4460–4471.
38. Ruefli, A. A., Ausserlechner, M. J., Bernhard, D., Sutton, V. R., Tainton, K. M., Kofler, R., Smyth, M. J., and Johnstone, R. W. (2001) *Proc Natl Acad Sci USA* 98(19), 10833–10838.
39. Rosato, R. R., Almenara, J. A., Dai, Y., and Grant, S. (2003) *Mol Cancer Ther* 2(12), 1273–1284.
40. Nakata, S., Yoshida, T., Horinaka, M., Shiraishi, T., Wakada, M., and Sakai, T. (2004) *Oncogene* 23(37), 6261–6271.
41. Singh, T. R., Shankar, S., and Srivastava, R. K. (2005) *Oncogene* 24(29), 4609–4623.
42. Glick, R. D., Swendeman, S. L., Coffey, D. C., Rifkind, R. A., Marks, P. A., Richon, V. M., and La Quaglia, M. P. (1999) *Cancer Res* 59(17), 4392–4399.
43. Mitsiades, N., Mitsiades, C. S., Richardson, P. G., McMullan, C., Poulaki, V., Fanoura-kis, G., Schlossman, R., Chauhan, D., Munshi, N. C., Hideshima, T., Richon, V. M., Marks, P. A., and Anderson, K. C. (2003) *Blood* 101(10), 4055–4062.
44. Zhang, X. D., Gillespie, S. K., Borrow, J. M., and Hersey, P. (2004) *Mol Cancer Ther* 3(4), 425–435.

45. Lucas, D. M., Davis, M. E., Parthun, M. R., Mone, A. P., Kitada, S., Cunningham, K. D., Flax, E. L., Wickham, J., Reed, J. C., Byrd, J. C., and Grever, M. R. (2004) *Leukemia* 18(7), 1207–1214.

46. Mitsiades, C. S., Mitsiades, N. S., McMullan, C. J., Poulaki, V., Shringarpure, R., Hideshima, T., Akiyama, M., Chauhan, D., Munshi, N., Gu, X., Bailey, C., Joseph, M., Libermann, T. A., Richon, V. M., Marks, P. A., and Anderson, K. C. (2004) *Proc Natl Acad Sci USA* 101(2), 540–545.

47. Peart, M. J., Smyth, G. K., van Laar, R. K., Bowtell, D. D., Richon, V. M., Marks, P. A., Holloway, A. J., and Johnstone, R. W. (2005) *Proc Natl Acad Sci USA* 102(10), 3697–3702.

48. Facchetti, F., Previdi, S., Ballarini, M., Minucci, S., Perego, P., and La Porta, C. A. (2004) *Apoptosis* 9(5), 573–582.

49. Moore, P. S., Barbi, S., Donadelli, M., Costanzo, C., Bassi, C., Palmieri, M., and Scarpa, A. (2004) *Biochim Biophys Acta* 1693(3), 167–176.

50. Duan, H., Heckman, C. A., and Boxer, L. M. (2005) *Mol Cell Biol* 25(5), 1608–1619.

51. Henderson, C., Mizzau, M., Paroni, G., Maestro, R., Schneider, C., and Brancolini, C. (2003) *J Biol Chem* 278(14), 12579–12589.

52. Zhao, Y., Tan, J., Zhuang, L., Jiang, X., Liu, E. T., and Yu, Q. (2005) *Proc Natl Acad Sci USA* 102(44), 16090–16095.

53. Tan, J., Zhuang, L., Jiang, X., Yang, K. K., Karuturi, K. M., and Yu, Q. (2006) *J Biol Chem*.

54. Zhang, Y., Adachi, M., Kawamura, R., and Imai, K. (2006) *Cell Death Differ* 13(1), 129–140.

55. Rosato, R. R., Almenara, J. A., and Grant, S. (2003) *Cancer Res* 63(13), 3637–3645.

56. Yu, C., Subler, M., Rahmani, M., Reese, E., Krystal, G., Conrad, D., Dent, P., and Grant, S. (2003) *Cancer Biol Ther* 2(5), 544–551.

57. Rahmani, M., Reese, E., Dai, Y., Bauer, C., Payne, S. G., Dent, P., Spiegel, S., and Grant, S. (2005) *Cancer Res* 65(6), 2422–2432.

58. Martirosyan, A., Leonard, S., Shi, X., Griffith, B., Gannett, P., and Strobl, J. (2006) *J Pharmacol Exp Ther*.

59. Rosato, R. R., Maggio, S. C., Almenara, J. A., Payne, S. G., Atadja, P., Spiegel, S., Dent, P., and Grant, S. (2006) *Mol Pharmacol* 69(1), 216–225.

60. Butler, L. M., Zhou, X., Xu, W. S., Scher, H. I., Rifkind, R. A., Marks, P. A., and Richon, V. M. (2002) *Proc Natl Acad Sci USA* 99(18), 11700–11705.

61. Ungerstedt, J. S., Sowa, Y., Xu, W. S., Shao, Y., Dokmanovic, M., Perez, G., Ngo, L., Holmgren, A., Jiang, X., and Marks, P. A. (2005) *Proc Natl Acad Sci USA* 102(3), 673–678.

62. Sade, H., and Sarin, A. (2004) *Cell Death Differ* 11(4), 416–423.

63. Sambucetti, L. C., Fischer, D. D., Zabludoff, S., Kwon, P. O., Chamberlin, H., Trogani, N., Xu, H., and Cohen, D. (1999) *J Biol Chem* 274(49), 34940–34947.

64. Vrana, J. A., Decker, R. H., Johnson, C. R., Wang, Z., Jarvis, W. D., Richon, V. M., Ehinger, M., Fisher, P. B., and Grant, S. (1999) *Oncogene* 18(50), 7016–7025.

65. Richon, V. M., Sandhoff, T. W., Rifkind, R. A., and Marks, P. A. (2000) *Proc Natl Acad Sci USA* 97(18), 10014–10019.

66. Sawa, H., Murakami, H., Kumagai, M., Nakasato, M., Yamauchi, S., Matsuyama, N., Tamura, Y., Satone, A., Ide, W., Hashimoto, I., and Kamada, H. (2004) *Acta Neuropathol* (Berl) 107(6), 523–531.

67. Sandor, V., Senderowicz, A., Mertins, S., Sackett, D., Sausville, E., Blagosklonny, M. V., and Bates, S. E. (2000) *Br J Cancer* 83(6), 817–825.

68. Rosato, R. R., Wang, Z., Gopalkrishnan, R. V., Fisher, P. B., and Grant, S. (2001) *Int J Oncol* 19(1), 181–191.

69. Gabrielli, B. G., Johnstone, R. W., and Saunders, N. A. (2002) *Curr Cancer Drug Targets* 2(4), 337–353.
70. Sasakawa, Y., Naoe, Y., Inoue, T., Sasakawa, T., Matsuo, M., Manda, T., and Mutoh, S. (2002) *Biochem Pharmacol* 64(7), 1079–1090.
71. Munster, P. N., Troso-Sandoval, T., Rosen, N., Rifkind, R., Marks, P. A., and Richon, V. M. (2001) *Cancer Res* 61(23), 8492–8497.
72. Klisovic, M. I., Maghraby, E. A., Parthun, M. R., Guimond, M., Sklenar, A. R., Whitman, S. P., Chan, K. K., Murphy, T., Anon, J., Archer, K. J., Rush, L. J., Plass, C., Grever, M. R., Byrd, J. C., and Marcucci, G. (2003) *Leukemia* 17(2), 350–358.
73. Grignani, F., De Matteis, S., Nervi, C., Tomassoni, L., Gelmetti, V., Cioce, M., Fanelli, M., Ruthardt, M., Ferrara, F. F., Zamir, I., Seiser, C., Lazar, M. A., Minucci, S., and Pelicci, P. G. (1998) *Nature* 391(6669), 815–818.
74. Qiu, L., Burgess, A., Fairlie, D. P., Leonard, H., Parsons, P. G., and Gabrielli, B. G. (2000) *Mol Biol Cell* 11(6), 2069–2083.
75. Burgess, A. J., Pavey, S., Warrener, R., Hunter, L. J., Piva, T. J., Musgrove, E. A., Saunders, N., Parsons, P. G., and Gabrielli, B. G. (2001) *Mol Pharmacol* 60(4), 828–837.
76. Taddei, A., Maison, C., Roche, D., and Almouzni, G. (2001) *Nat Cell Biol* 3(2), 114–120.
77. Taddei, A., Roche, D., Bickmore, W. A., and Almouzni, G. (2005) *EMBO Rep* 6(6), 520–524.
78. Williams, R. J. (2001) *Expert Opin Investig Drugs* 10(8), 1571–1573.
79. Zgouras, D., Becker, U., Loitsch, S., and Stein, J. (2004) *Biochem Biophys Res Commun* 316(3), 693–697.
80. Sasakawa, Y., Naoe, Y., Noto, T., Inoue, T., Sasakawa, T., Matsuo, M., Manda, T., and Mutoh, S. (2003) *Biochem Pharmacol* 66(6), 897–906.
81. Michaelis, M., Michaelis, U. R., Fleming, I., Suhan, T., Cinatl, J., Blaheta, R. A., Hoffmann, K., Kotchetkov, R., Busse, R., Nau, H., and Cinatl, J., Jr. (2004) *Mol Pharmacol* 65(3), 520–527.
82. Rossig, L., Li, H., Fisslthaler, B., Urbich, C., Fleming, I., Forstermann, U., Zeiher, A. M., and Dimmeler, S. (2002) *Circ Res* 91(9), 837–844.
83. Kim, M. S., Kwon, H. J., Lee, Y. M., Baek, J. H., Jang, J. E., Lee, S. W., Moon, E. J., Kim, H. S., Lee, S. K., Chung, H. Y., Kim, C. W., and Kim, K. W. (2001) *Nat Med* 7(4), 437–443.
84. Kwon, H. J., Kim, M. S., Kim, M. J., Nakajima, H., and Kim, K. W. (2002) *Int J Cancer* 97(3), 290–296.
85. Deroanne, C. F., Bonjean, K., Servotte, S., Devy, L., Colige, A., Clausse, N., Blacher, S., Verdin, E., Foidart, J. M., Nusgens, B. V., and Castronovo, V. (2002) *Oncogene* 21(3), 427–436.
86. Crazzolara, R., Johrer, K., Johnstone, R. W., Greil, R., Kofler, R., Meister, B., and Bernhard, D. (2002) *Br J Haematol* 119(4), 965–969.
87. Qian, D. Z., Kato, Y., Shabbeer, S., Wei, Y., Verheul, H. M., Salumbides, B., Sanni, T., Atadja, P., and Pili, R. (2006) *Clin Cancer Res* 12(2), 634–642.
88. Avecilla, S. T., Hattori, K., Heissig, B., Tejada, R., Liao, F., Shido, K., Jin, D. K., Dias, S., Zhang, F., Hartman, T. E., Hackett, N. R., Crystal, R. G., Witte, L., Hicklin, D. J., Bohlen, P., Eaton, D., Lyden, D., de Sauvage, F., and Rafii, S. (2004) *Nat Med* 10(1), 64–71.
89. Rossig, L., Urbich, C., Bruhl, T., Dernbach, E., Heeschen, C., Chavakis, E., Sasaki, K., Aicher, D., Diehl, F., Seeger, F., Potente, M., Aicher, A., Zanetta, L., Dejana, E., Zeiher, A. M., and Dimmeler, S. (2005) *J Exp Med* 201(11), 1825–1835.
90. Dunn, G. P., Bruce, A. T., Ikeda, H., Old, L. J., and Schreiber, R. D. (2002) *Nat Immunol* 3(11), 991–998.
91. Maeda, T., Towatari, M., Kosugi, H., and Saito, H. (2000) *Blood* 96(12), 3847–3856.

92. Magner, W. J., Kazim, A. L., Stewart, C., Romano, M. A., Catalano, G., Grande, C., Keiser, N., Santaniello, F., and Tomasi, T. B. (2000) *J Immunol* 165(12), 7017–7024.

93. Reddy, P., Maeda, Y., Hotary, K., Liu, C., Reznikov, L. L., Dinarello, C. A., and Ferrara, J. L. (2004) *Proc Natl Acad Sci USA* 101(11), 3921–3926.

94. Kruhlak, M. J., Hendzel, M. J., Fischle, W., Bertos, N. R., Hameed, S., Yang, X. J., Verdin, E., and Bazett-Jones, D. P. (2001) *J Biol Chem* 276(41), 38307–38319.

95. Cimini, D., Mattiuzzo, M., Torosantucci, L., and Degrassi, F. (2003) *Mol Biol Cell* 14(9), 3821–3833.

96. Ekwall, K., Olsson, T., Turner, B. M., Cranston, G., and Allshire, R. C. (1997) *Cell* 91(7), 1021–1032.

97. Shiloh, Y. (2003) *Nat Rev Cancer* 3(3), 155–168.

98. Bakkenist, C. J., and Kastan, M. B. (2003) *Nature* 421(6922), 499–506.

99. Camphausen, K., Burgan, W., Cerra, M., Oswald, K. A., Trepel, J. B., Lee, M. J., and Tofilon, P. J. (2004) *Cancer Res* 64(1), 316–321.

100. Zhang, Y., Jung, M., and Dritschilo, A. (2004) *Radiat Res* 161(6), 667–674.

101. Gaymes, T. J., Padua, R. A., Pla, M., Orr, S., Omidvar, N., Chomienne, C., Mufti, G. J., and Rassool, F. V. (2006) *Mol Cancer Res* 4(8), 563–573.

102. Munshi, A., Kurland, J. F., Nishikawa, T., Tanaka, T., Hobbs, M. L., Tucker, S. L., Ismail, S., Stevens, C., and Meyn, R. E. (2005) *Clin Cancer Res* 11(13), 4912–4922.

103. Subramanian, C., Opipari, A. W., Jr., Bian, X., Castle, V. P., and Kwok, R. P. (2005) *Proc Natl Acad Sci USA* 102(13), 4842–4847.

104. Yaneva, M., Li, H., Marple, T., and Hasty, P. (2005) *Nucleic Acids Res* 33(16), 5320–5330.

105. Cohen, H. Y., Lavu, S., Bitterman, K. J., KHekking, B., Imahiyerobo, T. A., Miller, C., Frye, R., Ploegh, H., Kessler, B. M., and Sinclair, D. A. (2004) *Mol Cell* 13, 627–638.

106. Lindemann, R. K., Gabrielli, B., and Johnstone, R. W. (2004) *Cell Cycle* 3(6), 779–788.

107. Brush, M. H., Guardiola, A., Connor, J. H., Yao, T. P., and Shenolikar, S. (2004) *J Biol Chem* 279(9), 7685–7691.

108. Canettieri, G., Morantte, I., Guzman, E., Asahara, H., Herzig, S., Anderson, S. D., Yates, J. R., 3rd, and Montminy, M. (2003) *Nat Struct Biol* 10(3), 175–181.

109. Whitesell, L., and Lindquist, S. L. (2005) *Nat Rev Cancer* 5(10), 761–772.

110. Budillon, A., Bruzzese, F., Di Gennaro, E., and Caraglia, M. (2005) *Curr Drug Targets* 6(3), 337–351.

111. Kovacs, J. J., Murphy, P. J., Gaillard, S., Zhao, X., Wu, J. T., Nicchitta, C. V., Yoshida, M., Toft, D. O., Pratt, W. B., and Yao, T. P. (2005) *Mol Cell* 18(5), 601–607.

112. Fuino, L., Bali, P., Wittmann, S., Donapaty, S., Guo, F., Yamaguchi, H., Wang, H. G., Atadja, P., and Bhalla, K. (2003) *Mol Cancer Ther* 2(10), 971–984.

113. Nimmanapalli, R., Fuino, L., Bali, P., Gasparetto, M., Glozak, M., Tao, J., Moscinski, L., Smith, C., Wu, J., Jove, R., Atadja, P., and Bhalla, K. (2003) *Cancer Res* 63(16), 5126–5135.

114. Bali, P., George, P., Cohen, P., Tao, J., Guo, F., Sigua, C., Vishvanath, A., Scuto, A., Annavarapu, S., Fiskus, W., Moscinski, L., Atadja, P., and Bhalla, K. (2004) *Clin Cancer Res* 10(15), 4991–4997.

115. Yu, X., Guo, Z. S., Marcu, M. G., Neckers, L., Nguyen, D. M., Chen, G. A., and Schrump, D. S. (2002) *J Natl Cancer Inst* 94(7), 504–513.

116. Olsen, E., Kim, Y. H., Kuzel, T., Pacheco, T. R., Foss, F., Parker, S., Wang, J. G., Frankel, S. R., Lis, J., and Duvic, M. (2006) *J Clin Oncol* (Meeting Abstracts) 24(18 suppl), 7500.

117. Duvic, M., Talpur, R., Ni, X., Zhang, C., Hazarika, P., Kelly, C., Chiao, J. H., Reilly, J. F., Ricker, J. L., Richon, V. N., and Frankel, S. (2006) *Blood.*

118. Piekarz, R. L., Frye, R., Turner, M., Wright, J., Allen, S., Kirschbaum, M. H., Zain, J., Prince, M., Bates, S. E., and for All Collaborators. (2005) *ASH Annu Meet Abstr* 106(11), 231.

119. Prince, H. M., George, D. J., Johnstone, R., Williams-Truax, R., Atadja, P., Zhao, C., Dugan, M., and Culver, K. (2006) *J Clin Oncol* (Meeting Abstracts) 24(18 suppl), 7501.

120. Carducci, M. A., Gilbert, J., Bowling, M. K., Noe, D., Eisenberger, M. A., Sinibaldi, V., Zabelina, Y., Chen, T. L., Grochow, L. B., and Donehower, R. C. (2001) *Clin Cancer Res* 7(10), 3047–3055.

121. Gilbert, J., Baker, S. D., Bowling, M. K., Grochow, L., Figg, W. D., Zabelina, Y., Donehower, R. C., and Carducci, M. A. (2001) *Clin Cancer Res* 7(8), 2292–2300.

122. Phuphanich, S., Baker, S. D., Grossman, S. A., Carson, K. A., Gilbert, M. R., Fisher, J. D., and Carducci, M. A. (2005) *Neurooncology* 7(2), 177–182.

123. Gore, S. D., Weng, L. J., Zhai, S., Figg, W. D., Donehower, R. C., Dover, G. J., Grever, M., Griffin, C. A., Grochow, L. B., Rowinsky, E. K., Zabalena, Y., Hawkins, A. L., Burks, K., and Miller, C. B. (2001) *Clin Cancer Res* 7(8), 2330–2339.

124. Gore, S. D., Weng, L. J., Figg, W. D., Zhai, S., Donehower, R. C., Dover, G., Grever, M. R., Griffin, C., Grochow, L. B., Hawkins, A., Burks, K., Zabelena, Y., and Miller, C. B. (2002) *Clin Cancer Res* 8(4), 963–970.

125. Kuendgen, A., Strupp, C., Aivado, M., Bernhardt, A., Hildebrandt, B., Haas, R., Germing, U., and Gattermann, N. (2004) *Blood* 104(5), 1266–1269.

126. Patnaik, A., Rowinsky, E. K., Villalona, M. A., Hammond, L. A., Britten, C. D., Siu, L. L., Goetz, A., Felton, S. A., Burton, S., Valone, F. H., and Eckhardt, S. G. (2002) *Clin Cancer Res* 8(7), 2142–2148.

127. Reid, T., Valone, F., Lipera, W., Irwin, D., Paroly, W., Natale, R., Sreedharan, S., Keer, H., Lum, B., Scappaticci, F., and Bhatnagar, A. (2004) *Lung Cancer* 45(3), 381–386.

128. Kelly, W. K., Richon, V. M., O'Connor, O., Curley, T., MacGregor-Curtelli, B., Tong, W., Klang, M., Schwartz, L., Richardson, S., Rosa, E., Drobnjak, M., Cordon-Cordo, C., Chiao, J. H., Rifkind, R., Marks, P. A., and Scher, H. (2003) *Clin Cancer Res* 9(10 Pt 1), 3578–3588.

129. Kelly, W. K., O'Connor, O. A., Krug, L. M., Chiao, J. H., Heaney, M., Curley, T., Mac-Gregore-Cortelli, B., Tong, W., Secrist, J. P., Schwartz, L., Richardson, S., Chu, E., Olgac, S., Marks, P. A., Scher, H., and Richon, V. M. (2005) *J Clin Oncol* 23(17), 3923–3931.

130. Garcia-Manero, G., Yang, H., Sanchez-Gonzalez, B., Verstovsek, S., Ferrajoli, A., Keating, M., Andreeff, M., O'Brien, S., Cortes, J., Wierda, W., Faderl, S., Koller, C., Davis, J., Morris, G., Issa, J.-P., Frankel, S. R., Richon, V., Fine, B., and Kantarjian, H. (2005) *ASH Annu Meet Abstr* 106(11), 2801.

131. O'Connor, O. A., Heaney, M. L., Schwartz, L., Richardson, S., Willim, R., MacGregor-Cortelli, B., Curly, T., Moskowitz, C., Portlock, C., Horwitz, S., Zelenetz, A. D., Frankel, S., Richon, V., Marks, P., and Kelly, W. K. (2006) *J Clin Oncol* 24(1), 166–173.

132. Krug, L. M., Curley, T., Schwartz, L., Richardson, S., Marks, P., Chiao, J., and Kelly, W. K. (2006) *Clin Lung Cancer* 7(4), 257–261.

133. Ottmann, O. G., Deangelo, D. J., Stone, R. M., Pfeifer, H., Lowenberg, B., Atadja, P., Peng, B., Scott, J. W., Dugan, M., and Sonneveld, P. (2004) *J Clin Oncol* (Meeting Abstracts) 22(14 suppl), 3024.

134. Rowinsky, E. K., Pacey, S., Patnaik, A., O'Donnell, A., Mita, M. M., Atadja, P., Peng, B., Dugan, M., Scott, J. W., and De Bono, S. (2004) *J Clin Oncol* (Meeting Abstracts) 22(14 suppl), 3022.

135. Beck, J., Fischer, T., Rowinsky, E., Huber, C., Mita, M., Atadja, P., Peng, B., Kwong, C., Dugan, M., and Patnaik, A. (2004) *J Clin Oncol* (Meeting Abstracts) 22(14 suppl), 3025.

136. Giles, F., Fischer, T., Cortes, J., Garcia-Manero, G., Beck, J., Ravandi, F., Masson, E., Rae, P., Laird, G., Sharma, S., Kantarjian, H., Dugan, M., Albitar, M., and Bhalla, K. (2006) *Clin Cancer Res* 12(15), 4628–4635.

137. Gimsing, P., Wu, F., Qian, X., Jeffers, M., Knudsen, L., Sehested, M., and Lichenstein, H. S. (2005) *ASH Annu Meet Abstr* 106(11), 3337.

138. Sandor, V., Bakke, S., Robey, R. W., Kang, M. H., Blagosklonny, M. V., Bender, J., Brooks, R., Piekarz, R. L., Tucker, E., Figg, W. D., Chan, K. K., Goldspiel, B., Fojo, A. T., Balcerzak, S. P., and Bates, S. E. (2002) *Clin Cancer Res* 8(3), 718–728.

139. Marshall, J. L., Rizvi, N., Kauh, J., Dahut, W., Figuera, M., Kang, M. H., Figg, W. D., Wainer, I., Chaissang, C., Li, M. Z., and Hawkins, M. J. (2002) *J Exp Ther Oncol* 2(6), 325–332.

140. Byrd, J. C., Marcucci, G., Parthun, M. R., Xiao, J. J., Klisovic, R. B., Moran, M., Lin, T. S., Liu, S., Sklenar, A. R., Davis, M. E., Lucas, D. M., Fischer, B., Shank, R., Tejaswi, S. L., Binkley, P., Wright, J., Chan, K. K., and Grever, M. R. (2005) *Blood* 105(3), 959–967.

141. Whittaker, S., McCulloch, W., Robak, T., Baran, E., and A. Prentice and all investigators. (2006) *J Clin Oncol* (Meeting Abstracts) 24(18 suppl), 3063.

142. Su, Y. B., Tuttle, R. M., Fury, M., Ghossein, R., Singh, B., Herman, K., Venkatraman, E. S., Stambuk, H., Robbins, R., and Pfister, D. G. (2006) *J Clin Oncol* (Meeting Abstracts) 24(18 suppl), 5554.

143. Molife, R., Patterson, S., Riggs, C., Higano, C., Stadler, W. M., Dearnaley, D., Parker, C., McCulloch, W., Shalaurov, A., and De-Bono, J. S. (2006) *J Clin Oncol* (Meeting Abstracts) 24(18 suppl), 14554.

144. Whitehead, R. P., McCoy, S., Wollner, I. S., Wong, L., Harker, W. G., Hoff, P. M., Gold, P. J., Billingsley, K. G., and Blanke, C. D. (2006) *J Clin Oncol* (Meeting Abstracts) 24(18 suppl), 3598.

145. Niesvizky, R., Ely, S., DiLiberto, M., Cho, H. J., Gelbshtein, U. Y., Jayabalan, D. S., Aggarwal, S., Gabrilove, J. L., Pearse, R. N., Pekle, K., Zafar, F., Goldberg, Z., Leonard, J. P., Wright, J. J., Chen-Kiang, S., and Coleman, M. (2005) *ASH Annu Meet Abstr* 106(11), 2574.

146. Fouladi, M., Furman, W. L., Chin, T., Freeman, B. B., III, Dudkin, L., Stewart, C. F., Krailo, M. D., Speights, R., Ingle, A. M., Houghton, P. J., Wright, J., Adamson, P. C., and Blaney, S. M. (2006) *J Clin Oncol* 24(22), 3678–3685.

147. Gore, L., Holden, S. N., Basche, M., Raj, S. K. S., Arnold, I., O'Bryant, C., Witta, S., Rohde, B., McCoy, C., and Eckhardt, S. G. (2004) *J Clin Oncol* (Meeting Abstracts) 22(14 suppl), 3026.

148. Gojo, I., Gore, S. D., Jiemjit, A., Greer, J., Tidwell, M. L., Sparreboom, A., Figg, D., Heyman, M. R., Rollins, S., Trepel, J., and Sausville, E. A. (2003) *Blood* 102(11), Abstract # 1408.

149. Hauschild, A., Trefzer, U., Garbe, C., Kaehler, K., Ugurel, S., Kiecker, F., Eigentler, T., Krissel, H., and Schadendorf, D. (2006) *J Clin Oncol* (Meeting Abstracts) 24(18 suppl), 8044.

150. Donovan, E. A., Sparreboom, A., Figg, W., Trepel, J., Maynard, K., Zwiebel, J., Melillo, G., Gutierrez, M., Doroshow, J., and Kummar, S. (2006) *J Clin Oncol* (Meeting Abstracts) 24(18 suppl), 13036.

151. Prakash, S., Foster, B. J., Meyer, M., Wozniak, A., Heilbrun, L. K., Flaherty, L., Zalupski, M., Radulovic, L., Valdivieso, M., and LoRusso, P. M. (2001) *Invest New Drugs* 19(1), 1–11.

152. Wozniak, A., O'Shaughnessy, J., Fiorica, J., and Grove, W. (1999) *J Clin Oncol* (Meeting Abstracts) Abstract # 1878.

153. O'Shaughnessy, J., Flaherty, L., Fiorica, J., and Grove, W. (1999) *J Clin Oncol* (Meeting Abstracts) Abstract # 1346).

154. Zalupski, M., O'Shaughnessy, J., Vukelja, S., Shields, A., Diener, K., and Grove, W. (2000) *J Clin Oncol* (Meeting Abstracts) Abstract # 1115.
155. Garcia-Manero, G., Minden, M., Estrov, Z., Verstovsek, S., Newsome, W. M., Reid, G., Besterman, J., Li, Z., Pearce, L., and Martell, R. (2006) *J Clin Oncol* (Meeting Abstracts) 24(18 suppl), 6500.
156. Carducci, M., Siu, L. L., Sullivan, R., Maclean, M., Kalita, A., Chen, E. X., Pili, R., Martell, R. E., Besterman, J., and Reid, G. K. (2006) *J Clin Oncol* (Meeting Abstracts) 24(18 suppl), 3007.

6 Sirtuins in Biology and Disease

Alejandro Vaquero and Danny Reinberg

CONTENTS

6.1   Introduction 74
6.2   The Sir2 Family 74
6.3   Prokaryotic Sir2 Homologs 75
6.4   Sir2 Homologs in Lower Eukaryotes 76
        6.4.1   Yeast 76
        6.4.2   *Plasmodium* 76
        6.4.3   *Trypanosoma* 77
6.5   *Drosophila* and *C. elegans* SIR2 Homologs 77
6.6   Mammalian Sirtuins 78
        6.6.1   Sirtuins and Cellular Localization 78
        6.6.2   Dual Activity of Sirtuins 79
        6.6.3   Sirtuins and H4K16Ac 79
        6.6.4   SirT1 80
                    6.6.4.1   SirT1 and Heterochromatin Regulation 80
                    6.6.4.2   SirT1 and Cell Survival 85
                    6.6.4.3   SirT1 and Metabolic Homeostasis 87
                    6.6.4.4   SirT1 and Cell Differentiation 87
        6.6.5   SirT2 88
                    6.6.5.1   SirT2 and Cytoskeleton 89
                    6.6.5.2   SirT2, H4K16Ac, and Cell Cycle Regulation 89
        6.6.6   SirT3 90
        6.6.7   SirT4 and SirT5 91
        6.6.8   SirT6 and SirT7 91
6.7   Sirtuins and Cancer 92
6.8   Sirtuins and Aging 94
6.9   Sirtuins and Neurological Diseases 96
6.10 Conclusions 96
References 97

6.1   INTRODUCTION

During the past 10 years, concerted efforts have been made to understand the mechanisms underlying chromatin dynamics and epigenetic phenomena, and these efforts have generated considerable advances in the field. This has been possible, in part, because of the identification of the factors involved and the characterization of mechanisms that alter chromatin.[1,2] Among the factors involved in the epigenetic regulation of chromatin, the Sir2 family, or sirtuins, stands out for several reasons. Sirtuins have an essential role in modulating histone H4 acetylation (H4K16Ac),[3] are sensors of the metabolic environment, and, in higher organisms, have diversified to target nonhistone proteins.[4] These features define sirtuins as major coordinators of the cellular response to environmental stimuli to promote viability of the cell/organism, including adaptation to metabolic changes and DNA damage, and activation of detoxifying machinery. As described below, the different roles these proteins exhibit throughout evolution have important medical implications, such as in the control of antigenic variation in response to invading pathogens in eukaryotes, lifespan regulation, cancer processes, neurological diseases, and hormone-related pathologies.

6.2   THE SIR2 FAMILY

The *SIR* (Silent Information Regulator) genes *(SIR1–4)* were originally identified by genetic screening in the budding yeast *Saccharomyces cerevisiae* based on their ability to rescue mating deficiency.[5] They are involved in the epigenetic silencing of three main loci: mating type loci, nucleolar rDNA, and telomeres.[3,6] Of the four *SIR2* gene products identified, only Sir2p is required for silencing all three loci, suggesting its role in these processes is critical.[7] Sir2p-dependent epigenetic silencing is associated with a compacted form of chromatin in which the N-terminal lysine residues of histones H3 and H4 are hypoacetylated.[8] Furthermore, hypoacetylation of one of these residues, H4K16, is a feature of Sir2p-dependent silencing.[7,9,10]

Although the exact role of Sir2p was unknown for a long time, several groups reported in 2000 that Sir2p is a histone deacetylase (HDAC) that requires NAD⁺ (nicotinamide adenine dinucleotide) as a cofactor for its activity. Given that NAD⁺ is involved in the transfer of electrons generated through intermediary metabolic pathways,[11,12] Sir2p appears to have a role in sensing the cell's metabolic redox state. Sir2p-dependent silencing not only affects the expression of genes located in the silenced regions, but it also has consequences for chromatin structure and genomic stability. Silencing of tandem rDNA copies (up to 100–200 copies) in the nucleolus inhibits genomic loci recombination.[13] Sir2p is involved in cell cycle control, DNA replication, meiosis progression, and double-strand DNA break (DSB) repair via a nonhomologous end-joining (NHEJ) mechanism (Figure 6.1).[7]

The Sir2 family is defined by its homology to the catalytic domain of the budding yeast factor, Sir2p, which spans approximately 250 residues. Despite the fact that all family members share an important degree of identity in this catalytic domain, they show remarkable differences in specificity and catalytic activity.[14,15] Some, such as Sir2p, seem to be strict HDACs, whereas others have developed wide-ranged specificity for nonhistone proteins.[3,4,6] Still others do not show any detectable deacetylase

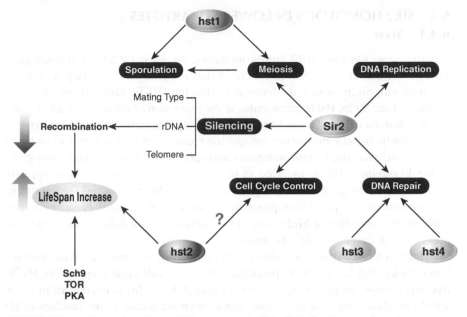

FIGURE 6.1 (See color figure following page 52.) Sir2 family members in yeast are involved in many biological processes. Hst1 is mainly linked to meiosis and regulation of the sporulation program, whereas Hst3 and Hst4 are implicated in DNA repair signaling through deacetylation of H3K56Ac. Sir2 is involved in epigenetic silencing, DNA repair, DNA replication, cell cycle control, meiosis regulation, and lifespan control through its inhibitory effect on rDNA recombination. The cellular role of Hst2 is not clear, although it might be involved in cell cycle control (indicated by "?"). An additional Sir2-independent pathway promoter of lifespan increase has been described and involves the nutrient-dependent kinases Sch9, TOR, and PKA.

activity but do display ADP-ribosyltransferase activity, an enzymatic activity present in all sirtuins (see Section 6.6.2). There are two Sir2 family members in prokaryotes, five in *S. cerevisiae*,[7] three in the fission yeast *Schizosaccharomyces pombe*,[15] four in *Caenorhabditis elegans*,[16] five in *Drosophila melanogaster*,[17] and seven in mammals.[14,15]

6.3 PROKARYOTIC SIR2 HOMOLOGS

Their prokaryotic homologs suggest that the Sir2p family evolved from phosphoribosyltransferases involved in metabolism. This explains the similarities between the catalytic mechanisms of Sir2 enzymes and mono- and poly-ADP-ribosyltransferases. One of the earliest known orthologs is CobB, a *Salmonella* enzyme that modulates acetyl-CoA synthetase (ACS) activity through NAD^+-dependent deacetylation.[18] Additionally, CobB has nicotinate mononucleotide (NaMN)-dependent phosphoribosyltransferase activity, which is important for cobalamin (vitamin B12) biosynthesis.[19] The regulation of structured forms of DNA is not a feature unique to eukaryotic SIR2 orthologs. The archaeobacteria *Sulfolobus* has a "chromatin-like" structure determined by Alba, a DNA-binding protein that organizes DNA in a stoichiometry of 5 bp DNA/Alba dimer.[20] The *Sulfolobus* Sir2 deacetylates Alba, which seems to increase its DNA-binding affinity.[21]

6.4 SIR2 HOMOLOGS IN LOWER EUKARYOTES

6.4.1 YEAST

Budding yeast has four *SIR2* family members, designated *HST1–4* (homologs of *SIR2*, 1–4), that exhibit specific functions that in some cases overlap with those involved with Sir2p-mediated silencing. Of the four HSTs, Hst1p is the closest to Sir2p.[15,22] Like Sir2p, Hst1p participates in the repression of a particular set of genes through histone deacetylation. Furthermore, Hst1p interacts with the transcription factor Sum1p, thereby promoting the specific recruitment of this factor to its target genes.[23] Hst1p has been found to repress mid-sporulation genes[23,24] and some genes of the kynureine pathway involved in de novo NAD⁺ synthesis.[25] As a repressor of mid-sporulation genes during vegetative growth, Hst1p is crucial for the proper development of the sporulation program (Figure 6.1). Hst1p activity seems to partially overlap with that of Sir2p, since its overexpression in *SIR2* mutants can suppress some of the genetic defects observed.[23]

Hst2p is located in the cytoplasm,[26] and from what we know about its mammalian ortholog SirT2, it seems to participate in mitotic cell cycle progression. Hst2p, like SirT2, shows an in vitro specificity for H4K16Ac.[27] This is in contrast to Sir2p, which, in addition to H4K16Ac, also targets other acetylated lysine residues of H3 and H4 histone tails.[27] Although not completely understood, this may reflect the tight regulation of H4K16Ac before mitotic entry in yeast cells, as this appears to be the case for SirT2.[27] Interestingly, *HST2* overexpression in *SIR2*-deficient yeast partially compensates for Sir2-dependent silencing defects, inducing the silencing of rDNA tandem copies while at the same time aggravating the derepression of subtelomeric domains upon *SIR2* loss.[26] Regarding Hst3p and Hst4p, these enzymes are involved in the DNA damage response through specific deacetylation of lysine 56 in the globular domain of histone H3.[28,29]

6.4.2 *PLASMODIUM*

The genus *Plasmodium* is responsible for malaria. Among the *Plasmodium* species, *Plasmodium falciparum* causes the deadliest form of the disease. These intracellular parasites have a complex life cycle with several stages including that of a vector (*Anopheles* mosquito) and a mammalian host.[30] One of the reasons for *Plasmodium*'s virulence is its capacity to avoid the host's immunological system. This is achieved through its alternative expression of variant antigens that avoid unequivocal identification by the host organism. The most important multigene family involved in antigenic variation is the *var* family, located in the subtelomeric regions of the *Plasmodium* genome and comprised of 60–70 genes that encode the main surface adhesin expressed on parasite-infected erythrocytes.[31] In normal blood-stage malaria, a single *var* gene is expressed at a certain moment while the rest of the *var* genes are epigenetically silenced through heterochromatin formation. A switching mechanism allows *Plasmodium* to alternatively express a given *var* gene at a given time.[32] Recent evidence suggests a key role for *P. falciparum* Sir2 protein (PfSir2) in malaria virulence through its regulation of *var* gene epigenetic silencing.[33,34] *Plasmodium* contains two members of the Sir2 family,[31] only one of which, PfSir2, has been characterized.

PfSir2 shows both histone deacetylase and ADP-ribosyltranferase activity.[35] Its loss in *Plasmodium* results in the simultaneous upregulation of many *var* genes.[33,34] All evidence suggests that PfSir2 silences *var* genes in a manner similar to that of yeast Sir2p, which silences mating type loci (i.e., through the establishment of heterochromatic regions by deacetylation of the N-termini of histones H3 and H4).[31] However, the silencing mechanism is not yet well understood and will require more in-depth study. One possibility that arises from this interesting link between the Sir2 family and malaria is that PfSir2 might be a therapeutic target. However, given the degree of conservation within the Sir2 family, the successful development of therapeutic agents would necessitate that PfSir2, and not the human sirtuins, be targeted.

6.4.3 *TRYPANOSOMA*

Trypanosomes are unicellular flagellated parasites responsible for several human diseases such as sleeping sickness, Chagas' disease, and leishmaniasis. *Tripanosoma brucei* has three Sir2 family members, designated TbSIR2RP1, TbSIR2RP2, and TbSIR2RP3.[36] TbSIR2RP1 is a nuclear protein that has both histone deacetylase and ADP-ribosyltransferase activity.[36] In contrast to *Plasmodium* sirtuins, TbSIR2RP1 does not participate in antigenic variation. Instead, it has a role in telomeric silencing and promotes the parasite's survival under genotoxic stress through an uncharacterized role in DNA repair.[36] TbSIR2RP2 and TbSIR2RP3 are mitochondrial proteins and seem to be phylogenetically related to SirT4 and SirT5.[37]

The Sir2 protein from *Leishmania infantum*, LmSir2, has also been characterized. It resides in the cytoplasm and promotes cell viability by inhibiting apoptosis.[38] Treatment with the Sir2 inhibitor, sirtinol, arrests growth during certain stages of the *Leishmania infantum* cell cycle, and sirtinol-induced cell cycle arrest can be rescued by increasing the number of copies of the LmSir2 gene.[39]

6.5 *DROSOPHILA* AND *C. ELEGANS* SIR2 HOMOLOGS

Among the four members of the *Drosophila* Sir2 family, only the closest homolog to *Saccharomyces* Sir2, DmSir2, has been studied in some depth. Different lines of evidence have supported a role for DmSir2 in heterochromatin formation and gene regulation. First, DmSir2 behaves like a suppressor of position effect variegation (PEV).[17,40] Second, DmSir2 interacts physically and genetically with Hairy and Deadpan, two transcriptional repressors that belong to the HES (Hairy/Deadpan/ Enhancer of split) family.[17] The HES family members contain a basic helix–loop– helix (bHLH) domain required for protein dimerization and DNA binding and play a basic role in different stages of *Drosophila* development by regulating the expression of certain key genes.[17,41-43] Hairy and Deadpan exert their function by recruiting repressors, such as the histone deacetylases DmSir2 and dRpd3, to the target genes. In keeping with this function, loss of DmSir2 causes upregulation of a subset of *Drosophila* genes, which includes some HES repressor targets[43] but also others such as genes involved in glycolysis.[44] Third, DmSir2 is part of a complex containing the polycomb factor E(Z), a histone methyltransferase involved in epigenetic silencing, which allows proper regulation of a spatially restricted pattern of homeotic gene expression during *Drosophila* development.[45] A considerable amount of data suggests that there is a dynamic involvement of DmSir2 in *Drosophila* development. DmSir2 mRNA levels peak in the early stages of embryogenesis then decrease progressively until stabilizing in adults. Cellular localization of the DmSir2 protein changes during the various stages of the differentiation program, being found in the nucleus, cytoplasm or both simultaneously.[17] Despite all of this, DmSir2 mutants are viable and fertile,[40,46] suggesting that there is a strong redundancy amongst Sir2 members. However, to date, this has not been completely demonstrated.

The nematode *Caenorhabditis elegans* contains four members of the Sir2 family: Sir-2.1, 2.2, 2.3, and 2.4. Sir-2.1 is the closest to DmSir2 and yeast Sir2p.[15,16] Our knowledge regarding the role of these proteins is quite limited. Most efforts have focused on understanding the relationship between Sir-2.1 and lifespan control (discussed further in Section 6.8). In this regard, transgenic organisms carrying extra copies of Sir-2.1 have an increased lifespan, an effect that seems to be dependent on the interaction between SIR-2.1 and the transcription factor DAF-16 in the insulin-like growth factor (IGF) pathway.[16] Of interest is that binding of SIR-2.1 to DAF-16 depends on the chaperone protein 14-3-3 through a mechanism not yet completely understood.[47]

6.6 MAMMALIAN SIRTUINS

Mammals harbor seven members of the Sir2 family, SirT1–SirT7, or sirtuins.[14,15] Despite being involved in a wide range of functions, their main role is to act as sensors of the redox state of the cell or organism, and their functions underlie four basic processes:[3,4,6] (1) regulation of chromatin structure and integrity, (2) promotion of cell viability under conditions of stress by inhibiting senescence and apoptosis, (3) regulation of metabolic homeostasis in close relationship with the endocrine system, and (4) development and cell differentiation.

As discussed below, sirtuin functions are sensitive to a wide range of environmental changes that may induce antagonistic roles between them. However, although mammalian sirtuins are, together with yeast Sir2, the best known of all members of the Sir2 family, we still have only a partial picture of their function overall.

6.6.1 SIRTUINS AND CELLULAR LOCALIZATION

Sirtuins exhibit a very diverse pattern of localization that includes the nucleus (SirT1, SirT6, SirT3), the cytoplasm (SirT2), mitochondria (SirT3, SirT4, SirT5), and the nucleolus (SirT7).[3,4,6,48] Some sirtuins can change their localization as a function of cell or tissue type, developmental stage, metabolic status, and certain stress conditions. This suggests that localization is important for the regulation of sirtuin function. For instance, in most cells SirT1 usually localizes to the nucleus,[48] but in certain cell types, such as pancreatic β cells, SirT1 is found in the cytoplasm.[49] In others, like rat cardiomyocytes, SirT1 localizes to the nucleus in the embryo, to the cytoplasm in newborns, and to both the nucleus and cytoplasm in adults.[50,51] In line with these observations, SirT1 has been found to relocalize from the nucleus to the cytoplasm during the differentiation of the myoblastic stable cell line C2C12.[51] This phenomenon is also conserved in SirT1 orthologs, such as DmSir2 (mentioned in

Section 6.5), which is usually located in the nucleus except during some developmental stages, such as the cellular blastoderm stage when it localizes exclusively in the cytoplasm or during the last stages of embryogenesis when DmSir2 is found in both the nucleus and the cytoplasm.[17] Also, in certain subpopulations of 293T cells, a human embryonic kidney cell line, SirT1, has been detected abundantly in the nucleolus[52] whereas in other cell lines it is either completely excluded from this cellular fraction or present in only discrete amounts.[53] Another interesting example of sirtuin dynamics is SirT3, which is present in both the mitochondria[54,55] and the nucleus.[56] Upon certain stress conditions (see below), the nuclear population of SirT3 translocates to the mitochondria. However, whether this occurs as part of a mechanism to avoid nuclear substrate deacetylation is currently unknown.

6.6.2 Dual Activity of Sirtuins

The reaction catalyzed by the Sir2 family is quite different from that of the rest of the HDACs and instead resembles the catalytic mechanism of ADP-ribosyltransferases. The reaction is accomplished basically in three stages.[57] The Sir2 enzyme (1) binds to NAD^+, then (2) catabolyzes NAD^+ in the presence of the acetylated substrate, releasing nicotinamide, with the resultant formation of the ternary ADP-ribose enzyme (acetylated substrate complex), followed by (3) transfer of the acetyl group from the substrate to ADP-ribose, releasing the product O-acetyl-ADP ribose. Much evidence suggests that O-acetyl-ADP ribose in turn performs a signaling role by binding to several factors. For instance, it binds to Sir3p in yeast, producing a conformational change that increases its capacity to recruit Sir2p.[58] Other binding partners include (1) macroH2A, a histone H2A variant involved in facultative heterochromatin formation, the structural function of which might be modulated upon binding to O-acetyl-ADP ribose,[59] and (2) the transient receptor potential melastatin-related channel 2 (TRPM2), a nonselective cation channel that is activated upon binding to O-acetyl-ADP ribose but whose function is poorly understood.[60]

The Sir2 mechanism of action allows for an alternative activity, a mono-ADP ribosyltransferase that all sirtuins seem to exhibit and is key for the function of some. Thus, in stage 2 the enzyme can transfer the ADP-ribose to a protein, including itself, instead of using it as an acetyl group acceptor.[3,5,57]

Based on what we know so far, only SirT1, SirT2, SirT3, and possibly SirT5[61] are deacetylases, whereas SirT4 and SirT6 are exclusively mono-ADP-ribosyltransferases.[62,63] In any case, it is still a matter of discussion as to whether these sirtuins are true deacetylases of targets yet to be identified.

6.6.3 Sirtuins and H4K16Ac

The NAD^+-dependent histone deacetylase activity of SirT1–SirT3 exhibits a preference for the acetylated K16 residue of histone H4 (H4K16Ac).[27,52,56] That preference is especially strong in the case of SirT2.[27] All three also exhibit a preference, albeit to a lesser extent, for H3K9Ac. However, only knockdown of SirT1 and SirT2 result in a global hyperacetylation of H4K16.[27,52] The link between sirtuins and H4K16Ac is conserved from yeast to humans.[3] H4K16 acetylation is involved in epigenetic phenomena throughout evolution, and studies in yeast suggest that it is the only residue of histone H4 targeted for acetylation that has specific functions in transcription and chromatin structure regulation.[64] Additionally, recent studies linking H4K16 hypoacetylation and cancer processes suggest a general role of this residue in proper regulation of genome integrity.[65]

6.6.4   SirT1

SirT1 is the closest homolog to yeast *SIR2* and the only sirtuin that has a clear role in silencing through heterochromatin formation. A SirT1 knockdown in mice produced smaller individuals at birth, which in most cases died during the postnatal period.[66] These mutants also showed defects in gametogenesis, sterility, eyelid opening problems, chronic lung infection, and atrophy of the pancreas.[66]

Considerable efforts during recent years have uncovered a role for SirT1 in several processes, such as chromatin regulation, survival under stress response, cell differentiation and development, regulation of metabolic homeostasis, cell cycle control, and neuronal protection (see below and Section 6.8). These functions, in turn, link SirT1 directly to cancer, diabetes, and lifespan control. The main biochemical form of SirT1 in the cell is a homotrimer,[52] but it is also found in certain multiprotein complexes like PRC4, where SirT1 colocalizes with Ezh2, the ortholog of *Drosophila* Polycomb group protein E(Z).[67] The remarkable diversity of SirT1-associated functions is achieved mainly through its direct interaction with a wide range of proteins, including transcription factors and other regulatory or structural proteins, and through its deacetylation of a myriad of histone and nonhistone protein substrates (Table 6.1).

6.6.4.1   SirT1 and Heterochromatin Regulation

SirT1 is the closest homolog to yeast *SIR2* not only from a phylogenetic point of view, but also from a functional perspective. Like *SIR2*, its function is closely related to heterochromatin formation in its two main forms, facultative and constitutive.[52,68] Facultative heterochromatin refers to those regions of chromatin where heterochromatin has been established epigenetically following a specific program or pattern, such as occurs during development, and differentiation. It can revert to open euchromatin when required.[69] This constitutes a rather local phenomenon that can span a single gene up to a whole cluster of genes. In limited cases, it may span a whole chromosome, as in the case of mammalian female X chromosome inactivation.[70] Other examples of facultative heterochromatin are developmentally related Polycomb-silenced regions in metazoans,[71] mating-type loci in yeast, and *var* gene silencing in *Plasmodium*.[31]

Constitutive heterochromatin refers to the regions of chromatin that never revert to euchromatin, spanning large portions of the chromosome and, in most cases, exhibiting a more structural function, like pericentromeric regions and telomeres.[72]

Both forms of heterochromatin are structurally quite similar, but they do show differences in certain components such as their histone modifications and other specific factors. For instance, facultative heterochromatin is dimethylated and trimethylated in H3K9 and H3K27, and monomethylated in histone H4K20. It contains facultative heterochromatin-specific proteins, such as the heterochromatin protein

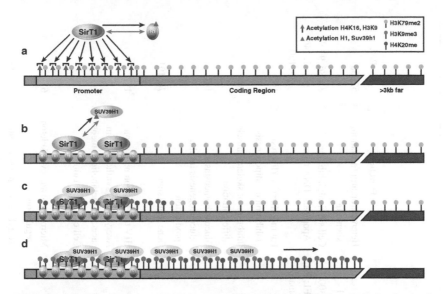

FIGURE 6.2 (See color figure following page 52.) Model of SirT1's role in facultative heterochromatin formation. SirT1 promotes the formation of facultative heterochromatin through several steps: (*a*) SirT1 arrival to chromatin is associated with deacetylation of H4K16Ac and H3K9Ac and recruitment and deacetylation of histone H1 (*red arrows* indicate interaction; *black arrows*, deacetylation). These changes are restricted to the promoter regions. (*b*) SirT1 recruits the H3K9 methyltransferase Suv39h1 (*magenta*) and induces Suv39h1 enzymatic activity (*yellow* Suv39h1 in *c, d*) through a conformational change and deacetylation of the SET domain. (*c,d*) The spreading of Suv39h1 throughout the coding region is independent of SirT1. Similar to Suv39h1-dependent spreading of H3K9me3, H4K20me also spreads from the promoter throughout the coding region, but the role of SirT1 in the arrival and spreading of this heterochromatin mark is unknown. In parallel to this heterochromatin spreading, a loss of an active mark, H3K79me2, is detected, a phenomenon that spans at least 3 kb away from the coding region.

1 isoform γ (HP1γ), and it is also enriched in the linker histone H1, involved in the establishment of high-order chromatin organization, but mainly in regulatory regions such as promoters. In contrast, constitutive heterochromatin seems to be trimethylated mainly in both H3K9 and H4K20, contains histone H1 throughout its structure, and is enriched in HP1α and, to a lesser extent, HP1β.[69] Although both forms are generally hypoacetylated in histones, there are some exceptions,[73,74] such as *Drosophila* and yeast heterochromatin, which is specifically enriched in histone H4K12Ac.[75]

6.6.4.1.1 SirT1 and Facultative Heterochromatin

SirT1 can be recruited to chromatin by transcription factors resulting in gene repression[52,76,77] through the formation of heterochromatin structures that can span several kilobases.[52] SirT1 promotes the formation of facultative heterochromatin through the coordination of different mechanisms (Figure 6.2).[52] First, SirT1 deacetylates H4K16Ac and H3K9Ac in promoter regions. Second, SirT1 recruits histone H1 through its N-terminal domain and deacetylates H1K26Ac. Third, SirT1 promotes the arrival of heterochromatin markers di- and trimethyl H3K9 and monomethylated

TABLE 6.1

Sirtuin General Features and Targets

	Localization	Ortholog Yeast	Activity	Interaction Partners	Targets	Function	Ref.
SIRT1	Nuclear	Sir2/Hst1	HDAC ADPRT	Histone H1; Suv39h1	H4K16Ac; H3K9Ac; H1K26; Suv39h1	Heterochromatin formation, silencing	52, 68
				FOXO 1, 3a, 4/p300p53; NF-κB; E2F1; Rb; Ku70; NJBS1; Smad7	FOXO 1, 3a, 4; p300; p53; NF-κB; E2F1; Rb; Ku70; NJBS1; Smad7	Cell survival; inhibition of Apoptosis; DNA repair	83–86, 90–93, 96, 97, 146, 189, 190
				PPAR-γ/ NCOR/SMRT	PPAR-γ	Fat mobilization (lipolysis); inhibition of adipogenesis	103
				PCAF/MyoD	PCAF; MyoD	Inhibition of skeletal muscle differentiation	108
				PGC-1α/4α	PGC-1α	Glucose homeostasis; gluconeogenesis insulin production	101
				CTIP2; HES1; HEY2; BCL11A	—	Gene repression	76, 191
				Tat protein (HIV)	Tat protein (HIV)	HIV transciption regulation	192
				—	AceCS1	Acetyl-CoA synthesis from acetate; metabolic homeostasis	
SIRT2	Cytoplasm/ nuclear (?)	Hst2	HDAC ADPRT	—	H4K16	Cell cycle regulation	27
				HDAC6/tubulin	α-tubulin	Microtubule and cytoskeleton regulation	61
SIRT3	Mitochondria/ nuclear	Hst2	HDAC ADPRT	—	AceCS2	Acetyl-CoA synthesis from acetate; metabolic homeostasis	131, 132

Sirtuin	Localization	Homolog	Activity	Target	Substrate	Function	Ref
					H4K16	Gene repression?	56
					Mitochondrial proteins	Mitochondrial activity?	130
SIRT4	Mitochondria	—	ADPRT		GDH	Amino acid-dependent insulin production	62
					—	thermogenesis?	61
SIRT5	Mitochondria	Hst3, Hst4?	HDAC? ADPRT?				
SIRT6	Nucleus	—	ADPRT	GCIP	SirT6 (itself)	Modulation of SirT6 ADPRT activity	63
					—	DNA repair? Tumor suppressor?	139
SIRT7	Nucleolus	—	ADPRT	RNA-polymerase I (pol I)	—	Activation of pol I transcription	140

HDAC: histone deacetylase; ADPRT: ADP-ribosyltransferase activity; ?: not demonstrated

H4K20. This initiates signaling that spreads beyond the promoter regions through the gene coding regions.

Recently, the link between SirT1 and H3K9 methylation has been uncovered.[68] SirT1 induces di-/trimethylation of H3K9 in heterochromatin regions through three simultaneous mechanisms: (1) SirT1 directly recruits Suv39h1, an H3K9 di- and trimethyltransferase involved in heterochromatin formation from *S. pombe* to humans;[69,72] (2) the N-terminus of SirT1 binds to Suv39h1 and induces its H3K9 methylation activity, most likely through a conformational change of the enzyme; and (3) SirT1 specifically targets K266Ac in the catalytic SET domain of Suv39h1. Evidence suggests that deacetylation of K266Ac renders a more active enzyme, although the role of this modification with regard to Suv39h1 function and heterochromatin formation is still not completely understood.[68] The relationship between sirtuins and K9 trimethylation is conserved through evolution. *S. pombe* Sir2 is required for proper H3K9 methylation in heterochromatin by the Suv39h1 ortholog, Clr4.[78]

Furthermore, SirT1 promotes demethylation of histone H3K79me, a modification located in the globular domain of H3 and involved in gene expression, through an unknown mechanism.[52] Of interest is that the demethylation of this residue is detected several kilobases away from the coding region (Figure 6.2).

Another fascinating aspect of SirT1 involvement in facultative heterochromatin formation is its presence in PRC4, one of the Ezh2-containing complexes.[67] The Polycomb group protein, Ezh2, is a histone methyltransferase (HMT) responsible for trimethylation of H3K27, a specific hallmark of Polycomb-dependent facultative heterochromatin, and can also target H1K26 in certain situations.[67,79]

Overall, the evidence highlights an interesting mechanistic difference between yeast Sir2p- and SirT1-mediated silencing. While Sir2p is involved in spreading silenced regions and is recruited throughout the whole silenced chromatin region,[80] the presence of SirT1 in the promoter region is sufficient for heterochromatin formation.[52] This reflects a key difference between heterochromatin in *S. cerevisiae* and higher eukaryotes. The silencing mechanism in *S. cerevisiae* relies exclusively on histone deacetylation for heterochromatin formation, as methylation of H3K9, H4K20, and H3K27 are absent. In contrast, heterochromatin in higher eukaryotes acquired additional specific modifications such as methylation along with specific methylation-reading proteins like HP1, as a consequence of their greater complexity and intricately regulated programs for development and differentiation. In addition, the role of acetylation adapted to being more regulatory than structural, as in yeast, which explains why gene acetylation is often detected in promoter regions in higher eukaryotes and does not span the whole coding region[81,82] as in yeast.

6.6.4.1.2 SirT1 and Constitutive Heterochromatin

Nuclear localization of SirT1 is restricted to euchromatic/facultative heterochromatin regions and seems to be excluded from constitutive heterochromatin.[48,52] However, analysis of SirT1 knockout mice strongly suggests an important role for SirT1 in constitutive heterochromatin regulation through Suv39h1 function.[68] Mouse embryonic fibroblasts (MEFs) derived from SirT1[-/-] mice showed a loss of H3K9me3 from heterochromatic foci in 50% of the cells analyzed. Trimethylation of H3K9 in these foci is exclusively dependent on Suv39h1 as well as its close ortholog Suv39h2 and is required to localize HP1α to heterochromatic foci.[68] Consistent with this, those SirT1$^{-/-}$ MEFs that lost H3K9me3 in these foci also lost HP1α localization in the same domains.

This phenotype is actually a direct consequence of SirT1 function, since transfection using the SirT1 expression vector of SirT1$^{-/-}$ MEF cells restored the H3K9me3 marker to the foci after 48 hours. Interestingly, transfection using either a catalytically inactive point mutant of SirT1 or SirT1 deleted of its N-terminal domain resulted in a partial restoration of marker localization, suggesting that SirT1 requires both its N-terminal domain and its catalytic domain for proper localization.[68] This also seems to be the case in facultative heterochromatin and suggests either that SirT1 might actually localize to constitutive heterochromatic regions during specific stages of the cell cycle, perhaps during replication, or that regulation of Suv39h1 activity by SirT1 is more global than anticipated. Further studies should help define the involvement of SirT1 in these processes and its relationship with Suv39h1.

6.6.4.2   SirT1 and Cell Survival

One of the most interesting functions of SirT1 links it directly to lifespan control and cancer, and that is its ability to promote cell survival through different mechanisms (Figure 6.3A). The most studied of these mechanisms is the modulation of FOXO factors involved in the cellular response to stress.[83-85] Upon oxidative or genotoxic stress, the forkhead-box (FOXO) family of transcription factors moves from cytoplasm to nucleus and activates a set of genes that encode for a wide range of proteins, from DNA repair and detoxifying enzymes to inducers of cell cycle arrest (p27KIP1) and apoptosis (BIM).[86] Acetylation of FOXO proteins is important for the modulation of their activity.[87] Under conditions of stress, SirT1 binds to the FOXO transcription factors and deacetylates them. In the case of FOXO-1 and -4, SirT1 downregulates their activity, while in contrast, SirT1 deacetylation of FOXO-3 has a dual effect on its transcriptional activity, promoting the upregulation of DNA repair and the expression/activity of detoxifying enzymes while inhibiting cell cycle arrest and apoptosis-related genes.[88] A very similar case is observed with the transcription factor NF-κB, which affects the expression of multiple genes involved in, among other things, cell cycle, apoptosis, cell adhesion, and angiogenesis.[89] SirT1-mediated deacetylation of the RelA/p65 subunit of NF-κB represses the transcriptional activation of its target genes, inducing cell survival.[90]

Another factor targeted by SirT1 in response to stress is the tumor suppressor p53. SirT1 interacts with and deacetylates p53, which results in the inhibition of p53-dependent apoptotic activity.[91,92] Although SirT1$^{-/-}$ mice show p53 hyperacetylation,[93] there is no evidence of p53-dependent defects,[94] suggesting that the functional relevance of p53 deacetylation might be more complicated than previously anticipated.

Another interesting target of SirT1 is Ku70, which is involved in the repair of DSBs through NHEJ. Ku70 also inhibits apoptosis by binding Bax, a protein involved in the regulation of stress-induced apoptosis, thereby sequestering it from the mitochondria.[95] This binding is abolished by Ku70 acetylation. Under stress, SirT1 interacts with and deacetylates Ku70, inhibiting Bax-dependent apoptosis and inducing Ku70-dependent DNA repair of DSBs.[96] This is similar to the case of

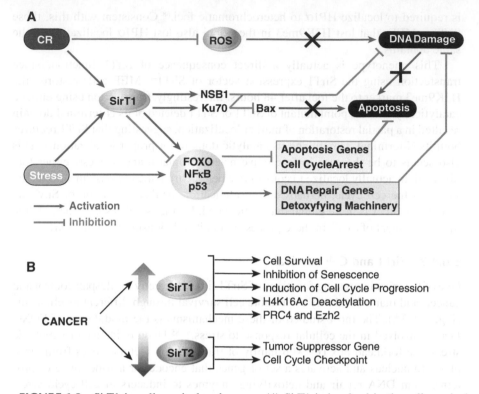

FIGURE 6.3 SirT1 in cell survival and cancer. (*A*) SirT1 is involved in the cell survival response to genotoxic and oxidative stress by interaction with and deacetylation of several factors (FOXO, NF-κB, p53, and others) that induce apoptosis inhibition, cell proliferation, DNA repair, and expression of detoxifying proteins. This modulatory effect is produced by either activation (FOXO1, 4) or inhibition (FOXO3a, NF-κB, p53) of the transcriptional activity of these factors. SirT1 also directly deacetylates factors of the DNA repair machinery (NBS1, Ku70). Caloric restriction (CR) upregulates SirT1 and therefore may activate this response in the absence of stress. This may constitute a protective mechanism and induce an increase in lifespan. (*B*) Of all the sirtuins, SirT1 and SirT2 are more clearly involved in cancer through antagonistic roles. SirT1 promotes cell viability (see *A*) and cancer progression by different mechanisms, some of which are not completely understood yet. SirT2 has a tumor suppressor gene effect through a checkpoint role in cell cycle progression not yet characterized.

Nijmegen breakage syndrome 1 (NBS1), a checkpoint factor that is involved in the detection and activation of DNA repair. NBS1 acetylation decreases cell survival, whereas deacetylation by SirT1 seems to promote it.[97] However, under certain conditions, such as chronic oxidative stress, extensive DNA damage, or acute oncogene overexpression, SirT1 does not promote cell survival but instead inhibits cell cycle progression and induces replicative senescence. This suggests that, under the conditions mentioned, other control or checkpoint signaling may target SirT1.[98]

6.6.4.3 SirT1 and Metabolic Homeostasis

The role of SirT1 as a metabolic sensor not only is important at the cellular level, but is key for the proper regulation of glucose homeostasis in the organism through endocrine signaling pathways.[3,6] SirT1 levels are upregulated upon nutrient deprivation or fasting, and this promotes a systemic coordinated response that involves at least three different organs.[99] In the liver, SirT1 deacetylates PGC-1α, a coactivator of the transcription factor HFN4α that is essential in the induction of gluconeogenesis (GNG).[100] SirT1 also positively modulates the inhibitory effect of PGC-1α in glycolysis during fasting conditions.[101] In agreement with these observations, high levels of glucose have been found to downregulate SirT1 levels in the liver.[101]

At the same time, SirT1 promotes insulin production by pancreatic β cells through the inhibition of uncoupling protein 2 (UCP2). UCP2 is involved in thermogenesis, controlling energy production by decoupling the electron transport chain (ETC) from ATP production.[49,102] SirT1 is located in the cytoplasm in these cells, whereas UCP2 is mitochondrial. Thus, the mechanism involved is not likely to be a direct one. Downregulation of UCP2 results in decreased ATP levels due to the ETC and increased insulin production.[49,99,102] Therefore, SirT1 not only induces de novo glucose synthesis by GNG in the liver, but also fosters glucose uptake in all tissues through increased insulin levels, a priority under fasting conditions. At the same time, SirT1 interferes with the response to insulin by white adipose tissue (WAT). Normally insulin would upregulate the processes of fat storage (adipogenesis) and WAT differentiation.[99] However, under fasting conditions, SirT1 impedes the transcriptional activator that is key to these processes, the peroxisome proliferator-activated receptor γ (PPARγ)[103] by recruiting the corepressors NCoR and SMRT to PPARγ-regulated genes.

In agreement with its role in systemic metabolic regulation and the endocrine system, SirT1 participates in the regulation of several hormone-signaling pathways in addition to insulin. SirT1 function is intimately related to that of the IGF pathway.[104] Moreover, SirT1 interacts with specific nuclear hormone receptors, such as androgen[105] or glucocorticoid receptors, which modulates their downstream signaling.[106] This suggests a wider role for SirT1 as a participant in the endocrine system response during metabolically compromised situations.[99]

6.6.4.4 SirT1 and Cell Differentiation

Certain tissues are highly responsive to the metabolic status of the organism. Such is the case for skeletal muscle and adipose tissue. In the case of muscle, its energetic requirements make it very sensitive to fluctuations reflected by the NAD^+/NADH ratio. In fact, skeletal muscle is the main tissue to uptake glucose in response to insulin and is particularly important in the regulation of systemic glucose levels in situations of diet or fasting.[107] Similar to the case of WAT, SirT1 inhibits muscle differentiation from myoblasts to myotubes through the silencing of certain muscle-specific genes such as *myogenin* or *MHC*.[108] SirT1 is recruited to these target genes through direct interaction with a complex formed by the transcription factor MyoD and the histone acetyltransferase (HAT) PCAF (p300/CBP associating factor). In normal situations, MyoD is acetylated by PCAF, promoting transcriptional activation of the muscle-specific genes.[109] In contrast, SirT1 deacetylates both PCAF and MyoD as well as histones in the promoter regions of their target genes, shutting down their expression. Interestingly, results from RNAi studies on cultured myoblasts suggest that SirT1 downregulation promotes a high expression of muscle-specific genes and differentiation.[108]

Thus when SirT1 is activated under energy-compromising conditions, it promotes the efficient use of available energy by inhibiting the differentiation of both adipose tissue and skeletal muscle, both avid consumers of blood glucose.[103,108] Another interesting aspect of SirT1's function in cellular differentiation comes from mouse studies. The observation that SirT1 levels are elevated in undifferentiated embryonic stem (ES) cells but decrease progressively throughout differentiation[67] suggests a general antagonistic relationship between SirT1 and cellular differentiation. On the other hand, both male and female SirT1$^{-/-}$ knockouts exhibit depletion of differentiating germ cells,[66] which suggests that SirT1 may be positively involved in differentiation in certain tissues.

6.6.5 SirT2

SirT2 is the eukaryotic ortholog of yeast *HST2*, and it is one of the few sirtuins that is clearly conserved through evolution from *S. cerevisiae* to humans.[14,15] Like Hst2p, SirT2 is cytoplasmic,[26] except during the G$_2$/M transition, when it is shuttled to nuclear chromatin.[27,110] A phosphorylation-dependent nuclear export mechanism has recently been defined for SirT2,[110-112] which suggests that SirT2 is actively kept out of the nucleus, except during the G$_2$/M transition, and that small amounts of SirT2 may temporally localize to the nucleus during certain conditions, as has been suggested for SirT2 and Hst2p.[113,114]

Although additional SirT2 functions have been reported (see below), current evidence strongly suggests that its main role is in cell cycle regulation. First, SirT2 overexpression delays mitosis, with resultant shortening of the G$_1$ phase.[110,113,115] Conversely, MEFs derived from SirT2$^{-/-}$ mice have a longer G$_1$ and a shorter S phase.[27] Second, overexpression of both SirT2 and Hst2 in starfish oocytes can delay cell division.[116] Third, SirT2 localization and protein levels seem to be regulated throughout the cell cycle through the phosphorylation state of SirT2 residues S368 and S372 mediated by cell-cycle-dependent kinases and phosphatases.[110,117,118]

In contrast to the general role of SirT1 in promoting cell survival, SirT2 seems to promote G$_2$/M arrest and cell death during stress or uncontrolled growth.[115,119] Interestingly, the overexpression of SirT2 seems more detrimental to cell viability than its loss in both mice (SirT2$^{-/-}$) and cultured cells.[27,113] This suggests that tight control of SirT2 protein levels may be key for its function and that there might exist an important redundancy for at least part of SirT2 functions.

Chromatin and the cytoskeleton have been defined as the main targets of SirT2 in the context of cell cycle regulation. However, at present the contribution of each target to the observed phenotypes and whether there are more SirT2 targets that account for part of the observed effects are largely unknown.

6.6.5.1 SirT2 and Cytoskeleton

The first function attributed to SirT2 was as regulator of microtubule organization involving deacetylation of α-tubulin through an interaction with the tubulin deacetylase HDAC6.[61] Tubulin acetylation was reported to stabilize microtubules after assembly,[120] whereas deacetylation was expected to negatively alter microtubule organization. In agreement with its role in microtubule dynamics, RNAi against SirT2 resulted in tubulin hyperacetylation in the stable 293 cell line.[61] Furthermore, SirT2 overexpression was found to induce microtubule network reorganization in glial cells.[115] However, whether these observations reflect a direct or indirect effect by SirT2 on tubulin and microtubules in vivo is currently not clear. In fact, unlike HDAC6, neither SirT2[-/-] mice nor MEFs derived from them show any defect in microtubule organization or acetyl-tubulin levels.[121] This suggests that either there is an important functional redundancy or SirT2 might target tubulin during very specific situations.

6.6.5.2 SirT2, H4K16Ac, and Cell Cycle Regulation

In contrast to SirT1, SirT2 does not participate directly in heterochromatin formation. Given its cytoplasmic localization, the targets of SirT2 were long assumed to be cytoplasmic as well. Indeed this seemed to be borne out when tubulin was identified as a SirT2 target. However, tubulin acetylation does not exist in lower eukaryotes,[122] and in light of SirT2 homology with yeast *HST2*, it seemed that there must be a more basic conserved function. In fact, despite the cytoplasmic localization of these two proteins, both SirT2 and Hst2p specifically target H4K16Ac and regulate global levels of this modification in vivo.[27] RNAi against SirT2 causes a global increase in H4K16Ac and yeast strains deficient in Hst2p (*hst2Δ*) display global H4K16 hyperacetylation.[27] In contrast, despite their homology/conservation, *sir2Δ* or *hst1Δ* mutant strains showed only very mild changes in global H4K16 hyperacetylation. This suggests that while Sir2 and Hst1 target H4K16Ac, their effect is local, whereas Hst2p and SirT2 regulate the global levels of the modification. H4K16Ac deacetylation specifically occurs during the G_2/M transition when cytoplasmic SirT2 is shuttled to the nucleus. SirT2[-/-] MEFs show consistently higher levels of H4K16Ac during mitosis, whereas SirT1[-/-] MEFs show no such change.[27] Despite that SirT2 overexpression causes delays in mitosis,[110,113,115] SirT2[-/-] MEFs do not show a clear change in mitosis duration. However, they do exhibit a delay in G_1 progression and a shortening of the S phase, suggesting that these cells encounter some problems when entering S-phase.[27] Although it is still unknown whether this phenotype is directly due to higher levels of H4K16Ac in the absence of SirT2 function, the facts that H4K16Ac levels peak in S phase and that this modification has been implicated in plant replication[123] as well as in yeast and in mammalian DNA repair[124,125] suggest this to be the case. Further work should elucidate the link between SirT2 function, H4K16 hyperacetylation, and cell cycle control.

SirT2 has also been recently linked to the inhibition of adipocyte and neural oligodendroglial differentiation.[126,127] Another interesting but not understood role in neurodegenerative diseases has been described (see Section 6.9).

6.6.6 SirT3

Human SirT3 is closely related to SirT2[15] and is the only sirtuin clearly linked to lifespan increase.[128,129] SirT3 localizes to the mitochondria after the first 142 residues of its N-terminus are proteolytically cleaved, rendering a 25-kDa protein from the 44-kDa full-length protein.[54,55] Its cellular role seems to be intimately linked to metabolism and mitochondrial function. For instance, its levels are upregulated by caloric restriction (CR) and cold exposure. SirT3 knockout mice show hyperacetylation of numerous mitochondrial proteins, suggesting that SirT3 might be a global deacetylase in this organelle.[130] However, only one substrate has been identified—the mitochondrial enzyme acetyl-CoA synthetase 2 (AceCS2). By deacetylating AceCS2 at Lys-642, SirT3 activates the synthetase, which then generates AcCoA from acetate thereby increasing the mitochondrial metabolic rate.[131,132] Recent studies suggest that, together with SirT4, SirT3 is involved in protecting cells from cell death induced by genotoxic agents such as methylmethane sulfonate (MMS). This function is dependent on the enzyme nicotinamide phosphorybosiltransferase (Nampt),[133] which, during stress and CR, can produce high levels of NAD^+ in the mitochondria; conditions expected to activate SirT3. The case of SirT4 is more complicated, since some evidence suggests it is downregulated by CR. In any case, the mechanism through which both enzymes may participate in these processes is completely unknown and will require further study.

Although SirT3 is expressed in numerous tissues in mice, its levels are particularly high in the brain, the kidneys, and brown adipose tissue (BAT).[134] SirT3 is actually poorly expressed in WAT, although CR upregulates SirT3 RNA levels in both brown and white adipose tissues.[134] In BAT, constitutive SirT3 expression induces the expression of certain mitochondrial factors, such as the transcription factor PGC1α, ATP synthetase, subunits of the cytochrome c oxidase complex, and the uncoupling protein 1 (UCP1), which, like UCP2, is involved in thermogenesis control.[134] Consistent with this, SirT3 overexpression in BAT increases cellular respiration and decreases mitochondrial membrane potential. In agreement with a functional link between CR and SirT3, the BAT of obese mice shows a significant drop in SirT3 RNA levels and in mitochondrial gene expression.[134]

Recent reports have shown that foci of unprocessed, full-length SirT3 exist in the nucleus, but, to date, they remain uncharacterized.[56] Under normal growth conditions, both nuclear and mitochondrial forms of SirT3 coexist in the cell. However, upon conditions eliciting cellular stress, such as DNA damage or SirT3 overexpression, the entire nuclear population can relocate to the mitochondria within 2 hours, via a leptomycin A-sensitive shuttling mechanism.[56] Nuclear SirT3 has been hypothesized to have a role in specific gene repression based on two observations. First, like SirT2, SirT3 has strong specificity for H4K16Ac.[27,56] Second, transient transfection experiments showed that a fusion protein containing full-length SirT3 and the Gal4 binding domain could deacetylate H4K16Ac and H3K9Ac on the promoter of a Gal4-dependent luciferase reporter gene, resulting in its repression.[56] Consistent with a role in the regulation of specific genes, a SirT3 knockdown, in contrast to a SirT2 knockdown, does not result in H4K16 hyperacetylation.[56] Another interesting possibility, not necessarily opposed to the previous one, is based on the observation that SirT3 overexpression causes cellular stress comparable to that of DNA damage. Perhaps unprocessed SirT3 nuclear foci serve as a reservoir to be used when required. Overall, the data suggests that, like SirT2 levels, SirT3 levels must be tightly regulated.

6.6.7 SiRT4 AND SiRT5

The other mitochondrial sirtuins, SirT4 and SirtT5, are phylogenetically related to the prokaryotic family members. Considerably less is known about SirT4 and SirT5 relative to SirT3, with SirT5 being virtually uncharacterized. Besides a weak deacetylase activity detected in vitro, additional targets/functions have not been identified for SirT5 to date.[61] SirT4, however, is known to be involved in the regulation of intermediary metabolism and insulin production in pancreatic β cells.[62] In contrast to SirT3, SirT4 does not show any detectable deacetylase activity, but instead shows strong ADP-ribosylase activity[130] that targets mitochondrial proteins. The only SirT4 target identified so far is glutamate dehydrogenase (GDH), an oxidoreductase involved in the catabolism of glutamic acid and glutamine. GDH catalyzes the conversion of glutamate to α-ketoglutarate, a citric acid cycle product. ADP-ribosylation inhibits GDH activity, depriving β cells of utilizing the amino acid pool for energetic purposes, which in turn inhibits amino acid-dependent insulin production during normal diet conditions.[62] Supporting evidence of SirT4's role in intermediary metabolism and β cell insulin production comes from SirT4 knockout mice that exhibit high GDH activity and glutamate-/glutamine-dependent insulin production. Contrary to what would be expected, CR increases GDH activity,[135] suggesting that unlike SirT1 and SirT3 CR negatively regulates SirT4 activity. Thus, under normal energetic conditions, SirT4 suppresses amino acid-dependent insulin production. If so, energetically compromised conditions would inhibit SirT4, which in turn would shift the energy source from that of carbohydrates to that of amino acids.[62] SirT4 also inhibits glucose-dependent insulin synthesis, but via an unknown mechanism. Interestingly, SirT1 stimulates glucose-dependent insulin production[49,102] and is positively regulated by CR,[96] suggesting an opposing role between SirT1 and SirT4 in this context. Further studies will be required to elucidate whether both sirtuins regulate insulin production through common or completely unrelated pathways.

6.6.8 SiRT6 AND SiRT7

SirT6 and SirT7 are highly homologous sirtuins present in different nucleolar compartments—SirT6 localizes to the nucleoplasm, whereas SirT7 localizes exclusively to the nucleolus. Both are ADP-ribosyl-transferases with no detectable deacetylase activity.[48,61,63] With the exception of ADP-ribosylation, which is a putative autoregulatory activity exhibited by SirT6, no clear targets have yet been identified for either enzyme.[63] SirT6 is a chromatin-bound protein that maintains genomic integrity through an undefined role during the base excision repair (BER) pathway, a mechanism that eliminates single-stranded breaks in DNA produced by different conditions like genotoxic and oxidative stress.[136] SirT6 is ubiquitously expressed from development through to adulthood. SirT6$^{-/-}$ mice exhibit higher sensitivity to ionizing radiations and both genotoxic and oxidative stress, and they show notable genomic instability resulting in metabolic defects and aberrant aging, a phenotype also ascribed to defects in the components of other DNA-repair pathways.[137,138] However, much evidence suggests that the effect of SirT6 on genomic integrity is indirect and that it might have additional, unidentified roles. First, in contrast to knockdowns of other BER factors, SirT6[-/-] deletion leads to lethality in mice.[138] Second, SirT6[-/-]-derived extracts can perform BER on naked DNA substrates in vitro to the same extent as wild type.[137] Third, DNA damage to SirT6[-/-]-derived MEFs causes the appearance of characteristic BER foci. Unlike the BER machinery under DNA-damaging conditions, SirT6 is diffused in the nucleoplasm.[137] Fourth, SirT6 has been found to interact with the factor GCIP (also known as CCNDBP1), a negative regulator of cell proliferation and a putative tumor suppressor.[139] The role of SirT6 in GCIP function is completely unknown but is consistent with that of a "guardian," poised to stimulate DNA repair or proliferation arrest whenever the cellular genomic integrity is compromised.

Nucleolar SirT7 is peculiar in that its role is unexpected, given its physical proximity to actively transcribed rDNA repeats as well as its activation of, and interaction with, RNA polymerase I (Pol I).[140] This latter stimulatory effect is mediated by direct interaction with Pol I and requires SirT7 catalytic activity. Yet Pol I does not seem to be targeted by SirT7 and no other target for its activity has been identified. SirT7 functions to stabilize Pol I binding to rDNA copies, promoting ribosomal gene expression. In keeping with this, loss of SirT7 causes both a decrease in the amount of Pol I bound to rDNA regions and a decrease in overall Pol I transcription, resulting in the inhibition of cell proliferation and apoptosis.[140] As mentioned above, SirT7 is not the only nucleolar sirtuin. Low, but detectable amounts of SirT1 localize to silenced rDNA regions due to recruitment by DNA methyltransferase 1 (Dnmt1).[53] Accordingly, RNAi against SirT1 increases the levels of H4K16Ac present in rDNA copies and promotes nucleolus reorganization. In all, the findings suggest that SirT1 and SirT7 have opposing functions when it comes to the regulation of rDNA expression. Whether they antagonize each other in this context is unknown and will require further investigation.

6.7 SIRTUINS AND CANCER

That the activity of sirtuins depends upon the cellular metabolic redox state and that active sirtuins inhibit senescence implicates them in the regulation of cellular proliferation and, by extension, cancer. Sirtuins are categorized into three groups depending on whether their levels in tumor cells are upregulated, downregulated, or unchanged.

SirT1, SirT3, SirT4, and SirT7 fall under the upregulated category and are thought to play a positive role in tumor formation. Within this group, SirT1 is most clearly linked to cancer due to its general role in promoting cell viability during DNA damage, cell cycle control defects, or oxidative stress (Figure 6.3B). SirT1 is overexpressed in many types of cancers, including pancreatic, breast, epithelial, skin, and lung cancer.[67,77,141,142] This overexpression is not completely understood, but it seems to occur at the level of transcription and mRNA stability. In the former case, SirT1 gene expression is tightly regulated by multiple factors known to play a role in cancer.

The best characterized among these factors is the tumor suppressor Hic1,[143] which inhibits SirT1 expression by directly binding the promoter.[144] Conversely, loss of Hic1 is associated with increased SirT1 levels.[144] Another SirT1 transcriptional repressor is Aiolos, a member of the Ikaros family that is involved in B-cell differentiation.[145] Aiolos seems to act as a tumor suppressor since its loss is associated with lymphoma development. Regarding SirT1 activation, the E2F1 transcription factor involved in cell cycle control and apoptosis regulation, promotes SirT1 expression during DNA damage. Interestingly, SirT1 binds to E2F1 and inhibits its proapoptotic activity.[146]

The second mechanism driving SirT1 upregulation is mRNA stabilization mediated by direct interaction between the RNA-binding protein HuR and SirT1 mRNA. This interaction is inhibited when HuR is phosphorylated by the cell cycle checkpoint kinase, Chk2,[147] implying a checkpoint-dependent control of SirT1 levels.

The mechanisms triggering SirT1 overexpression are not completely understood. Nevertheless, one obvious prediction is that in certain cases of tumor formation, hypermethylation of the Hic1 promoter results in SirT1 overproduction.[144] Another relevant finding is that SirT1 associates with the Polycomb group protein Ezh2, specifically in the PRC complex designated PRC4.[67] Overexpression of several PRC4 complex components, such as Ezh2, Suz12, Eed, and SirT1, has been reported in colon, pancreatic, and breast cancer tissue.[67,148,149] Interestingly, overexpression of Ezh2 results in the overexpression of all the other PRC4 components through an unknown mechanism. This suggests that SirT1 protein upregulation might also occur by indirect mechanisms in certain types of cancer.[67]

SirT1 might contribute to cancer progression through its inhibition of senescence and its promotion of viability under genotoxic or oxidative stress (see Section 6.6.4.2). Adding to this complexity is the close relationship between SirT1 and cell cycle progression through FOXO and other factors. Some evidence suggests that the levels of SirT1 protein, but apparently not SirT1 mRNA levels, decrease upon serial cell passage and correlate with an increase in senescence phenotypes.[150] In fact, SirT1 levels are low in MEFs that exhibit a premature senescence phenotype but are high in MEFs with delayed senescence.[150] In cells exiting the cell cycle, SirT1 levels decrease dramatically and increase only when the cells resume it. In agreement with this, SirT1 levels decrease with age in mouse tissues like testes or thymus in which mitotic activity decreases with age, but not in tissues in which mitotic activity remains unchanged, such as the brain.[150] The mechanism underlying these observations is unknown but suggests a direct role of SirT1 in cell cycle maintenance.

SirT1 is also involved in cancer through chromatin modification. This has been shown at two different levels. First, SirT1 is responsible for the direct silencing of certain tumor suppressor genes (TSG). For instance, SirT1 is detected in tumor suppressor promoter regions, which correlates with H4K16 and H3K9 hypoacetylation. Its absence from the generally hypermethylated promoters of these genes, reactivates them without affecting DNA methylation levels.[77] Second, cancer phenomena have been associated with loss of global H4K16Ac and H4K20me3, which suggests that a sirtuin may be involved.[65] The facts that SirT1 and SirT2 are the only sirtuins whose knockdowns produce global H4K16 hyperacetylation[27,52] and that SirT1 is specifically upregulated in tumors[67] suggest that SirT1's effect on chromatin could be involved in regulating cancer.

SirT3, SirT4, and SirT7 have also been reported as being overexpressed in certain types of cancers.[151-153] While SirT7 is upregulated in breast and thyroid cancers,[151,153] SirT3 is upregulated in node-positive breast cancer, and SirT4 is overexpressed in acute myeloid leukemia (AML).[152] However, whether the upregulation of these other sirtuins is directly related to the cancers per se or is just an indirect effect is not known. One could speculate that the increased proliferative activity of tumor cells associated with cancer processes is coupled with increased mitochondrial metabolic activity (SirT3 and maybe SirT4) and ribosomal gene expression (SirT7). However, more work is needed to elucidate the actual role of these sirtuins in cancer.

So far only one sirtuin, SirT2, has been reported to be downregulated in cancer. This sirtuin is downregulated in gliomas and gastric carcinomas.[115,119] Meanwhile, in melanomas, SirT2 has been found to carry a P199L mutation in its catalytic domain that eliminates its enzymatic activity.[154] Overall, the evidence points to a tumor suppressor role for SirT2. This is supported by studies showing that in situations of uncontrolled proliferation or mitotic stress, SirT2 overexpression induces cell cycle arrest before mitotic entry (Figure 6.3B).[119]

The third category refers to those sirtuins that have not been clearly linked to cancer, SirT5 and SirT6. In the case of SirT6, a direct interaction with GCIP, a candidate tumor suppressor, has been reported.[139] The gene encoding GCIP is located in a chromosome region frequently deleted in breast, colon, and prostate cancer.[155] Overexpression of GCIP inhibits the proliferation of tumorigenesis.[155] This evidence, together with the role of SirT6 as a guardian of genomic integrity,[136] suggests that it has much in common with SirT2. However, further work is needed to understand the significance of SirT6 interaction with GCIP as well as the possible role of SirT6 in cancer.

6.8   SIRTUINS AND AGING

Aging is associated with a group of degenerative diseases and with cancer. Different factors have been shown to determine lifespan increase in several organisms. Some examples are the IGF and growth hormone (GH) pathways, telomere length, CR, and reactive oxygen species (ROS).[156-160]

The mechanism best known to delay aging is mitochondrial ROS reduction. Reactive oxygen species are a group of molecules generated in the electron transport chain that very toxic to a cell because, among other things, they cause DNA damage and lipid peroxidation.[157] These oxidants can be decreased by either reducing the cellular metabolism rate, as during CR,[156] or expressing detoxifying enzymes to eliminate them. This close link between metabolism and lifespan control suggests a role for metabolic sensors such as sirtuins in lifespan mechanisms. This premise has actually been demonstrated in yeast,[161] C. elegans,[16] and Drosophila.[162] Moreover, some evidence suggests that in mammals, sirtuins might also be involved in lifespan mechanisms.[156,163]

In yeast, extra copies of SIR2 increase replicative lifespan, while a SIR2 deletion causes the opposite effect.[4,7,13,156] Replicative lifespan refers to the number of divisions a cell undergoes during its lifetime, while chronological lifespan refers to the length of time a cell lives.[164] Sir2p seems to affect lifespan by inhibiting the accumulation of extrachromosomal rDNA circles (ERC) produced from the recombination events of nucleolar rDNA copies. ERCs seem to induce replicative senescence in the mother cell.[13,165,166] Sir2p-mediated silencing of rDNA copies inhibits recombination and therefore ERC formation, whereas a *SIR2* deletion induces the opposite effect (Figure 6.1). Hst2p has also been reported to promote lifespan increase through a currently uncharacterized Sir2p-independent mechanism.[167]

Drosophila Sir2 also promotes lifespan increase through an undefined pathway that seems to be repressed by the histone deacetylase Rpd3.[162] As discussed above, *C. elegans* SIR-2.1 increases lifespan via an IGF pathway-dependent mechanism and, in particular, through its interaction with the FOXO ortholog DAF-16.[16] Under normal conditions, DAF-16 is cytoplasmic, but during stress it becomes phosphorylated and is transported to the nucleus. SIR-2.1 binds to phosphorylated DAF-16 in a 14-3-3-dependent manner and promotes expression of DAF-16 target genes, thereby inducing stress resistance and senescence inhibition.[47]

In mammals, SirT1 is also closely involved in the regulation of the IGF pathway.[104] Findings presented in Section 6.6.3.2 suggest that SirT1 could induce lifespan increase in mammals in a manner similar to that of the *C. elegans* SIR-2.1 gene.[16] However, this remains to be demonstrated. Another mammalian sirtuin that has been clearly linked to survival in older humans is SirT3.[128,129] A polymorphism residing in intron 5 of SirT3 has been found to enhance SirT3 expression. Interestingly, SirT3 alleles lacking this enhancer are completely absent in human males 90 years or older, suggesting that increased SirT3 expression might be required for a long life.[129]

A controversial aspect of the role of *SIR2* in aging is whether CR-dependent lifespan, shown to occur from yeast to mammals, is actually mediated by *SIR2* homologs.[164] The rationale behind this hypothesis is that CR increases NAD^+ levels that in turn have been shown to activate sirtuin activity. Although some initial studies support this idea, it is currently controversial.[164]

In yeast, CR increases both replicative and chronological lifespans, whereas Sir2p and probably Hst2p affect only replicative lifespan.[164,167] This suggests that the effect of CR on cellular lifespan may be mediated via pathways besides those involving Sir2p. In fact, a Sir2p-independent lifespan mechanism has been proposed and involves CR-induced inhibition of the nutrient-dependent kinases TOR, Sch9p, and PKA.[168] Interestingly, TOR and Sch9 knockouts increase lifespan in *C. elegans* and *Drosophila*.[169–171]

This link to Sir2 members has been demonstrated only in *Drosophila*.[162] SIR-2.1 is clearly not involved in CR-induced lifespan increases in *C. elegans*,[172,173] while in mammals some evidence links SirT1 with CR. In addition to its role in cell survival mechanisms, SirT1 is overexpressed in the brain, fat, and liver of rodents during CR.[96,101] Moreover, the IGF signaling pathway, in which SirT1 function is implied, is involved in mammalian lifespan increase.[174] Finally, loss of SirT1 in mice eliminates the CR-induced increase in physical activity, strongly suggesting that, at least in part, the effects produced by CR in mammals are dependent on sirtuins.[175] Given its metabolic function in the mitochondria and its link to survival in humans (see Section 6.6.5), SirT3 is another possible mediator in the regulation of mammalian CR-dependent lifespan, although this has not yet been determined.

6.9   SIRTUINS AND NEUROLOGICAL DISEASES

Recent reports have identified a protective role for SirT1 in neurodegenerative diseases such as Alzheimer's disease (AD), amyotrophic lateral sclerosis (ALS), Parkinson's disease (PD), and Huntington's disease (HD).[176–179] SirT1 seems to be overexpressed in mouse models of several of these diseases.[176] Furthermore, activation of SirT1 activity by resveratrol or SirT1 overexpression in the hippocampus results in neuron survival and protection against learning impairment in mouse Alzheimer's models.[176] SirT1 is responsible for the previously described axonal protection via increased NAD[+] biosynthesis.[180] Axonal degeneration is a critical feature of many neurodegenerative diseases, preceding the death of neurons in PD and AD.[181] In the case of AD, some findings suggest that SirT1 might also act through its inhibition of NF-κB in microglias, downregulating amyloid-β peptide (Aβ) levels. Upon accumulation of Aβ, microglial cells induce neurodegeneration, resulting in neuronal cell death.[177,182]

PD is characterized by the presence of cytotoxic, α-synuclein-enriched Lewy bodies within dopaminergic neurons.[183] SirT1 activation by resveratrol inhibits neurotoxin-induced accumulation of α-synuclein in dopaminergic neurons.[184] The inhibitory mechanism is unknown, but it may involve p53. Recently, an antagonistic role for SirT2 has been described in PD. Specific inhibition of SirT2 activity or knockdowns of SirT2 levels prevent α-synuclein accumulation and toxicity through an unknown mechanism in mice and *Drosophila* PD models.[185]

Finally, some evidence suggests that SirT1 function might be protective in polyglutamine-related neurodegenerative diseases such as HD.[179,186] These diseases are caused by alterations in certain genes that produce abnormal glutamine tract expansions (polyQ). Once expressed, these aberrant proteins aggregate and seem to be cytotoxic. Recent reports show that polyQ tract expansion in ataxin-3, a protein involved in proteosome-dependent ubiquitination and gene repression, induces Ku70 acetylation.[186] As described above, acetylated Ku70 can no longer sequester Bax, which is then free to move to the mitochondria and induce apoptosis. Consistent with what is seen during SirT1-mediated Ku70 deacetylation, resveratrol, a SirT1 activator, has been shown to reduce polyQ-ataxin-3 toxicity in neuronal cell lines.[186]

6.10   CONCLUSIONS

Recent developments in the study of the Sir2 family of proteins have provided a new perspective regarding the mechanisms in which they participate and their roles in multiple human pathologies. This new perspective, and the fact that Sir2 catalytic activity is so radically different from the rest of the HDACs, has prompted considerable efforts from the academic and private sectors to develop novel drugs capable of specifically modulating sirtuin functions. Studies of the sirtuin activator resveratrol,[187] the sirtuin inhibitor sirtinol,[188] and many others suggest that this class of pharmaceuticals may have the potential to drastically change the medical landscape in the near future.

REFERENCES

1. Allis, C.D., Jenuwein, T., and Reinberg, D., *Epigenetics*, 1st ed., Cold Spring Harbor Press, Cold Spring Harbor, 2006.
2. Vaquero, A., Loyola, A., and Reinberg, D., The constantly changing face of chromatin, *Sci. Aging Knowledge Environ.*, 2003(14), RE4, 2003.
3. Vaquero, A., Sternglanz, R., and Reinberg, D., NAD$^+$-Dependent Deacetylation of H4 Lysine 16 by Class III HDACs, *Oncogene*, 26, 5505, 2007.
4. Haigis, M.P. and Guarente, L., Mammalian sirtuins—emerging roles in physiology, aging, and calorie restriction, *Genes Dev.*, 20, 2913, 2006.
5. Shore, D., Squire, M., and Nasmyth, K.A., Characterization of two genes required for the position-effect control of yeast mating-type genes, *EMBO J.*, 3, 2817, 1984.
6. Saunders, L.R., Verdin, E., Sirtuins: critical regulators at the crossroads between cancer and aging, *Oncogene*, 26, 5489, 2007.
7. Guarente, L., Diverse and dynamic functions of the Sir silencing complex, *Nat. Genet.*, 23, 281, 1999.
8. Braunstein, M. et al., Transcriptional silencing in yeast is associated with reduced nucleosome acetylation, *Genes Dev.*, 7, 592, 1993.
9. Suka, N. et al., Highly specific antibodies determine histone acetylation site usage in yeast heterochromatin and euchromatin, *Mol. Cell.*, 8, 473, 2001.
10. Robyr, D. et al., Microarray deacetylation maps determine genome-wide functions for yeast histone deacetylases, *Cell*, 109, 437, 2002.
11. Imai, S. et al., Transcriptional silencing and longevity protein Sir2 is an NAD-dependent histone deacetylase, *Nature*, 403, 795, 2000.
12. Landry, J. et al., The silencing protein SIR2 and its homologs are NAD-dependent protein deacetylases, *Proc. Natl. Acad. Sci. U.S.A.*, 97, 5807, 2000.
13. Sinclair, D. and Guarente, L., The SIR2/3/4 complex and SIR2 alone promote longevity in Saccharomyces cerevisiae by two different mechanisms, *Genes Dev.*, 13, 2570, 1999.
14. Frye, R.A., Characterization of five human cDNAs with homology to the yeast SIR2 gene: Sir2-like proteins (sirtuins) metabolize NAD and may have protein ADP-ribosyltransferase activity, *Biochem. Biophys. Res. Commun.*, 260, 273, 1999.
15. Frye, R.A., Phylogenetic classification of prokaryotic and eukaryotic Sir2-like proteins, *Biochem. Biophys. Res. Commun.*, 273, 793, 2000.
16. Tissenbaum, H.A. and Guarente, L., Increased dosage of a sir-2 gene extends lifespan in Caenorhabditis elegans, *Nature*, 410, 227, 2001.
17. Rosenberg, M.I. and Parkhurst, S.M., Drosophila Sir2 is required for heterochromatic silencing and by euchromatic Hairy/E(Spl) bHLH repressors in segmentation and sex determination, *Cell*, 109, 447, 2002.
18. Starai, V.J. et al., Sir2-dependent activation of acetyl-CoA synthetase by deacetylation of active lysine, *Science*, 298, 2390, 2002.
19. Tsang, A.W. and Escalante-Semerena, J.C., CobB, a new member of the SIR2 family of eucaryotic regulatory proteins, is required to compensate for the lack of nicotinate mononucleotide: 5,6-dimethylbenzimidazole phosphoribosyltransferase activity in cobT mutants during cobalamin biosynthesis in Salmonella typhimurium LT2, *J. Biol. Chem.*, 273, 31788, 1998.
20. Wardleworth, B.N. et al., Structure of Alba: an archaeal chromatin protein modulated by acetylation, *EMBO J.*, 21, 4654, 2002.
21. Bell, S.D. et al., The interaction of Alba, a conserved archaeal chromatin protein, with Sir2 and its regulation by acetylation, *Science*, 296, 148, 2002.
22. Derbyshire, M.K., Weinstock, K.G., and Strathern, J.N., HST1, a new member of the SIR2 family of genes, *Yeast*, 12, 631, 1996.

23. Xie, J. et al., Sum1 and Hst1 repress middle sporulation-specific gene expression during mitosis in Saccharomyces cerevisiae, *EMBO J.*, 18, 6448, 1999.

24. Sutton, A. et al., A novel form of transcriptional silencing by Sum1-1 requires Hst1 and the origin recognition complex, *Mol. Cell. Biol.*, 21, 3514, 2001.

25. Robert, F. et al., Global position and recruitment of HATs and HDACs in the yeast genome, *Mol. Cell.*, 16, 199, 2004.

26. Perrod, S. et al., A cytosolic NAD-dependent deacetylase, Hst2p, can modulate nucleolar and telomeric silencing in yeast, *EMBO J.*, 20, 197, 2001.

27. Vaquero, A. et al., SirT2 is a histone deacetylase with preference for histone H4 Lys 16 during mitosis, *Genes Dev.*, 20, 1256, 2006.

28. Maas, N.L. et al., Cell cycle and checkpoint regulation of histone H3 K56 acetylation by Hst3 and Hst4, *Mol. Cell*, 23, 109, 2006.

29. Celic, I. et al., The sirtuins hst3 and Hst4p preserve genome integrity by controlling histone h3 lysine 56 deacetylation, *Curr. Biol.*, 16, 1280, 2006.

30. Prudencio, M., Rodriguez, A., and Mota, M.M., The silent path to thousands of merozoites: the Plasmodium liver stage, *Nat. Rev. Microbiol.*, 4, 849, 2006.

31. Merrick, C.J. and Duraisingh, M.T., Heterochromatin-mediated control of virulence gene expression, *Mol. Microbiol.*, 62, 612, 2006.

32. Scherf, A. et al., Antigenic variation in malaria: in situ switching, relaxed and mutually exclusive transcription of var genes during intra-erythrocytic development in Plasmodium falciparum, *EMBO J.*, 17, 5418, 1998.

33. Duraisingh, M.T. et al., Heterochromatin silencing and locus repositioning linked to regulation of virulence genes in Plasmodium falciparum, *Cell*, 121, 13, 2005.

34. Freitas-Junior, L.H. et al., Telomeric heterochromatin propagation and histone acetylation control mutually exclusive expression of antigenic variation genes in malaria parasites, *Cell*, 121, 25, 2005.

35. Merrick, C.J. and Duraisingh, M.T., Plasmodium falciparum sir2: an unusual sirtuin with dual histone deacetylase and ADP-ribosyltransferase activity, *Eukaryot. Cell*, EC.00114-07, 2007.

36. García-Salcedo, J.A. et al., A chromosomal SIR2 homologue with both histone NAD-dependent ADP-ribosyltransferase and deacetylase activities is involved in DNA repair in Trypanosoma brucei, *EMBO J.*, 22, 5851, 2003.

37. Alsford, S., et al., A sirtuin in the African trypanosome is involved in both DNA repair and telomeric gene silencing but is not required for antigenic variation, *Mol. Microbiol.*, 63, 724, 2007.

38. Vergnes, B. et al., Cytoplasmic SIR2 homologue overexpression promotes survival of Leishmania parasites by preventing programmed cell death, *Gene*, 296, 139, 2002.

39. Vergnes, B. et al., Targeted disruption of cytosolic SIR2 deacetylase discloses its essential role in Leishmania survival and proliferation, *Gene*, 363, 85, 2005.

40. Newman B.L. et al., A Drosophila homologue of Sir2 modifies position-effect variegation but does not affect life span, *Genetics*, 162, 1675, 2002.

41. Younger-Sepherd, S. et al., Deadpan, an essential pan-neural gene encoding an HLH protein, acts as a denominator in Drosophila sex determination, *Cell*, 18, 70, 911, 1992.

42. Bier, E. et al., Deadpan, an essential pan-neural gene in Drosophila, encodes a helix-loop-helix protein similar to the hairy gene product, *Genes Dev.*, 6, 2137, 1992.

43. Bianchi-Frias, D. et al., Hairy transcriptional repression targets and cofactor recruitment in Drosophila, *PLoS Biol.*, 2, E178, 2004.

44. Cho, Y. et al., Individual histone deacetylases in Drosophila modulate transcription of distinct genes, *Genomics*, 86, 606, 2005.

45. Furuyama, T. et al., SIR2 is required for polycomb silencing and is associated with an E(Z) histone methyltransferase complex, *Curr. Biol.*, 14, 1812, 2004.

46. Astrom, S.U., Cline, T.W., and Rine, J., The Drosophila melanogaster sir2+ gene is nonessential and has only minor effects on position-effect variegation, *Genetics*, 163, 931, 2003.
47. Berdichevsky, A. et al., C. elegans SIR-2.1 interacts with 14-3-3 proteins to activate DAF-16 and extend life span, *Cell*, 125, 1165, 2006.
48. Michishita, E. et al., Evolutionarily conserved and nonconserved cellular localizations and functions of human SIRT proteins, *Mol. Biol. Cell*, 16, 4623, 2005.
49. Moynihan, K.A., et al., Increased dosage of mammalian Sir2 in pancreatic beta cells enhances glucose-stimulated insulin secretion in mice, *Cell Metab.*, 2, 105, 2005.
50. Chen, I.Y. et al., Histone H2A.z is essential for cardiac myocyte hypertrophy but opposed by silent information regulator 2α, *J. Biol. Chem.*, 281, 19369, 2006.
51. Tanno, M., et al., Nucleocytoplasmic shuttling of the NAD+-dependent histone deacetylase SIRT1, *J. Biol. Chem.*, 282, 6823, 2007.
52. Vaquero, A., et al., Human SirT1 interacts with histone H1 and promotes formation of facultative heterochromatin, *Mol. Cell.*, 16, 93, 2004.
53. Espada, J. et al., Epigenetic disruption of ribosomal RNA genes and nucleolar architecture in DNA methyltransferase 1 (Dnmt1) deficient cells, *Nucleic Acids Res.*, 35, 2191, 2007.
54. Onyango, P., et al., SIRT3, a human SIR2 homologue, is an NAD-dependent deacetylase localized to mitochondria, *Proc. Natl. Acad. Sci. U.S.A.*, 99, 13653, 2002.
55. Schwer, B. et al., The human silent information regulator (Sir)2 homologue hSIRT3 is a mitochondrial nicotinamide adenine dinucleotide-dependent deacetylase, *J. Cell Biol.*, 158, 647, 2002.
56. Scher, M.B., Vaquero, A., and Reinberg D., SirT3 is a nuclear NAD+-dependent histone deacetylase that translocates to the mitochondria upon cellular stress, *Genes Dev.*, 21, 920, 2007.
57. Sauve, A.A. and Schramm, V.L., Sir2 regulation by nicotinamide results from switching between base exchange and deacetylation chemistry, *Biochemistry*, 42, 9249, 2003.
58. Liou, G.G. et al., Assembly of the SIR complex and its regulation by O-acetyl-ADP-ribose, a product of NAD-dependent histone deacetylation, *Cell*, 121, 515, 2005.
59. Kustatscher, G. et al., Splicing regulates NAD metabolite binding to histone macroH2A, *Nat. Struct. Mol. Biol.*, 12, 624, 2005.
60. Grubisha, O. et al., Metabolite of SIR2 reaction modulates TRPM2 ion channel, *J. Biol. Chem.*, 281, 14057, 2006.
61. North, B.J. et al., The human Sir2 ortholog, SIRT2, is an NAD+-dependent tubulin deacetylase, *Mol. Cell*, 11, 437, 2003.
62. Haigis, M.C. et al., SIRT4 inhibits glutamate dehydrogenase and opposes the effects of calorie restriction in pancreatic beta cells, *Cell*, 126, 941, 2006.
63. Liszt, G. et al., Mouse Sir2 homolog SIRT6 is a nuclear ADP-ribosyltransferase, *J. Biol. Chem.*, 280, 21313, 2005.
64. Dion, M.F. et al., Genomic characterization reveals a simple histone H4 acetylation code, *Proc. Natl. Acad. Sci. U.S.A.*, 102, 5501, 2005.
65. Fraga, M.F. et al., Loss of acetylation at Lys16 and trimethylation at Lys20 of histone H4 is a common hallmark of human cancer, *Nat Genet.*, 37, 391, 2005.
66. McBurney, M.W. et al., The mammalian SIR2α protein has a role in embryogenesis and gametogenesis, *Mol. Cell. Biol.*, 23, 38, 2003.
67. Kuzmichev, A. et al., Composition and histone substrates of polycomb repressive group complexes change during cellular differentiation, *Proc. Natl. Acad. Sci. U.S.A.*, 102, 1859, 2005.
68. Vaquero, A. et al., SirT1 regulates the histone methyl-transferase Suv39h1 during heterochromatin formation, *Nature*, 450, 440, 2007.
69. Trojer, P. and Reinberg, D., Facultative heterochromatin: is there a distinctive molecular signature?, *Mol. Cell*, 28, 1, 2007.
70. Whitelaw, E., Unravelling the X in sex, *Dev. Cell*, 11, 759, 2006.

71. Schwartz, Y.B. and Pirrotta, V., Polycomb silencing mechanisms and the management of genomic programmes, *Nat. Rev. Genet.*, 8, 9, 2007.
72. Craig, J.M., Heterochromatin—many flavours, common themes, *Bioessays*, 27, 17, 2005.
73. Braunstein, M. et al., Efficient transcriptional silencing in Saccharomyces cerevisiae requires a heterochromatin histone acetylation pattern, *Mol. Cell. Biol.*, 16, 4349, 1996.
74. Turner, B.M., Birley, A.J., and Lavender, J., Histone H4 isoforms acetylated at specific lysine residues define individual chromosomes and chromatin domains in Drosophila polytene nuclei, *Cell*, 69, 375, 1992.
75. Frankel, S. et al., An actin-related protein in Drosophila colocalizes with heterochromatin protein 1 in pericentric heterochromatin, *J. Cell. Sci.*, 110, 1999, 1997.
76. Senawong, T. et al., Involvement of the histone deacetylase SIRT1 in chicken ovalbumin upstream promoter transcription factor (COUP-TF)-interacting protein 2-mediated transcriptional repression, *J. Biol. Chem.*, 278, 43041, 2003.
77. Pruitt, K. et al., Inhibition of SIRT1 reactivates silenced cancer genes without loss of promoter DNA hypermethylation, *PLoS Genet.*, 2, e40, 2006.
78. Shankaranarayana, G.D. et al., Sir2 regulates histone H3 lysine 9 methylation and heterochromatin assembly in fission yeast, *Curr. Biol.*, 13, 1240, 2003.
79. Kuzmichev, A. et al., Different EZH2-containing complexes target methylation of histone H1 or nucleosomal histone H3, *Mol. Cell*, 14, 183, 2004.
80. Grunstein, M., Molecular model for telomeric heterochromatin in yeast, *Curr. Opin. Cell. Biol.*, 9, 383, 1997.
81. Calestagne-Morelli, A. and Ausio, J., Long-range histone acetylation: biological significance, structural implications, and mechanisms, *Biochem. Cell. Biol.*, 84, 518, 2006.
82. Roh, T.Y., Cuddapah, S., and Zhao, K., Active chromatin domains are defined by acetylation islands revealed by genome-wide mapping, *Genes Dev.*, 19, 542, 2005.
83. Brunet, A. et al., Stress-dependent regulation of FOXO transcription factors by the SIRT1 deacetylase, *Science*, 303, 2011, 2004.
84. Motta, M.C. et al., Mammalian SIRT1 represses forkhead transcription factors, *Cell*, 116, 551, 2004.
85. van der Horst, A. et al., FOXO4 is acetylated upon peroxide stress and deacetylated by the longevity protein hSir2(SIRT1), *J. Biol. Chem.*, 279, 28873, 2004.
86. Giannakou, M.E. and Partridge, L.,The interaction between FOXO and SIRT1: tipping the balance towards survival, *Trends Cell Biol.*, 14, 408, 2004.
87. Van der Heide, L.P. et al., Regulation of FoxO activity by CBP/p300-mediated acetylation, *Trends Biochem. Sci.*, 30, 81, 2005.
88. Giannakou, M.E. et al., Long-lived Drosophila with overexpressed dFOXO in adult fat body, *Science*, 305, 361, 2004.
89. Piva, R., Belardo, G., and Santero, M.G., NF-κB: a stress-regulated switch for cell survival, *Antioxid. Redox Signal.*, 8, 478, 2006.
90. Yeung, F. et al., Modulation of NF-κB-dependent transcription and cell survival by the SIRT1 deacetylase, *EMBO J.*, 23, 2369, 2004.
91. Luo, J. et al., Negative control of p53 by Sir2α promotes cell survival under stress, *Cell*, 107, 137, 2001.
92. Vaziri, H. et al., hSIR2(SIRT1) functions as an NAD-dependent p53 deacetylase, *Cell*, 107, 149, 2001.
93. Cheng, H.L. et al., Developmental defects and p53 hyperacetylation in Sir2 homolog (SIRT1)-deficient mice, *Proc. Natl. Acad. Sci. U.S.A.*, 100, 10794, 2003.
94. Kamel, C. et al., SirT1 fails to affect p53-mediated biological functions, *Aging Cell*, 5, 81, 2006.
95. Subramanian, C. et al., Ku70 acetylation mediates neuroblastoma cell death induced by histone deacetylase inhibitors, *Proc. Natl. Acad. Sci. U.S.A.*, 102, 4842, 2005.

96. Cohen, H.Y. et al., Calorie restriction promotes mammalian cell survival by inducing the SIRT1 deacetylase, *Science*, 305, 390, 2004.

97. Yuan, Z. et al., SIRT1 regulates the function of the Nijmegen breakage syndrome protein, *Mol. Cell*, 27, 149, 2007.

98. Chua, K.F. et al., Mammalian SIRT1 limits replicative life span in response to chronic genotoxic stress, *Cell Metab.*, 2, 67, 2005.

99. Yang, T. et al., SIRT1 and endocrine signaling, *Trends Endocrinol. Metab.*, 17, 186, 2006.

100. Yoon, J.C. et al., Control of hepatic gluconeogenesis through the transcriptional coactivator PGC-1, *Nature*, 413, 131, 2001.

101. Rodgers, J.T. et al., Nutrient control of glucose homeostasis through a complex of PGC-1alpha and SIRT1, *Nature*, 434, 113, 2005.

102. Bordone, L. et al., Sirt1 regulates insulin secretion by repressing UCP2 in pancreatic beta cells, *PLoS Biol.*, 4, e31, 2005.

103. Picard, F. et al., Sirt1 promotes fat mobilization in white adipocytes by repressing PPAR-gamma, *Nature*, 429, 771, 2004.

104. Lemieux, M.E. et al., The Sirt1 deacetylase modulates the insulin-like growth factor signaling pathway in mammals, *Mech. Ageing Dev.*, 126, 1097, 2005.

105. Fu, M. et al., Hormonal control of androgen receptor function through SIRT1, *Mol. Cell Biol.*, 26, 8122, 2006.

106. Amat, R. et al., Sirt1 is involved in glucocorticoid-mediated control of uncoupling protein-3 gene transcription, *J. Biol. Chem.*, 10.1074/jbc.M707114200, 2007.

107. Freyssenet, D., Energy sensing and regulation of gene expression in skeletal muscle, *J. Appl. Physiol.*, 102, 529, 2007.

108. Fulco, M. et al., Sir2 regulates skeletal muscle differentiation as a potential sensor of the redox state, *Mol. Cell*, 12, 51, 2003.

109. Sartorelli, V. et al., Acetylation of MyoD directed by PCAF is necessary for the execution of the muscle program, *Mol. Cell*, 4, 725, 1999.

110. Dryden, S.C. et al., Role for human SIRT2 NAD-dependent deacetylase activity in control of mitotic exit in the cell cycle, *Mol. Cell. Biol.*, 23, 3173, 2003.

111. Wilson, J.M. et al., Nuclear export modulates the cytoplasmic Sir2 homologue Hst2, *EMBO Rep.*, 7, 1247, 2006.

112. North, B.J. and Verdin, E., Interphase nucleo-cytoplasmic shuttling and localization of SIRT2 during mitosis, *PLoS ONE*, 2, e784, 2007.

113. Bae, N.S. et al., Human histone deacetylase SIRT2 interacts with the homeobox transcription factor HOXA10, *J. Biochem. (Tokyo)*, 135, 695, 2004.

114. Halme, A. et al., Genetic and epigenetic regulation of the FLO gene family generates cell-surface variation in yeast, *Cell*, 116, 405, 2004.

115. Hiratsuka, M. et al., Proteomics-based identification of differentially expressed genes in human gliomas: down-regulation of SIRT2 gene, *Biochem. Biophys. Res. Commun.*, 309, 558, 2003.

116. Borra, M.T. et al., Conserved enzymatic production and biological effect of O-acetyl-ADP-ribose by silent information regulator 2-like NAD+-dependent deacetylases, *J. Biol. Chem.*, 277, 12632, 2002.

117. Nahhas, F. et al., Mutations in SIRT2 deacetylase which regulate enzymatic activity but not its interaction with HDAC6 and tubulin, *Mol. Cell. Biochem.*, 303, 221, 2007.

118. North, B.J. and Verdin E., Mitotic regulation of SIRT2 by cyclin-dependent kinase 1-dependent phosphorylation, *J. Biol. Chem.*, 282, 19546, 2007.

119. Inoue, T. et al., SIRT2, a tubulin deacetylase, acts to block the entry to chromosome condensation in response to mitotic stress, *Oncogene*, 26, 945, 2007.

120. Piperno, G., LeDizet, M., and Chang, X.J., Microtubules containing acetylated α-tubulin in mammalian cells in culture, *J. Cell Biol.*, 104, 289, 1987.

121. Vaquero, A. and Reinberg, D., unpublished data, 2005.
122. Polevoda, B. and Sherman, F., The diversity of acetylated proteins, *Genome Biol.,* 3, reviews0006, 2002.
123. Belyaev, N.D. et al., Histone H4 acetylation in plant heterochromatin is altered during the cell cycle, *Cromosoma*, 106, 193, 1997.
124. Jazayeri, A. et al., Saccharomyces cerevisiae Sin3p facilitates DNA double-strand break repair, *Proc. Natl. Acad. Sci. U.S.A.*, 101, 1644, 2004.
125. Gupta, A., et al., Involvement of human MOF in ATM function, *Mol. Cell. Biol.*, 25, 5292, 2005.
126. Jing, E., Gesta, S., and Kahn, C.R., SIRT2 regulates adipocyte differentiation through FoxO1 acetylation/deacetylation. *Cell Metab.*, 6, 105, 2007.
127. Li, W. et al., Sirtuin 2, a mammalian homolog of yeast silent information regulator-2 longevity regulator, is an oligodendroglial protein that decelerates cell differentiation through deacetylating α-tubulin, *J. Neurosci.*, 27, 2606, 2007.
128. Rose, G. et al., Variability of the SIRT3 gene, human silent information regulator Sir2 homologue, and survivorship in the elderly, *Exp. Gerontol.*, 38, 1065, 2003.
129. Bellizzi, D. et al., A novel VNTR enhancer within the SIRT3 gene, a human homologue of SIR2, is associated with survival at oldest ages, *Genomics*, 85, 258, 2005.
130. Lombard, D.B. et al., Mammalian Sir2 Homolog SIRT3 Regulates Global Mitochondrial Lysine Acetylation, *Mol. Cell. Biol.*, 0: MCB.01636-07, 2007.
131. Hallows, W.C., Lee, S., and Denu, J.M., Sirtuins deacetylate and activate mammalian acetyl-CoA synthetases, *Proc. Natl. Acad. Sci. U.S.A.*, 103, 10230, 2006.
132. Schwer, B. et al., Reversible lysine acetylation controls the activity of the mitochondrial enzyme acetyl-CoA synthetase 2, *Proc. Natl. Acad. Sci. U.S.A.*, 103, 10224, 2006.
133. Yang, H. et al., Nutrient-sensitive mitochondrial NAD+ levels dictate cell survival, *Cell*, 130, 1095, 2007.
134. Shi, T. et al., SIRT3, a mitochondrial sirtuin deacetylase, regulates mitochondrial function and thermogenesis in brown adipocytes, *J. Biol. Chem.*, 280, 13560, 2005.
135. Hagopian, K., Ramsey, J.J., and Weindruch, R., Caloric restriction increases gluconeogenic and transaminase enzyme activities in mouse liver, *Exp. Gerontol.*, 38, 267, 2003.
136. Mostoslavsky, R. et al., Genomic instability and aging-like phenotype in the absence of mammalian SIRT6, *Cell*, 124, 315, 2006.
137. de Boer, J. et al., Premature aging in mice deficient in DNA repair and transcription, *Science*, 296, 1276, 2002.
138. Lombard, D.B. et al., DNA repair, genome stability, and aging, *Cell*, 120, 497, 2005.
139. Ma, W. et al., GCIP/CCNDBP1, a helix-loop-helix protein, suppresses tumorigenesis, *J. Cell. Biochem.*, 100, 1376, 2007.
140. Ford, E. et al., Mammalian Sir2 homolog SIRT7 is an activator of RNA polymerase I transcription, *Genes Dev.*, 20, 1075, 2006.
141. Ford, J., Jiang, M., and Milner, J., Cancer-specific functions of SIRT1 enable human epithelial cancer cell growth and survival, *Cancer Res.*, 65, 10457, 2005.
142. Yamamoto, H., Schoonjans, K., and Auwerx J., Sirtuin functions in health and disease, *Mol. Endocrinol.*, 21, 1745, 2007.
143. Wales, M.M. et al., p53 activates expression of HIC-1, a new candidate tumour suppressor gene on 17p13.3, *Nat. Med.*, 1, 570, 1995.
144. Chen, W.Y. et al., Tumor suppressor HIC1 directly regulates SIRT1 to modulate p53-dependent DNA-damage responses, *Cell*, 123, 437, 2005.
145. Caballero, R. et al., Combinatorial effects of splice variants modulate function of Aiolos, *J. Cell Sci.*, 120, 2619, 2007.
146. Wang, C. et al., Interactions between E2F1 and SirT1 regulate apoptotic response to DNA damage, *Nat. Cell. Biol.*, 8, 1025, 2006.

147. Abdelmohsen, K. et al., Phosphorylation of HuR by Chk2 regulates SIRT1 expression, *Mol. Cell*, 25, 543, 2007.
148. Varambally, S. et al., The polycomb group protein EZH2 is involved in progression of prostate cancer, *Nature*, 419, 624, 2002.
149. Kleer, C.G. et al., EZH2 is a marker of aggressive breast cancer and promotes neoplastic transformation of breast epithelial cells, *Proc. Natl. Acad. Sci. U.S.A.*, 100, 11606, 2003.
150. Sasaki, T. et al., Progressive loss of SIRT1 with cell cycle withdrawal, *Aging Cell*, 5, 413, 2006.
151. Ashraf, N. et al., Altered sirtuin expression is associated with node-positive breast cancer, *Br. J. Cancer*, 95, 1056, 2006.
152. Bradbury, C.A. et al., Histone deacetylases in acute myeloid leukaemia show a distinctive pattern of expression that changes selectively in response to deacetylase inhibitors, *Leucemia*, 19, 1751, 2005.
153. De Nigris, F. et al., Isolation of a SIR-like gene, SIR-T8, that is overexpressed in thyroid carcinoma cell lines and tissues, *Br. J. Cancer*, 87, 1479, 2002.
154. Lennerz, V. et al., The response of autologous T cells to a human melanoma is dominated by mutated neoantigens, *Proc. Natl. Acad. Sci. U.S.A.*, 102, 16013, 2005.
155. Natrajan, R. et al., High-resolution deletion mapping of 15q13.2- q21.1 in transitional cell carcinoma of the bladder, *Cancer Res.*, 63, 7657, 2003.
156. Bishop, N.A. and Guarente, L., Genetic links between diet and lifespan: shared mechanisms from yeast to humans, *Nat. Rev. Genet.*, 8, 835, 2007.
157. Balaban, R.S., Remoto, S. and Finkel, T., Mitochondria, oxidants, and aging, *Cell*, 120, 483, 2005.
158. Aguilaniu, H., Durieux, J., and Dillin, A., Metabolism, ubiquinone synthesis, and longevity, *Genes Dev.*, 19, 2399, 2005.
159. Kenyon, C., The plasticity of aging: insights from long-lived mutants, *Cell*, 120, 449, 2005.
160. Blasco, M.A., Telomere length, stem cells and aging, *Nat. Chem. Biol.*, 3, 640, 2007.
161. Kennedy, B.K. et al., Redistribution of silencing proteins from telomeres to the nucleolus is associated with extension of life span in S. cerevisiae, *Cell*, 89, 381, 1997.
162. Rogina, B. and Helfland, S.L., Sir2 mediates longevity in the fly through a pathway related to calorie restriction, *Proc. Natl. Acad. Sci. U.S.A.*, 101, 15998, 2004.
163. Baur, J.A. et al., Resveratrol improves health and survival of mice on a high-calorie diet, *Nature*, 444, 337, 2006.
164. Longo, V.D. and Kennedy, B.K., Sirtuins in aging and age-related disease, *Cell*, 126, 257, 2006.
165. Sinclair, D.A. and Guarente, L., Extrachromosomal rDNA circles—a cause of aging in yeast, *Cell*, 91, 1033, 1997.
166. Kaeberlein, M., McVey, M., and Guarente, L., The SIR2/3/4 complex and SIR2 alone promote longevity in Saccharomyces cerevisiae by two different mechanisms, *Genes Dev.*, 13, 2570, 1999.
167. Lamming, D.W. et al., HST2 mediates SIR2-independent life-span extension by calorie restriction, *Science*, 309, 1861, 2005.
168. Kaeberlein, M. et al., Regulation of yeast replicative life span by TOR and Sch9 in response to nutrients, *Science*, 310, 1193, 2005.
169. Vellai, T. et al., Genetics: influence of TOR kinase on lifespan in C. elegans, *Nature*, 426, 620, 2003.
170. Hertweck, M., Gobel, C., and Baumeister R., C. elegans SGK-1 is the critical component in the Akt/PKB kinase complex to control stress response and life span, *Dev. Cell*, 577, 2004.
171. Kapahi, P. et al., Regulation of lifespan in Drosophila by modulation of genes in the TOR signaling pathway, *Curr. Biol.*, 14, 885, 2004.

172. Kaeberlein, T.L. et al., Lifespan extension in Caenorhabditis elegans by complete removal of food, *Aging Cell*, 5, 487, 2006.
173. Lee, G.D. et al., Dietary deprivation extends lifespan in Caenorhabditis elegans, *Aging Cell*, 5, 515–524, 2006.
174. Carter, C.S., Ramsey, M.M., and Sonntag, W.E., A critical analysis of the role of growth hormone and IGF-1 in aging and lifespan, *Trends Genet.*, 18, 295, 2002.
175. Chen, D. et al., Increase in activity during calorie restriction requires Sirt1, *Science*, 310, 1641, 2005.
176. Kim, D. et al., SIRT1 deacetylase protects against neurodegeneration in models for Alzheimer's disease and amyotrophic lateral sclerosis, *EMBO J.*, 26, 3169, 2007.
177. Chen, J. et al., SIRT1 protects against microglia-dependent amyloid-beta toxicity through inhibiting NF-κB signalling, *J. Biol. Chem.*, 280, 40364, 2005.
178. Dillin, A. and Kelly, J.W., Medicine. The yin-yang of sirtuins, *Science*, 317, 461, 2007.
179. Parker, J.A. et al., Resveratrol rescues mutant polyglutamine cytotoxicity in nematode and mammalian neurons, *Nat. Genet.*, 37, 349, 2005.
180. Araki, T., Sasaki, Y., and Milbrandt, J., Increased nuclear NAD biosynthesis and SIRT1 activation prevent axonal degeneration, *Science*, 305, 1010, 2004.
181. Raff, M.C., Whitmore, A.V., and Finn, J.T., Axonal self-destruction and neurodegeneration, *Science*, 296, 868, 2002.
182. Qin, W. et al., Neuronal SIRT1 activation as a novel mechanism underlying the prevention of Alzheimer disease amyloid neuropathology by calorie restriction, *J. Biol. Chem.*, 281, 21745, 2006.
183. Garske, A.L., Smith, B.C., and Denu, J.M., Linking SIRT2 to Parkinson's disease, *A.C.S. Chem. Biol.*, 2, 529, 2007.
184. Okawara, M. et al., Resveratrol protects dopaminergic neurons in midbrain slice culture from multiple insults, *Biochem. Pharmacol.*, 73, 550, 2007.
185. Outeiro, T.F. et al., Sirtuin 2 inhibitors rescue α-synuclein-mediated toxicity in models of Parkinson's disease, *Science*, 317, 516, 2007.
186. Li, Y. et al., Bax-inhibiting peptide protects cells from polyglutamine toxicity caused by Ku70 acetylation, *Cell Death Differ.*, 14, 2058, 2007.
187. Baur, J.A. and Sinclair, D.A., Therapeutic potential of resveratrol: the in vivo evidence, *Nat. Rev. Drug Discov.*, 5, 493, 2006.
188. Grozinger, C.M. et al., Identification of a class of small molecule inhibitors of the sirtuin family of NAD-dependent deacetylases by phenotypic screening, *J. Biol. Chem.*, 276, 38837, 2001.
189. Wong, S. and Weber, J.D., Deacetylation of the retinoblastoma tumour suppressor protein by SIRT1, *Biochem. J.*, 407, 451, 2007.
190. Kume, S. et al., SIRT1 inhibits transforming growth factor beta-induced apoptosis in glomerular mesangial cells via Smad7 deacetylation, *J. Biol. Chem.*, 282, 151, 2007.
191. Takata, T. and Ishikawa, F., Human Sir2-related protein SIRT1 associates with the bHLH repressors HES1 and HEY2 and is involved in HES1- and HEY2-mediated transcriptional repression, *Biochem. Biophys. Res. Commun.*, 301, 250, 2003.
192. Pagans, S. et al., SIRT1 regulates HIV transcription via Tat deacetylation, *PLoS Biol.*, 3, e41, 2005.

7 microRNAs in Cell Biology and Diseases

Muller Fabbri, Carlo M. Croce,
and George A. Calin

CONTENTS

7.1   Physiology of microRNAs 105
7.2   microRNAs in Cell Biology.. 107
    7.2.1   microRNAs as Differentiating Genes.. 107
    7.2.2   microRNAs as Pro-/Anti-Apoptotic Genes 110
7.3   microRNAs in Human Diseases 111
    7.3.1   microRNAs and Cancer.. 111
        7.3.1.1   microRNAs are Frequently Located in Cancer-
                Associated Genomic Regions (CAGRs)........................... 111
        7.3.1.2   High-Throughput miRNA Profiling Methods Reveal
                Tissue-Specific Signatures 114
        7.3.1.3   miRNAs as Oncogenes 115
        7.3.1.4   miRNAs as Tumor Suppressor Genes............................... 118
    7.3.2   microRNAs and Neurological Diseases 120
    7.3.3   microRNAs and Immunity 121
    7.3.4   microRNAs and Viral Diseases.. 122
    7.3.5   microRNAs and Endocrine Diseases 124
7.4   Conclusions 124
Acknowledgments... 125
References... 125

7.1   PHYSIOLOGY OF microRNAs

microRNAs (miRNAs) are 18–24 nucleotide (nt) small, noncoding RNAs that target cognate mRNAs at a posttranscriptional level by degradation or at a translational level by repression through base pairing. The perfectly (in plants) or partially (in mammals) complementarity sites are located mainly, but not exclusively, in the 3′-untranslated region (3′-UTR) of the target mRNA [1,2]. The founding members of the miRNA family, *lin-4* and *let-7*, were first identified as fundamental development regulators in *Caenorhabditis elegans* [3,4]. Subsequently, miRNAs have been discovered in worms, flies, and human cells as well [5-7]. The majority of animal miRNAs

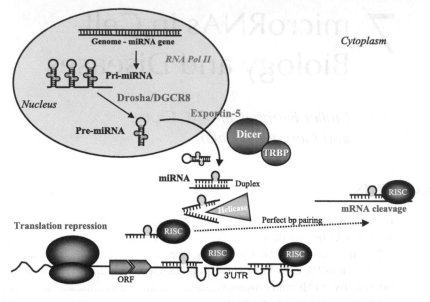

FIGURE 7.1   (See color figure following page 52.) Mechanisms of biogenesis and action of microRNAs. For details see the text. (Modified with permission from Ref. 126.)

have been discovered by cloning and by bioinformatic analyses [8,9], and their implications in basic cellular functions such as differentiation, proliferation, development, and apoptosis have been clearly documented in the last few years [2,10-12].

miRNAs are first transcribed by an RNA polymerase II as long primary transcripts called pri-miRNAs (Figure 7.1). These precursors are then processed by a ribonuclease III called Drosha [in conjunction with its binding partner, DiGeorge syndrome critical region gene 8 (DGCR8, or Pasha in *Drosophila* and *C. elegans*)], which is able to recognize double-stranded RNA and is responsible for the cleavage of pri-miRNAs into hairpin RNAs, called pre-miRNAs (70–100 nt in length). These direct precursors of the mature form of miRNAs are transported to the cytoplasm by the nuclear export factor exportin 5. Once in the cytoplasm, another double-stranded specific ribonuclease III, called Dicer, which works in concert with human immunodeficiency virus (HIV-1) transactivating response (TAR) RNA-binding protein (TRBP, or Loquacious in *Drosophila*), processes pre-miRNAs in an 18–24 nt miRNA duplex. This duplex is incorporated into a large protein complex called RISC (RNA-induced silencing complex), whose core includes components of the Argonaute protein family (Ago 1–4 in humans). While one strand of the miRNA duplex remains stably associated with RISC and represents the mature miRNA, the complementary strand, called passenger strand or miRNA*, is degraded through two different mechanisms, driven by the subtype of Ago protein present in RISC. If Ago2 is part of RISC, the miRNA may be cleaved; should a different Ago be involved, RISC may remove the passenger strand through a bypass mechanism, which is presumably based on duplex unwinding more than on their cleavage [13-15]. The mature miRNA drives RISC to the target mRNAs, inducing its cleavage in case of perfect miRNA–mRNA base pair complementarity (mechanism described as predominant in plants), or to its translational silencing (predominant mechanism in *C. elegans, Drosophila melanogaster*, and mammals), although also in the case of imperfect base pairing, a reduction of the target mRNA has been observed [1]. A match of seven consecutive base pairs between the target mRNA and nucleotides 2 to 8 at the 5′ end of the mature miRNA is sufficient to reduce the protein levels or the target [16]. These perfectly matching nucleotides are referred to as "seed" sites.

Recently it has been demonstrated that the mature miRNA sequence not only determines which mRNAs are targeted but also affects the subcellular localization and function of the miRNA itself. Among the miR-29 family (composed of miR-29a, -29b, and -29c), only miR-29b is predominantly localized to the nucleus, because it ends with a distinctive hexanucleotide terminal motif that acts as a transferable nuclear localization element [17]. The functional meaning of this subcellular miRNA compartmentalization is still under investigation, but it is well established that miRNAs negatively regulate the expression of target mRNAs. Microarray analyses showed that each miRNA downregulates more than 100 mRNAs [18], impacting many different biologic functions. Aberrant expression of the miRNome (defined as the full complement of miRNAs in a genome) has been associated with many different human diseases including cancer, diabetes, immuno, or neurodegenerative disorders [19].

7.2 microRNAs IN CELL BIOLOGY

7.2.1 MICRORNAs AS DIFFERENTIATING GENES

Among the cell functions regulated by miRNAs are differentiation and maintenance of stemness [20,21]. Embryonic stem cells (ES) are totipotent cell lines derived from the inner cell mass of the mammalian blastocyst. First derived from mice, ES are now available from a variety of mammalian systems, including human [22]. Stem cells are able to self-renew in an undifferentiated status under determined culture conditions, while retaining differentiation capacity [22-24]. Under in vitro specific differentiating conditions, ES cells are able to develop into different kinds of specialized somatic cells, recapitulating the different steps that normally occur in early embryonic development. In particular, they recapitulate some of the global genomic methylation that occurs shortly after implantation, allowing study of the epigenetic events responsible for X chromosome inactivation during midblastula [25]. Because of their high differentiation capacity, stem cells represent very attractive investigative targets in clinical and regenerative medicine [26,27]. Stem cells have been described in many different types of cancer, and cancer stem cells are considered responsible for recurrences of the disease and failure to respond to antitumoral treatments [28]; therefore, a better comprehension of the mechanisms that regulate both their self-renewal and differentiating properties is crucially needed.

miRNAs are involved in the regulation of stem cell function. Using Dicer-1 mutants (*dcr-1*), researchers observed that miRNA maturation and proper mRNA targeting were compromised (as expected), but also that ES cells had division and proliferation defects [29], leading to mice death at 7.5 days (embryonic stage) and to a complete loss of pluripotent stem cells [30]. Mutants for *dcr-1* in the germline stem cells of the fruitfly *D. melanogaster* show a sensibly decreased cyst production due to a delayed transition from the G_1 to the S phase [31]. This study indicates that miRNAs allow stem cells to overcome the G_1/S checkpoint of the cell cycle, despite the environmental signals that normally induce a cell cycle stop at the G_1/S transition in all other cell types. It can be assumed that cancer cells skip the G_1/S checkpoint through a similar mechanism and undergo higher proliferation and mutation rates, which are well known aspects of the malignant phenotype.

miRNAs also regulate stem cell differentiation. Mutations of the *dcr-1* gene in mouse ES cells induce globally reduced expression of miRNAs and severely alter the pattern of ES cell differentiation both in vitro and in vivo, with a reversibility of effects observed after reexpression of Dicer [32]. While the complete Dicer mutant failed to express both *oct-4* (a POU family transcription factor, which is a classical marker of all pluripotent cells and plays a critical role in establishing and maintaining cells in a pluripotent state) and *brachyury* (a primitive streak marker), with consequent cell death at an early stage, Dicer-deficient ES cells, expressed stem-cell-specific markers (*oct-4* and the short *α6-integrin* isoform), showed a normal ES cell morphology and preserved the ability to continually divide, albeit to a slower proliferation rate than the normal counterpart [32]. All these studies clearly demonstrate an involvement of miRNAs in stem cell physiology, but being based on Dicer's mutants (therefore, on a very general processing mechanism, shared by all miRNAs), they do not answer the question of which miRNAs are predominantly or exclusively expressed in stem cells and which miRNAs affect the commitment of a stem cell in one differentiation pathway or another. Using a direct cloning approach in undifferentiated, moderately differentiated, and differentiated mouse ES cells, Houbaviy et al. [33] were able to identify 15 miRNAs specifically expressed in mouse stem cells, 6 of which (miR-290, -291, -292, -293, -294, and -295) were clustered together, had similar sequences, and were specific for ES or early embryos. This early embryonic microRNA cluster (EEmiRC) can be identified only in placental (eutherian) mammals and only in trophoblastic stem cells, suggesting a possible important role in maintaining the pluripotent cell state [33,34]. Some of the miRNAs expressed in ES cell cultures were also found in adult tissues and should be considered as regulating general aspects of cell physiology (miR-15a, -16, -19b, -92, -93, -96, -130, and -130b) [33]. Finally, a group of miRNAs cloned from undifferentiated ES cells, whose expression dramatically increased upon differentiation (mir-21, and -22 among this set), indicates that miRNA expression timing is strictly regulated and that specific miRNAs are upregulated when the cell starts differentiation [33]. In accordance with what had been described in mice, in human embryonic stem cells an additional 36 miRNAs (mainly stem cell specific) were identified by cDNA cloning [35]. Interestingly, the expression of these miRNAs is significantly reduced as the embryonic stem cell differentiates into embryoid bodies, whereas miRNAs expressed in adult organs or tissues are poorly or not expressed in ES cells [35]. Three miRNAs (miR-296, -301, and -302) were common between the novel miRNA data sets from mouse and human ES cells [33, 35].

Many studies have addressed which miRNAs drive the commitment of stem cells in different lineages. (The main findings are summarized in Figure 7.2.) The whole picture has been particularly well investigated in hematopoiesis. Ectopic expression

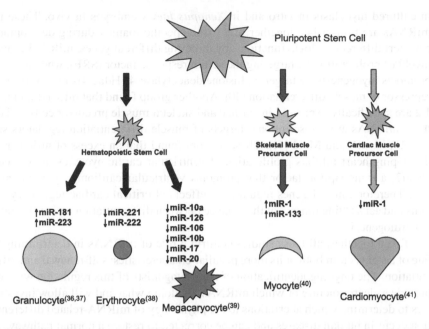

Pluripotent Stem Cell

Hematopoietic Stem Cell

Skeletal Muscle
Precursor Cell

Cardiac Muscle
Precursor Cell

↑miR-181   ↓miR-221   ↓miR-10a   ↑miR-1   ↓miR-1
↑miR-223   ↓miR-222   ↓miR-126   ↑miR-133
                                ↓miR-106
                                ↓miR-10b
                                ↓miR-17
                                ↓miR-20

Granulocyte[36,37]   Erythrocyte[38]   Myocyte[40]   Cardiomyocyte[41]
                              Megacaryocyte[39]

FIGURE 7.2   (See color figure following page 52.) Examples of microRNAs involved in differentiation. For details see the text.

of miR-181 increases the fraction of B lymphocytes both in vitro and in vivo. It has not been determined yet if this effect is due to an actual increased commitment of stem cells in the B lineage or to a higher survival rate of B cells [36]. Interestingly, Fazi et al. [37] have identified a minicircuitry which involves a miRNA (miR-223) and two transcription factors (NFI-A and C/EBPα) as regulators of human granulopoiesis. NFI-A binds to the promoter region of miR-223 and keeps it at low levels of expression. Retinoic acid (RA) is a differentiating agent that induces replacement of NFI-A with C/EBPα at the same promoter region of miR-223, resulting in an induced higher expression of this miRNA. In turn, miR-223 targets NFI-A mRNA, therefore completing the regulatory minicircuitry. Both silencing of NFI-A by silencing RNA techniques (siRNA) and ectopic expression of miR-223 potentiate differentiation of acute promyelocytic leukemia cells; conversely, silencing of miR-223 (with miRNA antisense molecules) prevents the differentiation response to RA [37]. The erythropoietic lineage has been investigated by Felli et al. [38]. In this study normal erythropoiesis was inhibited by miR-221 and -222. These two miRNAs were also able to contrast the growth of the erythroleukemic cell line TF-1, by targeting the KIT oncogene [38]. Garzon et al. [39] demonstrated that the megakaryocytic differentiation from CD34+ hematopoietic precursors is associated with a characteristic miRNA signature, which includes a down-modulation of miR-10a, -126, -106, -10b, -17, and -20. Interestingly, it was observed that miR-130a (downregulated itself in the megakaryocytic committed cells) targets MAFB, a transcription factor involved in the activation of the GPIIB promoter, important for platelet physiology.

miRNAs are also involved in stem cell differentiation in nonhematologic lineages. miR-1 and -133 regulate skeletal muscle differentiation, as demonstrated in cultured myoblasts in vitro and in *Xenopus laevis* embryos in vivo. These two miRNAs are transcribed together in a tissue-specific manner during development, but exert different biologic functions, by targeting different genes. miR-133 induces myoblast proliferation by targeting the serum response factor (SRF), whereas miR-1 promotes myogenesis by targeting histone deacetylase 4 (Hdac 4), a transcriptional repressor of muscle differentiation [40]. Another group found that miR-1-1 and miR-1-2 are specifically expressed in cardiac and skeletal muscle precursor cells and that these miRNAs are transcriptional targets of muscle differentiation regulators such as SRF, MyoD, and Mef2 [41]. These authors found that an excess of miR-1 in the developing heart inhibits proliferation of ventricular cardiomyocytes, by targeting Hand2, a transcription factor that promotes ventricular cardiomyocyte expansion [41]. Therefore, miR-1 genes regulate the effects of critical cardiac regulatory proteins and act as "fine tuners" of the balance between differentiation and proliferation in cardiogenesis.

Taken together, all these studies confirm a role of miRNAs in determining the fate of a stem cell in both of its more peculiar characteristics: self-renewal and differentiation. The ongoing identification of the protagonists of this regulation will soon provide a clearer picture of which miRNAs are doing what and will allow investigators to determine which aberrations to the physiology of miRNA-related differentiation occur in human diseases and can be corrected to restore a normal pathway.

7.2.2 MICRORNAS AS PRO-/ANTI-APOPTOTIC GENES

We have previously discussed the effects of miRNAs on stem cell proliferation. miRNAs can affect cell growth and apoptosis of different cell lineages in a cell-specific manner. Initially subdivided into pro-apoptotic miRNAs (e.g., miR-15a, miR-16, let-7, etc.) or inducers of cell proliferation (e.g., miR-21, miR-155, etc.), it is now becoming more and more evident the dual nature (both as pro- and anti-apoptotic) of at least some miRNAs, in different cell types and probably under different environmental conditions also in the same cell line [for extensive review, see Ref. 42]. In order to determine which miRNAs are involved in cell growth and apoptosis, Cheng et al. [43] employed a library of more than 90 antisense anti-miRNA inhibitors in HeLa (cervical cancer) and A549 (lung cancer) cell lines. In HeLa cells the silencing of 19 miRNAs (miR-95, -124, -125, -133, -134, -144, -150, -152, -187, -190, -191, -192, -193, -204, -211, -218, -220, -296, and -299) was associated with a decreased cell growth, whereas silencing of miR-21 and -24 produced increased cell growth. In A549 cells antagonists of 9 miRNAs (miR-7, -19a, -23, -24, -134, -140, -150, -192, and -193) down-regulated cell growth, whereas the silencing of miR-107, -132, -155, -181, -191, -194, -203, -215, and -301 increased cell growth. From this analysis emerges how the profile of cell growth up- and down-regulating miRNAs is different in two different cell lines and in some cases (e.g., miR-24, -191) with opposite effects. This study also identified miRNAs with anti-apoptotic effects (miR-1d, -7, -148, -204, -210, -216, and -296), whereas miR-214 was the only one to show a pro-apoptotic effect in HeLa cells [43]. Altogether, these data confirm a role for miRNAs in regulating cell growth and apoptosis and identify some of the culprits involved in these fundamental biologic events. We will discuss later for which miRNAs and

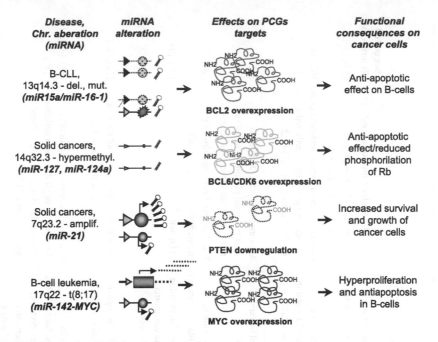

Disease, Chr. aberation (miRNA)	miRNA alteration	Effects on PCGs targets	Functional consequences on cancer cells
B-CLL, 13q14.3 - del., mut. (miR15a/miR-16-1)		BCL2 overexpression	Anti-apoptotic effect on B-cells
Solid cancers, 14q32.3 - hypermethyl. (miR-127, miR-124a)		BCL6/CDK6 overexpression	Anti-apoptotic effect/reduced phosphorilation of Rb
Solid cancers, 7q23.2 - amplif. (miR-21)		PTEN downregulation	Increased survival and growth of cancer cells
B-cell leukemia, 17q22 - t(8;17) (miR-142-MYC)		MYC overexpression	Hyperproliferation and antiapoptosis in B-cells

FIGURE 7.3   microRNA involvement in human cancers as oncogenes or tumor suppressors. For details see the text. (Modified with permission from Ref. 47.)

to what extent these effects on cell proliferation justify a role as oncogenes and/or tumor suppressor genes.

7.3   microRNAs IN HUMAN DISEASES

7.3.1   MICRORNAS AND CANCER

The study of the role of miRNAs in a complex disease such as cancer inevitably implies investigating the possible function of these noncoding RNAs in many different fields, from genetic to epigenetic modifications of cancer cells. Cancer is the genetic disease with the most complex mechanism. More recently, this postulate has been modified by adding that cancer is an epigenetic disease as well [for a comprehensive review, see Ref. 44]. We will discuss the role of miRNAs in both the genetics and epigenetics of cancer in the following paragraphs (Figure 7.3 and Table 7.1).

7.3.1.1   microRNAs are Frequently Located in Cancer-Associated Genomic Regions (CAGRs)

Some of the first evidence of a correlation between miRNAs and cancer came from the observation that miRNAs are frequently located in CAGRs [45]. Using a database of 186 human miRNAs, the authors generated several databases containing the exact genomic positions of markers used for cloning CAGRs, including 157 minimal regions of loss of heterozygosity (LOH), possible clues to the presence of tumor suppressor genes (TSGs); 37 minimal regions of amplification, suggestive of the

TABLE 7.1

Oncogenic and Suppressor microRNAs

Putative Function	miRNA	Location	Chromosomal Rearrangements	Expression in Cancer	Molecular Consequences*	Suggested References
	let-7 family	Various	LOH in lung cancers	Reduced expression in lung	let-7 regulates RAS oncogene expression in lung tumors	[54], [90], [91]
Suppressors	miR-16-1/miR-15a cluster	13q14.3, intron 4 of DLEU2	Deletion of 13q143 band or LOH in hematopoietic and solid cancers	Downregulated in the majority of B-CLLs and in the majority of DLBCLs	Exogenous restoration in leukemia cells induces apoptosis by directly reducing levels of anti-apoptotic BCL2	[62], [82], [88]
	miR-124a-1	8p23.1 intergenic		Downregulated in various solid cancers	Activation of cyclin D kinase 6 oncogene and phosphorylation of retinoblastoma suppressor gene	[97]
	miR-127	14q32.31	LOH in solid cancers	Reduced expression in prostate and bladder cancers	Translational downregulation of the transcriptional repressor BCL6	[96]
	miR-145/ miR-143 cluster	5q32, intergenic	Deletion of 5q32 band or LOH in MDS (5q-syndrome)	Reduced expression in colon adenomas and carcinomas and in breast cancers	Unknown	[56], [94]
Oncogenic	miR-17/ 92 cluster	13q313, intron 3 C13orf25	AMPLIF in follicular lymphomas	Overexpressed in malignant lymphomas and lung cancers	The miRNAs cluster, but not the host C13orf25 gene, enhances cell proliferation	[71], [73]
	miR-21	7q23.2,3'UTR VMP1	AMPLIF in neuroblastoma and breast, colon, and lung cancers	Elevated levels in glioblastomas, breast, colon, lung, pancreas, prostate, and stomach cancers, and cholangiocarcinomas	Increased apoptotic cell death after knockdown in glioblastoma cells;miR-21 modulated gemcitabine induced apoptosis by PTEN	[53], [56], [59]

| miR-155 | 21q21.3, exon 3 of ncRNA BIC | Not reported | Unknown | High expression in pediatric BL, in Hodgkin's, primary mediastinal and DLBCL lymphomas; overexpressed in breast, colon, and lung cancers | [53], [54], [56], [61], [62], [64] |
| miR-372, miR-373 | 19q1342, intergenic | Not reported | Proliferation of p53 wild-type cancer cells sensible to DNA-damaging agents | High expression in testicular germ cell tumors | [81] |

Note: *Molecular consequences with proven cancer relevance, as reported by the references in the last column, were presented; B-CLL: B-cell chronic lymphocytic leukemia; BIC: noncoding RNA gene; BL: Burkitt lymphoma; DLBCL: diffuse large B-cell lymphoma; DLEU2: noncoding RNA gene; MDS: myelodysplastic syndrome; VMP1: vacuole membrane protein 1. Modified with permission from Ref. 47.

presence of oncogenes (OGs); 45 common breakpoint regions in or near possible OGs or TSGs; and 29 fragile sites (FRA). Overall, more than half of miRNAs are located in CAGRs, 19% are inside or close to FRA, including FRA in which no known TSGs map (e.g., FRA7H and miR-29a and -29b-1). About half of miRNAs are in LOH areas or in regions of amplification. Interestingly, many miRNAs are located close to breakpoint regions. This is the case for miR-142, which is 50 nt from the t(8;17) breakpoint regions, which involves chromosome 17 and the onco-gene MYC. The translocation juxtaposes a truncated version of MYC (without the first exon) downstream of the miR-142 promoter, therefore inducing an abnormally increased MYC expression, which leads to a very aggressive B-CLL (B-cell chronic lymphocytic leukemia) phenotype [45]. Another example involves miR-180, located at 1 kb from the MN1 gene, which is involved in a t(4;22) translocation in menin-gioma. As a consequence, the chromosomal aberrancy inactivates both MN1 and the miRNA gene [45]. In a patient with B-ALL, a rearranged miR-125b-1, juxtaposed to the immunoglobulin heavy-chain locus, was described as a possible early step in leukemogenesis [47]. miR-122a maps in the minimal amplicon around MALT1 in aggressive marginal zone lymphoma (MZL) and about 160 kb from the breakpoint region of translocation t(11;18), in mucosa-associated lymphoid tissue lymphoma (MALT) [45]. Other miRNAs are located in target regions for viral integration (e.g., miR-142 at FRA17B, a target for HPV16 integration in cervical tumors). [For a review on chromosomal abnormalities and miRNAs, see Ref. 47.] Strengthening the importance of these findings, an extensive study of high-resolution array-based comparative genomic hybridization on 227 human ovarian cancer, breast cancer, and melanoma specimens clearly proved that regions hosting miRNAs exhibit high-frequency genomic alterations in human cancer [48].

Recently, a statistically significant association has been reported between the chromosomal location of miRNAs and mouse cancer susceptibility loci that are involved in the development of solid tumors, and for many of those miRNAs, spe-cific patterns of flanking DNA sequences are present in inbred strains with different tumor susceptibilities. This indicates that miRNAs could represent genes involved in the development and penetrance of solid tumors [49].

7.3.1.2   High-Throughput miRNA Profiling Methods Reveal Tissue-Specific Signatures

The expression of miRNAs in different tissues and species has been detected by two different methods which allow high-throughput miRNA profiling. The first method is based on oligonucleotide miRNA microarray chips, containing hundreds of human precursor and mature miRNA probes [50]. The method has been vali-dated by demonstrating its ability to identify different and tissue-specific patterns of miRNA expression in human and mouse tissues [50]. Therefore, it has been pos-sible to identify miRNAs highly expressed only in one or a few tissues: miR-133s (skeletal and cardiac muscle, prostate), miR-223 (spleen), and miR-1-b-2, -99b, -125, and -128 (brain). Interestingly, the microarray data confirm that each tissue has its specific miRNAs expression profile, that this specificity is maintained among differ-ent individuals, and that there is a different pattern of expression among fetal or adult origin of the same tissue (e.g., brain) [50]. The microarrays have many advantages: they require less RNA than a normal Northern blotting with no need for radioactive isotopes, they detect both mature and precursor miRNA molecules, and they analyze the global expression of hundreds of genes in the same sample at one time point.

The second high-throughput method was developed more recently and is a bead-based flow cytometric miRNA method [51], confirming the existence of a tissue-specific miRNA signature. The results of both techniques (microarrays and flow cytometric assay) have been confirmed by Northern blotting and quantitative real time (RT)-PCR. Quantitative RT-PCR is based on the quantitative relationship between the amount of starting target sample and the amount of PCR product at any given cycle number. The development of looped primers which anneal to specific miRNAs and are then extended by reverse transcriptase has allowed researchers to use this technique for the detection of miRNA levels in a very specific and sensitive manner [52]. Both the microarrays and the flow cytometric approach have been used by many investigators to study the different levels of expression of miRNAs in tumors. The involvement of miRNAs in cancer has been further confirmed by showing that their expression is significantly different between tumors and the normal tissue counterpart. These findings allow classification of some miRNAs as mainly oncogenetic and others as mainly oncosuppressors, although there is ever-increasing evidence of a dual role (both as oncogenes and as TSGs), at least for some miRNAs.

7.3.1.3   miRNAs as Oncogenes

Many studies have been conducted both in solid tumors and in hematologic malignancies, showing that some miRNAs are expressed at higher levels in cancers, with respect to the normal counterpart, suggesting a possible role as oncogenes (Figure 7.3 and Table 7.1). For some of these miRNAs the oncogenic pathway has been clarified. Volinia et al. [53] conducted a large-scale miRNome analysis on 540 samples including lung, breast, stomach, prostate, colon, and pancreatic tumors. The comparison of tumor samples with the normal counterpart showed a total of 43 aberrantly expressed miRNAs (26 up- and 17 downregulated). miR-21 resulted upregulated in all six types of solid cancers considered, miR-17-5p in 5 out of 6 tumors (not upregulated in gastric and breast cancer, respectively). Some of the miRNAs upregulated in this study have been well characterized for a strong cancer association by other groups (e.g., miR-17-5p, -20a, -21, -92, -106a, -155). Interestingly, this study shows the existence of a tumor-specific miRNA signature, which differs from the expression profile of noncancerous tissues and could have diagnostic and prognostic implications in the near future. In 32 lung cancer patients, high levels of miR-155 precursor were found to correlate significantly with a poor survival [54], whereas another study identified the miR-17-92 cluster as overexpressed, expecially in the small cell histotype [55], confirming a role of these miRNAs as oncogenes. In breast cancer the specific miRNA signature is composed of 29 deregulated miRNAs [56]. Among the most abnormally upmodulated again are miR-21 and miR-155 [56]. In an effort to explain the oncogenic effect of miR-21 in breast cancer, researchers demonstrated that downregulation of miR-21 with an antisense oligonucleotide suppresses both cell growth in vivo and tumor growth in the xenograft mouse model, by interfering with the BCL-2 pathway and inducing apoptosis [57]. These same two miRNAs were described as upregulated in papillary thyroid carcinoma (PTC) [58]. In PTC a set of five overexpressed miRNAs (including miR-221, -222, and -146b) distinguish between tumoral and normal thyroid tissue. In this study, polymorphisms in the 3'-UTR region of the KIT gene (in sites of interaction with the targeting miRNAs) were described [58]. In glioblastomas miR-21 has been implicated in the acquisition of the malignant phenotype [59], and miR-221 has emerged (out of 245 analyzed miRNAs, 9 of which strongly upregulated versus normal peritumoral tissues) as a tumor-specific marker [60].

   In hematologic malignancies an abnormally high expression of miR-155 and of its host gene BIC has been described in Burkitt lymphoma (BL) and several types of B-cell lymphoma [diffuse large B-cell lymphoma (DLBCL), primary mediastinal B-cell lymphoma (PMBL), and Hodgkin's lymphoma (HL)] [61]. Interestingly, in DLBCL, high levels of miR-155 are associated with the variant with the activated B-cell phenotype (poor prognosis), with respect to the germinal center phenotype (better prognosis) [62,63]. In addition, miR-21 is expressed at higher levels in the activated B-cell variant of DLBCL and is an independent prognostic indicator in de novo DLBCLs [63]. Since a 100-fold upregulation of the precursor of miR-155 has been described in pediatric BLs [64], whereas a lack of both BIC (B-cell integration cluster) and miR-155 expression has been observed in primary cases of BLs [65], a specific age-related role of this oncogenic miR substantially dependent on the age of onset of BL may be postulated. Recently, mir-155 both transgenic and knockout mice models have been developed. Our group observed the development of a preleukemic pre-B-cell proliferation evident in spleen and bone marrow of an Eμ transgenic mouse (able to express the miR-155 selectively in B-lymphocytes), followed by a frank B-cell malignancy [66]. Two different groups have studied the effects of miR-155 in a knockout mice model and they both found that miR-155 plays a critical role in the immune response [67,68]. In particular, one group found that miR-155 plays a specific role in the control of the germinal center (GC) reaction in the context of a T-cell-dependent antibody response. The control of the GC response is at least in part at the level of cytokine production, since bic/miR-155$^{-/-}$ cells produce more interleukin-4 and less interferon-γ, both hallmarks of a T_H2 differentiation [67]. The second group identified abnormalities in the protective immunity of bic-deficient mice. This effect was a consequence of diminished B- and T-cell responses at least in part mediated by a compromised ability of dendritic cells (DCs) to efficiently activate T cells because of defective antigen presentation or co-stimulatory functions [68]. Among the predicted targets of miR-155, the transcription factor c-MAF was identified in the knockout model and could explain why bic-deficient T_H cells are intrinsically biased toward T_H2 differentiation [68]. The regulation of miR-155 expression is still under investigation, since an imbalance between levels of BIC and miR-155 has been described in BL cell lines [69]. In particular, the BL cell line Ramos shows a yet unclear block of BIC processing to mature miR-155 after strong inducing stimuli of the transcription of BIC [such as protein kinase C (PKC) and NF-κB] [69]. The recent discovery that miR-155 expression is induced in primary murine macrophages by mediators of the inflammatory response (such as the polyriboinosinic:polyribocytidylic acid, or the cytokine interferon-β) indicates that miR-155-inducing signals use the JNK pathway and that this miR could represent a link between inflammation and cancer [70].

Another important cluster of miRNAs that plays an established oncogenic role in hematologic malignancies is the miR-17-92 cluster (which includes six miRNAs: miR-17-5p, -18a, -19a, -20a, -19b-1, and -92). They are located in the chromosomal region 13q31-32, which is well known to undergo amplification in malignant B-cell lymphomas [71]. The gene c13orf25 is the pri-miRNA encoding the cluster of miR-17-92, and its amplification in lymphomas explains the overexpression of the miR-17-92 cluster in this disease [71,72]. A strong indication of these clustered miRNAs as oncogenes comes from the observation that they are overexpressed in a variety of human cancers, including colon, lung, breast, pancreas, and prostate tumors [53,72,73]. In B-cell lymphoma-prone Eµ-myc transgenic mice (overexpressing c-myc specifically in B cells), the induced overexpression of the miR-17 cluster (not including miR-92), resulted in a dramatically accelerated lymphomagenesis and anticipated mice death [72]. Other studies have specifically addressed the oncogenic mechanism of these miRNAs. The transcription of miR-17-92 is directly activated by c-Myc [74]. Among the identified targets of miR-17-5p and miR-20a there is E2F1 [74], a transcription factor whose expression promotes both G_1 to S phase progression in mammalian cells (pro-proliferation effect [75]) and apoptosis in some settings [76]. C-myc and E2F1 reciprocally activate transcription of one another [77]. In normal cells, an overexpression of E2F1 leading to apoptosis might be expected, but in cancer cells, the amplification of the miR-17-92 cluster could promote oncogenesis through both increased transcription of c-myc and reduced levels of E2F1.

The mechanisms by which oncogenic miRNAs may become abnormally increased in cancers are chromosomal translocations, amplifications, and aberrant methylation. [For a detailed review, see Ref. 42.] An example of chromosomal translocation inducing a miR-dependent expression of a well-known oncogene is the t(8;17) translocation, which juxtaposes the coding region of the oncogene c-Myc under the control of the regulatory elements of miR-142, located only 50 nt from the translocation breakpoint [45]. As a result of this mechanism, a clinically aggressive acute prolymphocytic leukemia develops. miR-142-3p and 142-5p are also within the 17q23 minimal amplicon described in breast cancer [78]. Another very interesting translocation at 12q15 (occurring in many tumors) truncates the opening reading frame and the 3'-UTR of the oncogene HMGA2 (High Mobility Group A2) and appends those regulatory regions to the 3' end of known TSGs. It has been recently demonstrated that the microRNA let-7 targets HMGA2 in seven different complementary sites on the 3'-UTR or the gene. As a consequence of the translocation, two main oncogenic effects are achieved: (1) HMGA2 is released by the let-7 inhibition, and (2) let-7 negatively regulates the TSGs to which the HMGA2-3'-UTR region has been juxtaposed [79]. Genomic amplifications as overexpression mechanisms for oncogenic miRNAs have been described for the previously discussed miR-17-92 cluster (located at 13q31-32, in the gene c13orf25, which undergoes amplification in malignant B-cell lymphomas [71]) and for miR-21 [located at 17q23.2 in the 3'-UTR of VMP1 (vacuole membrane protein 1), a region frequently amplified in neuroblastomas as well as breast, colon, and lung cancers]. Silencing by promoter methylation has been reported for let-7a-3, located at 22q13.31 and associated with a CpG island region. In lung adenocarcinomas, hypomethylation of this miR promoter, a reactivation of the miR, and a pro-oncogenic effect as a consequence of the derepression have been described [80]. Finally, some miRNAs have a clearly established oncogenic role, but their regulatory mechanisms are not yet fully understood. As examples, miR-372 and -373 in primary cells can substitute for p53 loss and potentiate Ras-dependent cellular proliferation [81]. In testicular germ cell tumors p53 is rarely abnormal but miR-372 and -373 are frequently overexpressed through not yet fully defined mechanisms, and this contributes to the tumoral proliferation in the presence of activated Ras [81].

Altogether, these data confirm that miRNAs can act as oncogenes and that their overexpression is a consequence of well-known genetic and epigenetic mechanisms.

7.3.1.4   miRNAs as Tumor Suppressor Genes

The first correlation between miRNAs and cancer was identified in CLL, the most frequent form of adult leukemia in the Western world [82]. Hemizygous and/or homozygous deletions at 13q14 are the most common chromosomal abnormality in CLL (occurring in more than half of cases) but are very frequent also in mantle cell lymphomas (about 50% of cases [83]), in multiple myeloma (about 40% of cases [84]), and in about 70% of prostate carcinomas [85], revealing possible TSG location in this site. By comparison of 60 B-CLL patients and 30 human cancer cell lines to a panel of normal tissues (including CD5+ B cells isolated from tonsils of normal individuals, which can be considered the normal cells with respect to the CLL malignant counterpart), our group was able to identify two clustered miRNAs (miR-15a and miR-16-1) within a 30-kb minimal region of deletion, which are deleted or downregulated in 68% of B-CLL patients [82]. The oncogenic protein Bcl2 (B-cell lymphoma 2) is consistently overexpressed in CLL as well as in many other human tumors, both solid and hematologic [86]. In some malignancies (such as follicular lymphomas and a fraction of diffuse B-cell lymphomas) the mechanism of Bcl2 overexpression is known: a translocation t(14;18) (q32;q21) juxtaposes the BCL2 gene under the control of immunoglobulin heavy-chain enhancers, resulting in a hyperexpression of the gene [87]. Our finding of miR-15a and -16-1 as located in a region frequently deleted in CLL and other cancers suggested a TSG nature for these two clustered miRNAs, but the direct demonstration of their oncosuppressor nature came from the identification of Bcl2 as a target gene of theirs [88]. MEG-01 is a leukemia-derived cell line in which miR-15a and miR-16-1 are not expressed because of the deletion of one allele and the alteration of the other locus. Interestingly, exogenous expression of miR-15a and -16-1 in this cell line induces apoptosis both in vitro and in nude mice [88]. This finding provides the rationale for potential therapeutic implications of these miRNAs in Bcl2-overexpressing cancers. In 94 CLL patients our group identified a unique miRNA expression signature associated with prognosis and progression [89]. Among the 13 miRNAs (out of the 190 analyzed) that are part of the signature, miR-15a and -16-1 were widely deregulated in CLL; moreover, we described a germline point mutation in the primary precursor of these two miRNAs that determines a reduction of miR-15a and -16-1 expression both in vitro and in vivo [89]. Altogether, these findings document the role of miR-15a and miR-16-1 as TSGs.

A well-defined role as TSGs has been shown also for the let-7 family of miRNAs. In the human genome are localized 12 let-7 homologs, organized in 8 different clusters, 4 of which are positioned within genomic regions that are frequently deleted in many different human malignancies [45]. In lung cancer, lower expression of members of the let-7 family is associated with a poor prognosis [54], whereas reexpression of these miRNAs inhibits tumor growth in lung adenocarcinoma cell lines [90]. The molecular bases of the oncosuppressor nature of let-7s rely on the patterns of oncogenes that are targeted by let-7s: K-RAS and N-RAS are a class of potent oncogenes that are silenced by let-7 family members [91], as well as the oncogene HMGA2 [92]. Interestingly, for the let-7 family members a role as oncogenes (see previous paragraph) has been claimed recently. In neuroblastoma cells, miR-34a exerts an antitumoral effect by inducing apoptosis through a caspase-dependent mechanism. In particular, miR-34a directly targets E2F3, a transcription factor that induces cell cycle progression [93]. Moreover, retinoic acid-induced differentiation of neuroblastoma cell lines is associated with increased expression of miR-34a and decreased levels of E2F3 protein [93]. For some miRNAs the molecular mechanisms determining their oncosuppressor nature are not well understood yet. This is the case with miR-143 and miR-145. These two miRNAs are in cluster and are downregulated in many human tumors, including breast cancer [56], colon adenomas, and colorectal adenocarcinomas [94]. They are located at 5q32, a region of LOH and deletion in myelodysplastic syndromes [45], and this observation is in favor of their TSG nature.

For miRNAs that act as oncosuppressor, many regulatory mechanisms responsible for their abnormally reduced expression in tumors (with respect to the normal counterpart) have been identified. Chromosomal rearrangements (deletions, LOH), mutations, and promoter methylation are among these mechanisms. We have already discussed the chromosomal deletions that occur in the locus of miR-15a and miR-16-1 in CLL and other tumors and the significance of the mutations in the precursor of these miRNAs, as well as the LOH frequently occurring in the site of miR-143 and -145. Recent reports underline the importance of epigenetic silencing of miRNAs. miR-127 is located at 14q32.31, in the largest miRNA cluster identified to date (with miR-136, -431, -432, and -433) [95]. In cancer cells the putative promoter region of miR-127 is strongly hypermethylated, resulting in a lower expression of the miRNA. The epigenetic silencing could be reversed by a combination of 5-aza-2'-deoxycytidine (a DNA demethylating agent) and 4-phenylbutyric acid (a histone deacetylase inhibitor), whereas no effect was observed after treatment with each single drug or on other members of the cluster [96]. The strong methylation-mediated downregulation of miR-127 in cancer cells suggests a TSG role, further confirmed by the oncogene BCL6 as target of miR-127 [96]. Reexpression of miRNAs after treatment with demethylating agents and/or deacetylase inhibitors has not been described in non-small cell lung cancer cells [54], suggesting a tissue-specific role of epigenetics in miRNA regulation. Among the most strongly methylation-regulated miRNAs there is miR-124a. In a very elegant study, Lujambio et al. [97] determined a miRNA expression profiling in cancer cells genetically deficient for DNMT3B (responsible for de novo methylation) and DNMT1 (involved in the *maintenance* methylation) and identified miR-124a as silenced by promoter's methylation. The epigenetic silencing of this miRNA was associated with the activation of the oncogene CDK6 (cyclin D kinase 6) and the phosphorylation of the oncosuppressor gene RB (retinoblastoma) [97] (Figure 7.3 and Table 7.1).

The exponentially increasing number of reports showing an involvement of miRNAs in human malignancies (as OG and/or as TSGs) demonstrates the effort of the scientific community to decipher the role of these noncoding RNAs in the complex biology of cancer and to constantly put new pieces in a puzzle made of target mRNAs, pathways, and regulatory elements. These efforts have already led to some important diagnostic and therapeutic theoretical implications. With the shape of the puzzle becoming clearer and clearer, new hopes for cancer patients will become more an option than an abstract biological concept.

7.3.2 MICRORNAS AND NEUROLOGICAL DISEASES

We have already discussed the role of miRNAs in the neurological committment of stem cells. There is increasing evidence of the involvement of miRNAs both in the development of the nervous system and in neuronal plasticity, as well as in neurological diseases. [For extensive review, see Ref. 98.] The neurological disorder most frequently associated with miRNAs is the fragile X syndrome, which is the most common inherited mental retardation disease, affecting approximately 1:4000 males and 1:6000 females of all races and ethnic groups. This syndrome is characterized by a group of genetic conditions including a range of cognitive or intellectual disabilities with different grades of severity and is caused by mutations in the FMR1 gene, located on the long arm of the X chromosome. The fragile X mental retardation protein (FMRP) is an RNA-binding protein that can function as a translational suppressor. By its ability to form ribonucleoproteic complexes, FMRP can interact with the RISC complex, directly with miRNAs, and with other components of the miRNA machinery such as Dicer and the mammalian ortholog of Argonaute 1 (AGO1) [99]. Therefore, diseases causing mutation in FMR1 are likely to determine secondary effects on many RNAs, including miRNAs associated with FMR1.

Spinal muscular atrophy (SMA) is an inherited neurodegenerative disease in which a role of miRNAs can be hypothesized. The survival of motor neurons protein (SMN), part of the SMN complex, has critical functions in the assembly and restructuring of various ribonucleoproteic complexes. Some of the components of this SMN complex (such as GEMIN3 and 4) are also part of another complex containing eIF2C2 (a member of the Argonaute family) as well as many miRNAs that form novel ribonucleoproteins called miRNPs [100]. This intriguing correlation between a devastating neurodegenerative disease and the world of miRNAs will prompt more extensive studies on the possible implications of miRNAs for SMA patients both in terms of diagnosis and therapy.

The DiGeorge syndrome is clinically characterized by cardiac malformations, facial deformities, and endocrine and immune abnormalities. In the discussion in Section 7.1 about the physiology of miRNAs we described that the DGCR8 protein interacts with Drosha in the conversion of pri-miRNA into pre-miRNA (Figure 7.1).

It has been shown that the knockdown of the DGCR8 gene in *Drosophila* results in 5–23-fold accumulation of some pri-miRNAs [101,102]; therefore, confirming a strong link between this neurological disease and miRNA pathways.

In trichotillomania, an obsessive-compulsive disorder (OCD) characterized by excessive grooming leading to hair removal and skin lesions, a role for miR-196 is currently being investigated. In fact the homeobox gene HOXB8 is expressed in regions of the central nervous system known as "the OCD circuit" and some of the behavioral characteristics of trichotillomania have also been attributed to HOXB8 [103]. miR-196 can negatively regulate the expression of HOXB8 by targeting the 3′-UTR of its mRNA [39] and could therefore play a role in the pathogenesis of this disease.

Single-nucleotide polymorphisms (SNPs) in miRNA loci can be responsible for neurological diseases. In Texel sheep the G-to-A mutation of the 3′-UTR of the myostatin/growth differentiation factor 8 gene creates a target site for miR-1 and -206, which are highly expressed in skeletal muscle. As a result, a translational inhibition of the myostatin gene occurs. A rare sequence variant of the gene SLITRK1 (SLIT and Trk-like 1) in the binding site of miR-189 has been described in patients with Tourette syndrome [104]; moreover, an overlapping expression pattern of SLITRK1 mRNA and miR-189 has been observed in brain regions involved in Tourette syndrome.

Altogether, these data show a role of miRNAs in neurological diseases, and in some cases specific miRNA have been identified as involved in the pathogenetic tree of these neurological affections, raising new theoretical hopes for curing these patients.

7.3.3 MICRORNAs AND IMMUNITY

miRNAs play a role in the mammalian response to microbial infection. In human monocytic cells miR-146, miR-132, and miR-155 are strongly upregulated in response to lipopolysaccharide (LPS) [for review, see Ref. [105]). The miR-146 family is composed of two members (miR-146a and -146b) that are encoded by separate chromosomes and differ only for two nucleotides at the 3′ end of the mature sequence. Both loci respond to LPS, but whereas mature miR-146a levels are increased by LPS, miR-146b expression remains unchanged, therefore suggesting a fine regulation of this family member's expression [106]. miR-146a is involved in a feedback loop that modulates the cascade of events following the binding of LPS to the toll-like receptor 4 (TLR4). After the binding, a signaling cascade starts involving many adaptor molecules, which include TRAF6 (TNF receptor-associated factor 6), IRAK1 (IL-1 receptor-associated kinase 1), and many other mediators, which ultimately mobilize downstream transcription factors (such as AP-1 and NF-κB) with a well-established role in the regulation of immune-response genes. Among the genes whose expression is induced by NF-κB, there is miR-146a [106], which in turn suppresses both TRAF6 and IRAK1 by binding to the 3′-UTR of their mRNAs, therefore creating a negative feedback loop that controls the NF-κB-mediated response to LPS. miR-146a silencing could have a profound impact on this pathway because the two targets work in a linear signaling cascade and their downregulation substantially affects both IL-1 and TLR receptor signaling. Since overactivation of the innate immunity could lead to dangerous conditions (such as septic shock and autoimmune diseases), the identification of miR-146a as a possible regulator of this overactivation could have clinical implications. Moreover, one of the mechanisms invoked to explain endotoxin tolerance is the block of the LPS/TLR4 signaling. Reduced expression of IRAK1 protein in response to the first bacterial challenge could be responsible for the cell's becoming insensitive to the next LPS exposure [107,108]. Interestingly, a reduced level of IRAK1 protein, but normal levels of IRAK1 mRNA, have been described in this condition, strongly suggesting a miR-mediated action.

Another LPS-inducible miR is miR-132, a CREB-responsive gene able to control the expression of p250GAP, a GTPase-activating protein involved in the regulation of neuronal outgrowth in the rat [109]. Since members of the CREB family play a role in LPS signaling [110], miR-132 involvement in this cascade is currently under investigation as well.

Finally, miR-155 is induced by both bacterial (e.g., LPS) and viral (e.g., double-stranded RNA) ligands, indicating a function in the regulation of antimicrobial defense [106]. In bone marrow-derived macrophages, TLR3 induces miR-155, whereas IFN-mediated upregulation of this miR requires TNFα autocrine/paracrine signaling [70]. We have already discussed the role of this miRNA in cancer and the recent findings on knockout mice which strongly establish a role for miR-155 in immunity. Even if we have just begun to understand the role of miR-155 in human diseases, we can safely conclude that this miRNA represents an intriguing link between cancer and inflammatory response.

7.3.4 MICRORNAs AND VIRAL DISEASES

Plants and insects have developed a sophisticated and highly conserved antiviral defense system that uses small, noncoding RNA species to direct the sequence-specific silencing of gene expression. This defense system is called RNA interference (RNAi) and shares many components of the miRNA machinery. Viruses have evolved a number of mechanisms to overcome the RNAi defense system and to take advantage of the regulatory potentials of this machinery [for a review, see Ref. 111]. Viruses can encode miRNAs. The Epstein–Barr virus (EBV) hosts two clusters of miRNAs in its genome [112]. The first contains 14 miRNAs (miR-1 to miR-14) mapped to the intronic regions of the BamHI-A region rightward transcript (BART) gene and therefore will be referred to as the BART cluster. The second cluster is composed of three miRNAs and is located within the 5′- and 3′-UTRs of the BamHI fragment H rightward open reading frame 1 (BHRF1) gene and thus will be called the BHRF1 cluster [112,113]. Although the functions of these clusters remain unknown, it has been observed that among the predicted target mRNAs in human genome there are transcription factors, components of signal transduction pathways, genes involved in B-cell proliferation and apoptosis, and B-cell-specific chemokines. The profile of miRNA expression in EBV-infected cells shows that the BART cluster is constantly expressed at high levels in all stages of latency as well as in lytic infections, whereas the BHRF1 miRNAs were highly expressed in the late stages of the latency but were almost undetectable in the first stages [113], therefore indicating an important yet not fully understood role of these miRNAs in the life cycle of EBV. miRNAs have been found also in the genome of other members of the

Herpesviridae family—ten miRNAs in Kaposi's sarcoma-associated virus (KSHV or HHV-8), nine in the murine herpesvirus-68 (MHV-68), and nine in human cytomegalovirus (HCMV or HHV-5) [114]. For some members of the family—such as the human varicella-zoster virus (HHV-3) or the herpesviruses HHV-6 and HHV-7—no miRNAs have been identified so far; moreover, the lack of conservation in both location and sequences among the miRNAs of different members of the family suggests that these miRNAs could be important more in the adaptation to the specific cell types in which the virus persists than in core functions of the virus (such as viral replication, gene expression, new particles packaging, etc.). In the complex relationship established between the virus and the host, the following events have been demonstrated to occur: viral miRNAs can regulate host genes, viral miRNAs can target viral genes, and host miRNAs can regulate viral genes.

An example of the first kind of interaction involves the herpes simplex virus-1 (HSV-1). During viral latency, the latency-associated transcript (LAT) is the only viral gene expressed [115]. It has been shown that a miRNA can be encoded by exon 1 of the LAT gene [116]. Among the target human mRNAs for this miRNA are two genes associated with apoptosis: TGF-β (transforming growth factor-β) and SMAD3 (mothers against decapentaplegic homolog 3). In cells infected by HSV-1, the levels of TGF-β and SMAD3 were reduced and cells were protected from apoptosis, an effect that the virus needs to achieve in order to maintain latency. Mutants of HSV-1 with a deletion of exon 1 of the LAT gene were unable to prevent TGF-β-induced apoptosis.

Viral miRNAs interact not only with host mRNAs but also with viral genes. This has been seen in simian virus 40 (SV40), which hosts 2 miRNAs in the 3'-UTR region of the late pre-mRNA, downstream of the polyadenylation cleavage site [117]. The SV miRNAs are completely complementary to the 3'-UTR of the early SV transcripts that encode the T antigens. Therefore, they contribute to the silencing of early viral gene products. Moreover, since the T viral antigens are strong targets of cytotoxic T lymphocytes (CTLs), the miRNA-mediated silencing of these antigens represents a strategy for the virus to evade immune detection. Mutants of SV40, lacking the miRNA encoding region, show no decrease in early mRNAs and have a significantly higher susceptibility to lysis by CTLs [117]. Also, in EBV one miRNA (miR-BART2) is completely complementary to the 3'-UTR of the viral gene BALF5 (encoding the viral DNA polymerase), although the functional meaning of this interaction has yet to be understood. Viruses also reveal new possible regulatory mechanisms of miRNAs. In Kaposi's sarcoma-associated virus, one miRNA (miR-K12-10) is located within the open-reading frame of the gene on Kaposin (a protein involved in cellular transformation) [114,118]. Because of its location, the processing of the pri-miRNA by Drosha cleaves the kaposin mRNA, causing its degradation with no actual target complementarity between miR-K12-10 and kaposin mRNA. The last of the described host–virus interactions (host miRNAs regulating viral genes) is exemplified by the primate foamy virus type 1 (PFV-1). Computational analyses have identified a potential binding site for miR-32 in a specific region of the viral genome [119]. Mutations of this binding site, as well as transfection with oligonucleotides complementary to miR-32, resulted in an increased accumulation of progeny virus, confirming a role for miR-32 in silencing viral gene expression and replication. In order to circumvent this inhibitory effect, the PFV-1 virus encodes the Tas protein, a transcriptional activator expressed early in the viral life cycle, which is able to inhibit miRNA activity in tissue culture cells, therefore protecting the virus from the silencing mediated by the host miR-32. Also, miR-146 (in addition to its central role in the immune response) may exert antiviral properties, since its binding site has been detected in the genome of the PFV-1 virus, Dengue virus, HCV, influenza B virus, and many others [120].

Finally, specific consideration must be given to the unique relationship between miR-122 and the hepatitis C virus (HCV). miR-122 is specifically expressed in liver and seems to be required for the viral RNA expression to occur properly [121]. HCV replicons can replicate only in miR-122-expressing liver cell lines, and the silencing of miR-122 by antisense oligonucleotides impairs the ability of HCV to replicate in normally permissive cells. The molecular basis of this interaction must still be clarified.

7.3.5 MICRORNAs AND ENDOCRINE DISEASES

miRNAs have important implications in endocrinology, and their deregulation could lead to aberrations in hormone secretion. In mice, miR-375 (a pancreatic islet-specific miRNA) inhibits insulin secretion by targeting the expression of myotrophin, a cytoplasmic protein responsible for the exocytosis of insulin granules [122]. Although it is not known yet if aberrant expression of miR-375 occurs in diabetic patients, a better understanding of the regulation and functions of miR-375 (and the other 67 miRNAs found to be expressed in β cells), and a better definition of miRNA's role in the development of β cells as well as in the regulation of insulin secretion, could aid in the development of new treatments for diabetes. Another miRNA involved in endocrine functions is miR-143, which is involved in adipocyte differentiation [123]. Silencing of miR-143 by antisense oligonucleotides inhibits human pre-adipocyte differentiation in vitro and its ability to accumulate triglycerides. In fact, by reducing the levels of miR-143 in pre-adipocytes, the expression of adipocyte-specific genes (such as GLUT4, HSL, fatty acid-binding protein aP2, and PPARγ2) decreases. Among the many predicted targets of miR-143, ERK5/BMK1 could partially explain the effects of this miRNA on adipocyte differentiation, since ERK5 promotes cell growth and proliferation in these cells. Also, miR-14 can regulate adipocyte droplet size and triacylglycerol levels in *Drosophila* [124]. Considerable evidence is converging to indicate a role for miRNAs in endocrine biology. A better understanding of their role in diabetes, obesity, hypertension, and atherosclerosis could lead to greater comprehension of the physiopathology of these very common diseases and have a terrific clinical impact.

7.4 CONCLUSIONS

Despite the relatively recent identification of miRNAs as new players in the well-known biological paradigm that leads from DNA to protein, an impressive and constantly growing number of studies on their roles in human pathology are currently being conducted worldwide. Starting with high-throughput studies, which have identified miRNAs that are differentially expressed in pathologic versus normal tissues, the next step has been to identify which messenger RNAs are targeted by the culprit miRNAs, allowing determination of the biological pathway impacted by the deregulated miRNAs. In the meantime, other groups have identified both regulatory mechanisms of miRNA expression and regulatory loops in which miRNAs determine a "fine tuning" effect. The real challenge, though, is the translation of all the preclinical results into clinical benefits for patients. Although we are at the dawn of these clinical implications, a role for miRNAs in the diagnosis, prognosis, and therapy of human diseases can safely be hypothesized. miRNAs could help in the diagnosis of different pathologic conditions otherwise challenging to be clearly delineated. This seems to be the case for schizophrenia, which is a common neuropsychiatric disorder for which well-determined biologic and affordable parameters are still not available for the diagnosis. A recent study has identified a differentially expressed pattern of miRNAs from the postmortem prefrontal cortex tissue of 15 individuals with schizophrenia (or schizoaffective disorders) versus 21 psychiatrically unaffected individuals [125]. From this comparison, 15 miRNAs downregulated and one (miR-106b) upregulated in the schizophrenic patients. These differences could be determined by abnormal processing of precursor miRNAs into the mature product, although other studies are needed to address this hypothesis. These findings (once further confirmed by other groups in a higher number of patients) could identify a miRNA signature for schizophrenia that, in addition to the already existing criteria, could help in the diagnosis of this disease, especially in mild to borderline manifestations. We have previously discussed how for different kinds of cancers a specific miRNA signature has been identified, therefore helping pathologists to determine the origin of tumors of different origins, including the challenging diagnosis of "cancer of unknown primary" origin [51]. In some cases this miRNA's signature has a clearly documented prognostic meaning, such as in CLL [89]. Finally, from a therapeutic point of view, miRNAs are currently being investigated in preclinical gene therapy protocols, and the near future will tell whether restoring normal levels of deregulated miRNAs will affect the outcome of some of the most devastating human diseases.

ACKNOWLEDGMENTS

Dr. Croce is supported by program project grants from the National Cancer Institute and Dr. Calin by a Kimmel Foundation Scholar award, by the CLL Global Research Foundation, and by an M.D. Anderson Trust grant. We apologize to the many colleagues whose work was not cited due to space limitations.

REFERENCES

1. He, L. and Hannon, G.J., MicroRNA: small RNAs with a big role in gene regulation, *Nat. Rev. Genet.*, 5, 522, 2004.
2. Bartel, D.P., MicroRNAs: genomics, biogenesis, mechanism, and function, *Cell*, 116, 281, 2004.
3. Lee, R.C., Feinbaum, R.L., and Ambros, V., The C. elegans heterochronic gene lin-4 encodes small RNAs with antisense complementarity to lin-14, *Cell*, 75, 843, 1993.
4. Reinhart, B.J. et al., The 21-nucleotide let-7 RNA regulates developmental timing in Caenorhabditis elegans, *Nature*, 403, 901, 2000.

5. Lagos-Quintana, M., Rauhaut, R., Lendeckel, W., and Tuschl, T., Identification of novel genes coding for small expressed RNAs, *Science*, 294, 853, 2001.
6. Lau, N.C., Lim, L.P., Weinstein, E.G., and Bartel, D.P., An abundant class of tiny RNAs with probable regulatory roles in Caenorhabditis elegans, *Science*, 294, 858, 2001.
7. Lee, R.C. and Ambros, V., An extensive class of small RNAs in Caenorhabditis elegans, *Science*, 294, 862, 2001.
8. Lim, L.P. et al., Vertebrate microRNA genes, *Science*, 299, 1540, 2003.
9. Lim, L.P. et al., The micro-RNAs of Caenorhabditis elegans, *Genes & Dev.*, 17, 991, 2003.
10. Harfe, B.D., MicroRNAs in vertebrate development, *Curr. Opin. Genet. Dev.*, 15, 410, 2005.
11. Lewis, B.P. et al., Prediction of mammalian microRNA targets, *Cell*, 115, 787, 2005.
12. Lewis, B.P., Burge, C.B., and Bartel, D.P., Conserved seed pairing, often flanked by adenosines, indicates that thousands of human genes are microRNA targets, *Cell*, 120, 15, 2005.
13. Gregory, R.I., Chendrimada, T.P., Cooch, N., and Shiekhattar, R., Human RISC couples microRNA biogenesis and posttranscriptional gene silencing, *Cell*, 123, 631, 2005.
14. Matranga, C. et al., Passenger-strand cleavage facilitates assembly of siRNA into Ago2-containing RNAi enzyme complexes, *Cell*, 123, 607, 2005.
15. Rand, T.A., Petersen, S., Du, F., and Wang, X., Argonaute 2 cleaves the anti-guide strand of siRNA during RISC activation, *Cell*, 123, 621, 2005.
16. Brennecke, J., Stark, A., Russell, R.B., and Cohen, S.M., Principles of micro-RNA-target recognition, *PloS Biol.*, 3, e85, 2005.
17. Hwang, H.W., Wentzel, E.A., and Mendell, J.T., A hexanucleotide element directs microRNA nuclear import, *Science*, 315, 97, 2007.
18. Lim, L.P. et al., Microarray analysis shows that some microRNAs downregulate large numbers of target mRNAs, *Nature*, 433, 338, 2005.
19. Sevignani, C., Calin, G.A., Siracusa, L., and Croce, C.M., Mammalian microRNAs: a small world for fine-tuning gene expression, *Mamm. Genome*, 17, 189, 2006.
20. Ambros, V., The functions of animal microRNAs, *Nature*, 431, 350, 2004.
21. Griffiths-Jones, S., The microRNA registry, *Nucleic Acids Res.*, 32, 109, 2004.
22. Smith, A., Embryonic stem cells, in *Stem Cell Biology*, Marshak, D.R., Gardner, R.L., and Gottlieb, D., Eds., Cold Spring Harbor Laboratory Press, Cold Spring Harbor, New York, 2001, 205.
23. Evans, M.J., and Kaufman, M.H., Establishment in culture of pluripotential cells from mouse embryos, *Nature*, 292, 154, 1981.
24. Martin, G.R., Isolation of a pluripotent cell line from early mouse embryos cultured in medium conditioned by teratocarcinoma stem cells, *Proc. Natl. Acad. Sci. USA*, 78, 7634, 1981.
25. Wutz, A. and Jaenisch, R., A shift from reversible to irreversible X inactivation is triggered during ES cell differentiation, *Mol. Cell*, 5, 695, 2000.
26. Zhao, Y.X., Clark, J., and Ding, S., Genomic studies in stem cell systems, *Curr. Opin. Mol. Ther.*, 7, 543, 2005.
27. Cedar, S.H., The function of stem cells and their future roles in healthcare, *Br. J. Nurs.*, 15, 104, 2006.
28. Reya, T., Morrison, S.J., Clarke, M.F., and Weissman, I.L., Stem cells, cancer, and cancer stem cells, *Nature*, 415, 105, 2001.
29. Murchison, E.P. et al., Characterization of Dicer-deficient murine embryonic stem cells, *Proc. Natl. Acad. Sci. USA*, 102, 12135, 2005.
30. Bernstein, E. et al., Dicer is essential for mouse development, *Nat. Genet.*, 35, 215, 2003.

31. Hatfield, S.D. et al., Stem cell division is regulated by the microRNA pathway, *Nature*, 435, 974, 2005.
32. Kanellopoulou, C. et al., Dicer-deficient mouse embryonic stem cells are defective in differentiation and centromeric silencing, *Genes Dev.*, 19, 489, 2005.
33. Houbaviy, H.B., Murray, M.F., and Sharp, P.A., Embryonic stem cell-specific micro-RNAs, *Dev. Cell*, 5, 351, 2003.
34. Houbaviy, H.B., Dennis, L., Jaenisch, R., and Sharp, P.A., Characterization of a highly variable eutherian microRNA gene, *RNA*, 11, 1245, 2005.
35. Suh, M.R. et al., Human embryonic stem cells express a unique set of microRNAs, *Dev. Biol.*, 270, 488, 2004.
36. Chen, C.Z., Li, L., Lodish, H.F., and Bartel, D.P., MicroRNAs modulate hematopoietic lineage differentiation, *Science*, 303, 83, 2004.
37. Fazi, F. et al., A minicircuitry comprised of microRNA-223 and transcription factors NFI-A and C/EBPα regulates human granulopoiesis, *Cell*, 123, 819, 2005.
38. Felli, N. et al., MicroRNAs 221 and 222 inhibit normal erythropoiesis and erythroleukemic cell growth via kit receptor down-modulation, *Proc. Natl. Acad. Sci. USA*, 102, 18081, 2005.
39. Garzon, R. et al., MicroRNA fingerprints during human megakaryocytopoiesis, *Proc. Natl. Acad. Sci. USA*, 103, 5078, 2006.
40. Chen, J.F. et al., The role of microRNA-1 and microRNA-133 in skeletal muscle proliferation and differentiation, *Nat. Genet.*, 38, 228, 2006.
41. Zhao, Y, Samal, E., and Srivastava, D., Serum response factor regulates a muscle-specific microRNA that targets Hand2 during cardiogenesis, *Nature*, 436, 181, 2005.
42. Fabbri, M. et al., Regulatory mechanisms of microRNAs involvement in cancer: the strange case of Dr. Jekyll and Mr. Hyde, *Expert Opin. Biol. Ther.*, 7, 1009, 2007.
43. Cheng, A.M. et al., Antisense inhibition of human miRNAs and indications for an involvement of miRNA in cell growth and apoptosis, *Nucleic Acids Res.*, 33, 1290, 2005.
44. Esteller, M., Cancer epigenomics: DNA methylomes and histone-modification maps. *Nat. Rev. Genetics*, 8:286, 2007
45. Calin, G.A. et al., Human microRNA genes are frequently located at fragile sites and genomic regions involved in cancers, *Proc. Natl. Acad. Sci. USA*, 101, 2999, 2004.
46. Sonoki, T. et al., Insertion of microRNA-125b-1, a human homologue of lin-4, into a rearranged immunoglobulin heavy chain gene locus in a patient with precursor B-cell acute lymphoblastic leukemia, *Leukemia*, 19, 2009, 2005.
47. Calin, G.A. and Croce, C.M., MicroRNAs and chromosomal abnormalities in cancer cells, *Oncogene*, 25, 6202, 2006.
48. Zhang, L. et al., MicroRNAs exhibit high frequency genomic alterations in human cancer, *Proc. Natl. Acad. Sci. USA*, 103, 9136, 2006.
49. Sevignani, C. et al., MicroRNA genes are frequently located near mouse cancer susceptibility loci, *Proc. Natl. Acad. Sci. USA*, 104, 8017, 2007.
50. Liu, C.G. et al., An oligonucleotide microchip for genome-wide microRNA profiling in human and mouse tissues, *Proc. Natl. Acad. Sci. USA*, 101, 9740, 2004.
51. Lu, J. et al., MicroRNA expression profiles classify human cancers, *Nature*, 435, 834, 2005.
52. Chen C. et al., Real-time quantification of microRNAs by stem-loop RT-PCR, *Nucleic Acids Res.*, 33, e179, 2005.
53. Volinia, S. et al., A microRNA expression signature of human solid tumors defines cancer gene targets, *Proc. Natl. Acad. Sci. USA*, 103, 2257, 2005.
54 Yanaihara, N. et al., Unique microRNA molecular profiles in lung cancer diagnosis and prognosis, *Cancer Cell*, 9, 189, 2006.

55. Hayashita, Y. et al., A polycistronic microRNA cluster, miR-17-92, is overexpressed in human lung cancers and enhances cell proliferation, *Cancer Res.*, 65, 9628, 2005.
56. Iorio, M.V. et al., MicroRNA gene expression deregulation in human breast cancer, *Cancer Res.*, 65, 7065, 2005.
57. Si, M.L. et al., miR-21-mediated tumor growth, *Oncogene*, 26, 2799, 2007.
58. He, H. et al., The role of microRNA genes in papillary thyroid carcinoma, *Proc. Natl. Acad. Sci. USA*, 102, 19075, 2005.
59. Chan, J.A., Krichevsky, A.M., and Kosik, K.S., MicroRNA-21 is an antiapoptotic factor in human glioblastoma cells, *Cancer Res.*, 65, 6029, 2005.
60. Ciafre, S.A. et al., Extensive modulation of a set of microRNAs in primary glioblastoma, *Biochem. Biophys. Res. Commun.*, 334, 1351, 2005.
61. Kluiver, J. et al., BIC and miR-155 are highly expressed in Hodgkin, primary mediastinal and diffuse large B cell lymphomas, *J. Pathol.*, 207, 243, 2005.
62. Eis, P.S. et al., Accumulation of miR-155 and BIC RNA in human B cell lymphomas, *Proc. Natl. Acad. Sci. USA*, 102, 3627, 2005.
63. Lawrie, C.H. et al., Microrna expression distinguishes between germinal center B cell-like and activated B cell-like subtypes of diffuse large B cell lymphoma, *Int. J. Cancer*, doi: 10.1002/ijc.22800, 2007.
64. Metzler, M. et al., High expression of precursor microRNA-155/BIC RNA in children with Burkitt lymphoma, *Genes Chromosomes Cancer*, 39, 167, 2004.
65. Kluiver, J. et al., Lack of BIC and microRNA miR-155 expression in primary cases of Burkitt lymphoma, *Genes Chromosomes Cancer*, 45, 147, 2006.
66. Costinean, S. et al., Pre B cell proliferation and lymphoblastic leukemia/high grade lymphoma in Eμ-miR 155 transgenic mice, *Proc. Natl. Acad. Sci. USA*, 103, 7024, 2006.
67. Thai, T.H. et al., Regulation of the germinal center response by microRNA-155, *Science*, 316, 604, 2007.
68. Rodriguez, A. et al., Requirement of bic/microRNA-155 for normal immune function, *Science*, 316, 608, 2007.
69. Kluiver, J. et al., Regulation of pri-microRNA BIC transcription and processing in Burkitt lymphoma, *Oncogene*, 26, 3769, 2007.
70. O'Connell, R.M. et al., MicroRNA-155 is induced during the macrophage inflammatory response, *Proc. Natl. Acad. Sci. USA*, 104, 1604, 2007.
71. Ota, A. et al., Identification and characterization of a novel gene, C13orf25, as a target for 13q31-q32 amplification in malignant lymphoma, *Cancer Res.*, 64, 3087, 2004.
72. He, L. et al., A microRNA polycistronic as a potential human oncogene, *Nature*, 435, 828, 2005.
73. Hayashita, Y. et al., A polycistronic microRNA cluster, miR-17-92, is overexpressed in human lung cancers and enhances cell proliferation, *Cancer Res.*, 65, 9628, 2005.
74. O'Donnell, K.A. et al., C-myc-regulated microRNAs modulate E2F1 expression, *Nature*, 435, 839, 2005.
75. Bracken, A.P. et al., E2F target genes : unraveling the biology, *Trends Biochem. Sci.*, 29, 409, 2004.
76. Matsumura, I., Tanaka, H., and Kanakura, Y., E2F1 and c-Myc in cell growth and death, *Cell Cycle*, 2, 333, 2003.
77. Leone, G. et al., Myc and Ras collaborate in inducing accumulation of active cyclin E/Cdk2 and E2F, *Nature*, 387, 422, 1997.
78. Barlund, M. et al., Multiple genes at 17q23 undergo amplification and overexpression in breast cancer, *Cancer Res.*, 60, 5340, 2000.
79. Mayr, C., Hemann, M.T., and Bartel, D.P., Disrupting the pairing between let-7 and HmgA2 enhances oncogenic transformation, *Science*, 315, 1576, 2007.

80. Brueckner, B. et al., The human let-7a-3 locus contains an epigenetically regulated microRNA gene with oncogenic function, *Cancer Res.*, 67, 1419, 2007.
81. Voorhoeve, P.M. et al., A genetic screen implicates miRNA-372 and miRNA-373 as oncogenes in testicular germ cell tumors, *Cell*, 124, 1169, 2006.
82. Calin, G.A. et al., Frequent deletions and down-regulation of micro-RNA genes miR15 and miR16 at 13q14 in chronic lymphocytic leukemia, *Proc. Natl. Acad. Sci. USA*, 99, 15524, 2002.
83. Stilgenbauer, S. et al., Expressed sequences as candidates for a novel tumor suppressor gene at band 13q14 in B-cell chronic lymphocytic leukemia and mantle cell lymphoma, *Oncogene*, 16, 1891, 1998.
84. Elnenaei, M.O. et al., Delineation of the minimal region of loss at 13q14 in multiple myeloma, *Genes Chromosomes Cancer*, 36, 99, 2003.
85. Dong, J.T., Boyd, J.C., and Frierson, H.F. Jr., Loss of heterozygosity at 13q14 and 13q21 in high grade, high stage prostate cancer, *Prostate*, 49, 166, 2001.
86. Sanchez-Beato, M., Sanchez-Aguilera, A., and Piris, M.A., Cell cycle deregulation in B-cell lymphomas, *Blood*, 101, 1220, 2003.
87. Tsujimoto, Y. et al., Cloning of the chromosome breakpoint of neoplastic B cells with the t(14;18) chromosome translocation, *Science*, 226, 1097, 1984.
88. Cimmino, A. et al., miR-15 and miR-16 induce apoptosis by targeting BCL2, *Proc. Natl. Acad. Sci. USA*, 102, 13944, 2005.
89. Calin, G.A. et al., A microRNA signature associated with prognosis and progression in chronic lymphocytic leukemia, *N. Engl. J. Med.*, 353, 1793, 2005.
90. Takamizawa, J. et al., Reduced expression of the let-7 microRNAs in human lung cancers in association with shortened postoperative survival, *Cancer Res.*, 64, 3753, 2004.
91. Johnson, S.M. et al., RAS is regulated by the let-7 microRNA family, *Cell*, 120, 635, 2005.
92. Lee, Y.S. and Dutta, A., The tumor suppressor microRNA let-7 represses the HMGA2 oncogene, *Genes Dev.*, 21, 1025, 2007.
93. Welch, C., Chen, Y., and Stallings, R.L., MicroRNA-34a functions as a potential tumor suppressor by inducing apoptosis in neuroblastoma cells, *Oncogene*, 26, 5107, 2007.
94. Michael, M.Z. et al., Reduced accumulation of specific microRNAs in colorectal neoplasia, *Mol. Cancer Res.*, 1, 882, 2003.
95. Altuvia, Y. et al., Clustering and conservation patterns of human microRNAs, *Nucleic Acids Res.*, 33, 2697, 2005.
96. Saito, Y. et al., Specific activation of microRNA-127 with downregulation of the proto-oncogene BCL6 by chromatin-modifying drugs in human cancer cells, *Cancer Cell*, 9, 435, 2006.
97. Lujambio, A. et al., Genetic unmasking of an epigenetically silenced microRNA in human cancer cells, *Cancer Res.*, 67, 1424, 2007.
98. Kosik, K.S., The neuronal microRNA system, *Nat. Rev. Neurosci.*, 7, 911, 2006.
99. Jin, P., Alisch, R.S., and Warren, S.T., RNA and microRNAs in fragile X mental retardation, *Nat. Cell Biol.*, 6, 1048, 2004.
100. Mourelatos, Z. et al., miRNPs: a novel class of ribonucleoproteins containing numerous microRNAs, *Genes Dev.*, 16, 720, 2002.
101. Landthaler, M., Yalcin, A., and Tuschl, T., The human DiGeorge syndrome critical region gene 8 and its D. Melanogaster homolog are required for miRNA biogenesis, *Curr. Biol.*, 14, 2162, 2004.
102. Gregory, R.I. et al., The microprocessor complex mediates the genesis of microRNAs, *Nature*, 432, 235, 2004.
103. Greer, J.M. and Capecchi, M.R., Hoxb8 is required for normal grooming behaviour in mice, *Neuron*, 33, 23, 2002.

104. Abelson, J.F. et al., Sequence variants in SLITRK1 are associated with Tourette's syndrome, *Science*, 310, 317, 2005.
105. Taganov, K.D., Boldin, M.P., and Baltimore, D., MicroRNAs and immunity: tiny players in a big field, *Immunity*, 26, 133, 2007.
106. Taganov, K.D. et al., NF-κB-dependent induction of microRNA-146, an inhibitor targeted to signaling proteins of innate immune responses, *Proc. Natl. Acad. Sci. USA*, 103, 12481, 2006.
107. Li, L. et al., Characterization of interleukin-1 receptor-associated kinase in normal and endotoxin-tolerant cells, *J. Biol. Chem.*, 275, 13340, 2000.
108. Siedlar, M. et al., Tolerance induced by the lipopeptide Pam3Cys is due to ablation of IL-1R-associated kinase-1, *J. Immunol.*, 173, 2736, 2004.
109. Vo, N. et al., A cAMP-response element binding protein-induced microRNA regulates neuronal morphogenesis, *Proc. Natl. Acad. Sci. USA*, 102, 16426, 2005.
110. Gilchrist, M. et al., Systems biology approaches identify ATF3 as a negative regulator of Toll-like receptor 4, *Nature*, 441, 173, 2006.
111. Dykxhoorn, D.M., MicroRNAs in viral replication and pathogenesis, *DNA Cell Biol.*, 26, 239, 2007.
112. Pfeffer, S. et al., Identification of virus-encoded microRNAs, *Science*, 304, 734, 2004.
113. Cai, X. et al. Epstein-Barr virus microRNAs are evolutionarily conserved and differentially expressed, *PLoS Pathog.*, 2, e23, 2006.
114. Pfeffer, S. et al., Identification of microRNAs of the herpesvirus family, *Nat. Methods*, 2, 269, 2005.
115. Jones, C., Herpes simplex virus type 1 and bovine herpesvirus 1 latency, *Clin. Microbiol. Rev.*, 16, 79, 2003.
116. Gupta, A. et al., Anti-apoptotic function of a microRNA encoded by the HSV-1 latency-associated transcript, *Nature*, 442, 82, 2006.
117. Sullivan, C.S. et al., SV40-encoded microRNAs regulate viral gene expression and reduce susceptibility to cytotoxic T cells, *Nature*, 435, 682, 2005.
118. Cai, X. et al., Kaposi's sarcoma-associated herpesvirus expresses an array of viral microRNAs in latently infected cells, *Proc. Natl. Acad. Sci. USA*, 102, 5570, 2005.
119. Lecellier, C.H. et al., A cellular microRNA mediates antiviral defense in human cells, *Science*, 308, 557, 2005.
120. Hsu, P.W. et al., ViTa: prediction of host microRNAs targets on viruses, *Nucleic Acids Res.*, 35, D381, 2007.
121. Jopling, C.L. et al., Modulation of hepatitis C virus RNA abundance by a liver-specific microRNA, *Science*, 309, 1577, 2005.
122. Poy, M.N. et al., A pancreatic islet-specific microRNA regulates insulin secretion, *Nature*, 432, 226, 2004.
123. Esau, C. et al., MicroRNA-143 regulates adipocyte differentiation, *J. Biol. Chem.*, 279, 52361, 2004.
124. Xu, P. et al., The Drosophila microRNA miR-14 suppresses cell death and is required for normal fat metabolism, *Curr. Biol.*, 13, 790, 2003.
125. Perkins, D.O. et al., MicroRNA expression in the prefrontal cortex of individuals with schizophrenia and schizoaffective disorder, *Genome Biol.*, 8, R27, 2007.
126. Di Leva, G., Calin, G.A., and Croce, C.M., MicroRNAs: fundamental facts and involvement in human diseases, *Birth Defects Res. C. Embryo Today*, 78, 180, 2006.

8 Chromatin Modifications by Polycomb Complexes

Miguel Vidal

CONTENTS

8.1 Introduction 131
8.2 Genome-Wide Maps of Epigenetic Landmarks Associated with
Protein-Encoding Genes 132
    8.2.1 Histone Modifications in Promoter Proximal Regions 132
    8.2.2 DNA Methylation and Gene Expression 134
8.3 PcG Repressive Complexes 135
    8.3.1 Polycomb Repressive Complex of Type 2: Marking
    Chromatin for Repression 136
        8.3.1.1 The Activity of PRC2 Complexes Depends on
                Subunits Without Histone Methyltransfease Activity 137
    8.3.2 Polycomb Repressive Complex I: Maintaining the
    Repressed State 138
    8.3.3 Other PcG Complexes 139
        8.3.3.1 Complexes Containing the PRC1 Subunits
                Ring1A/Ring1B 139
8.4 Reversal of PcG-Dependent Epigenetic Marks 141
    8.4.1 Histone H3K27 Demethylases 141
    8.4.2 Histone H2A Deubiquitinases 142
8.5 Intersection of PcG and DNA Methylation Pathways 143
8.5 Recruiting of PcG Complexes and Mechanisms of Repression 145
8.6 Perspectives 145
Acknowledgments 146
References 146

8.1 INTRODUCTION

The Polycomb group (PcG) of genes is one of the constituents of the epigenetic regulatory machinery of pluricellular organisms. PcG genes were discovered in genetic screens for developmental regulators in the fruit fly *Drosophila melanogaster* (reviewed by Kennison, 1995).[1] They were identified as a set of genes that when mutated individually resulted in phenotypes similar to, or that enhanced, that of the *Polycomb* (*Pc*) gene (named after the supernumerary sex combs associated with Pc mutant male flies) after which the group is named. PcG are also found in plants, worms, and vertebrates.[2,3] Genetic analysis in *Drosophila* also uncovered the trithorax group (trxG) of genes, which counteract the activity of the PcG genes.[1] The trxG encode a complex mix of products with roles in transcriptional regulation that will not be considered here.

The diversity of specialized cell types in multicellular organisms is determined by cell-type-specific expression patterns arising from a differential reading of the same genetic information. The relative stability of these patterns and their maintenance through cell divisions are said to be the domains of epigenetic regulation, a term whose appropriate use it is still often debated.[4] The establishment and maintenance of patterns of gene expression involve chromatin modifications that often are equaled to epigenetic marks. Some of these modifications, of which DNA methylation and posttranslational modifications of histones are the best studied, influence transcription initiation through structural alterations in the chromatin fiber and through the recruitment of chromatin-binding proteins.[5-7]

PcG products assemble into chromatin-associated multiprotein complexes with a role in transcriptional repression. Their activity as transcriptional regulators depends on their ability to act as posttranslational modifiers of histones and possibly of other chromatin-associated proteins. The PcG system contributes to the ordered generation of cellular diversity that is intrinsic to developmental processes and to the regulated renewal of adult tissues. They ensure the maintenance of pluripotency programs in embryonic stem (ES) cells and also take part in processes that determine long-term silencing events, such as the monoallelic expression of imprinted genes or the inactivation of the X chromosome in mammals. PcG products are also active regulators of cell proliferation, and the genesis and expansion of tumor cells is often associated with PcG deregulation. Many of these aspects of PcG function in the biology of stem cells, the targets they regulate during early developmental stages, and their funcion in cell proliferation and cancer have been recently reviewed.[8-11]

This review examines recent developments in mammalian PcG complexes, the histone modifications associated with their activity together with their reversal, and the links between the PcG and DNA methylation regulatory systems.

8.2 GENOME-WIDE MAPS OF EPIGENETIC LANDMARKS ASSOCIATED WITH PROTEIN-ENCODING GENES

Conventionally, it has been expected that specific chromatin modifications would correlate with either activation or repression of transcription. However, recent work defining high-resolution maps of DNA methylation, RNA polymerase II (RNApol II) occupancy, and histone marks on a genome-wide scale shows that chromatin modifications do not follow a simple code,[12-15] suggesting that epigenetic states result from regulatory interactions between complexes that interpret these modifications and the enzymes that catalyze them.

8.2.1 HISTONE MODIFICATIONS IN PROMOTER PROXIMAL REGIONS

Nucleosomes around the promoter regions of most protein-encoding genes (75%) are associated with the histone H3 lysine 4 trimethyl mark (H3K4me3), whether they are transcriptionally active or inactive.[13] Together with this mark, commonly considered an indication of transcriptional activity, other "activating" marks such as acetylation of histone H3 lysines 9 and 14 (H3K9,14Ac) are also found on many inactive genes. Moreover, the form of RNApol II typically associated with the initiation of gene transcription (Ser5 phosphorylated) is found associated with a large number of genes that do not produce detectable transcripts. Therefore, transcription initiation, whether productive or not, appears as a more general phenomenon than previously anticipated.

These observations occur regardless of the cell type studied: undifferentiated pluripotent ES cells, hepatocytes, or B-lymphocytes. Therefore, it is likely they reflect general regulatory principles in transcriptional control. Thus most genes would fall in one of three categories.[13]

The first category would include those genes thought to be regulated mostly by specific mechanisms directed to prevent transcription initiation. Their promoters have no H3K4me3 marks and typically have a low content in CpG dinucleotides. It is likely that histone H3 in nucleosomes at these loci is (asymmetrically) dimethylated at arginine 2 (H3R2me2a), a mark present throughout the body and 3' region of genes, regardless of their activity, but that when present at promoter proximal areas prevents association of histone H3K4 methyltransferases.[16]

Another category, containing about 50% of the genes, is that of actively transcribing genes (i.e., producing detectable transcripts) whose nucleosomes contain H3K4me3 and H3K9,14Ac. A major feature of this large group of genes is the presence of histone H3 trimethylated at lysine 36 (HeK36me3) in nucleosomes downstream of the transcription initiation site, a transcription elongation-specific mark.[17] Their promoters are occupied by the initiating form of RNApolII. This category includes most of the ubiquitously expressed genes with housekeeping functions, and their promoters are enriched in the CpG dinucleotide.

The third category contains genes controlled mostly through posttranscription initiation steps, since their nucleosomes contain H3K4me3 and H3K9,14Ac but lack the elongation marks H3K36me3 or H3K79me2. This group of genes, about one third of the total, includes many developmental regulators, and although they seem to initiate transcription, their transcripts do not accumulate. Mechanisms involved in posttranscriptional regulation of these genes could also include mRNA degradation events. The promoters of these genes are also CpG rich. It is likely that, at least in ES cells, this set of genes overlaps with another group of genes characterized by the concurrent presence of H3K4me3 and H3K27me3 marks.[14] Nucleosomes marked in such a manner define the so-called bivalent domains, because when they were initially identified[18,19] these histone marks were associated with opposite transcriptional outcomes. Overall, these domains are transcriptionally silent and are occupied by PcG products.[14]

During differentiation of ES cells, the histone marks on these bivalent domains evolve. Thus, in lineage-committed cells, such as neuronal precursor cells derived from ES cells or embryonic fibroblasts, the genes whose expression increases only show H3K4me3 marks, whereas those that remain silent lose them both, become marked only with H3K27me3, or maintain their bivalent marks.[14] Bivalent sites, then, are not specific to pluripotent cells.[12] What sets of genes retain both marks or resolve the bivalent structure (into only H3K4me3 or only H3K27me3) is determined in a lineage-specific manner. The results confirm, on a genome-wide scale, the initial proposition that bivalent domains reflect a mechanism that prevents developmentally relevant genes from being inappropriately expressed in pluripotent cells and in their differentiating descendants.[14,19] The contribution of the PcG system to this regulatory strategy is essential since the H3K27me3 mark is strictly dependent on the activity of a subset of PcG complexes.

8.2.2 DNA Methylation and Gene Expression

In mammals, DNA methylation is restricted to the CpG dinucleotide, which is underrepresented except for short sequence stretches called CpG islands, often associated with promoters. Typically, DNA methylation is used to silence promoters in transposons, imprinted genes, and the inactive X chromosome, whereas its contribution to tissue-specific expression in somatic cells has been controversial (reviewed by Goll and Bestor, 2005).[20] Citosine methylation is the only chromatin modification with a known mechanism for propagation through cell generations.

DNA methylation maps have been generated by hybridizing methylated DNA (immunoprecipitated with antibodies specific for methylated cytosine) to microarrays containing sequences corresponding to >20,000 human promoters. The comparison between methylated DNA maps and those of RNApol II occupancy and histone marks has been used to evaluate the impact of DNA methylation on gene expression.[15] The results show that promoters with a high CpG content, typically marked by methylated H3K4, have their DNA always unmethylated regardless of their transcriptional state, except for two exceptions. One is that of a large subset of germline-specific promoters, including those containing CpG islands, which are methylated in somatic cells. Another is a small number of CpG islands corresponding to transcription factors that do not represent gene ontology groups and that are methylated in somatic tissues. Contrasting with the hypomethylated state of CpG-rich promoters, those with an intermediate content in CpG are methylated even if they are transcriptionally active.

The limited methylation of DNA in regulatory sequences contrasts with that of repeats and intergenic sequences that are consistently methylated, perhaps because of their importance in the maintenance of genome stability.[21] Overall, the data support the notion that differential methylation is not a general mechanism involved in tissue-specific gene expression.

What determines the unmethylated state of CpG islands is not known. The recent finding that a substantial number of conserved CpG islands lie in the proximity of PcG target loci defined in ES cells suggests that PcG products may contribute to the unmethylated state.[22] However, considering that the mechanisms mediating PcG recruiting are largely unknown, it would also be possible that PcG products are associated with these sites following an uncharacterized mechanism for the recognition of stretches of unmethylated sequences. The interplay between PcG products and DNA methylation is further underlined by the increased likelihood (by 10-fold) of PcG target genes being methylated in tumor cells compared to that of nontargets.[23–25]

8.3 PCG REPRESSIVE COMPLEXES

PcG genes encode a diverse set of proteins that display a number of functionally relevant protein motifs conserved in vertebrate and invertebrate PcG proteins. Vertebrates have several PcG gene paralogs, from two to six, for each of the PcG gene homologs in flies. For instance, whereas *D. melanogaster* has single *Pc* or *Posterior sex comb* (*Psc*) genes, mammals have four and six paralogs, respectively. A summary of products encoded by PcG genes is shown in Table 8.1.

TABLE 8.1
Mammalian PcG Products

PcG Complex	Subunits	Domains	Function	Fly Homolog
PRC1	M33/Cbx2	Chromodomain	Binding to H3K27me3	PC
	Pc2/Cbx4*	Chromodomain	Binding to H3K27me3	
	Pc3/Cbx8	Chromodomain	Binding to H3K27me3	
	Cbx7	Chromodomain	Binding to H3K27me3	
	Cbx6	Chromodomain		
	Ring1A/Ring1	RING finger	E3 ubiquitin ligase	SCE
	Ring1B/Rnf2	RING finger	(histone H2AK119 Ub1)	
	Bmi1/Pcgf4	RING finger	E3 ubiquitin ligase cofactors	PSC
	Mel18/Pcgf2	RING finger	E3 ubiquitin ligase cofactors	
	Ph1/Phc1	SAM, Zn finger	Oligomerization,	PH
	Ph2/Phc2	SAM, Zn finger	Heterodimerization,	
	Ph3/Phc3	SAM, Zn finger	RNA binding?	
	Scmh1	SAM, MBT	Oligomerization	SCMH
	Scmh2	SAM, MBT	Heterodimerization	
PRC2	Ezh2	SET	Histone methyl transferase (H3K27me3)	E(Z)
	Eed	WD40		ESC
	Suz12	Zn finger		SU(Z)12
Other complexes	Pcgf1	RING finger	E3 ubiquitin ligase cofactors	
	Pcgf3	RING finger		
	Pcgf5	RING finger		
	Pcgf6	RING finger		
	L3mbtl1	SAM, MBT	Oligomerization,	L3MBTL
	L3mtbtl2	MBT	Heterodimerization	
	L3mbtl3	SAM, MBT		
	Sfmbt1	SAM, MBT		
	Mbtd1	MBT		
	YY1	Zn finger	DNA binding	PHO, PHOL

*Pc2/Cbx4 has a E3 SUMO ligase activity

Early evidence from genetic interactions between PcG genes and from immunoprecipitation studies suggested that the PcG products work as multiprotein complexes. Searches for PcG subunits identified separate PcG complexes on the basis of the lack of associations among sets of PcG products in yeast two hybrid interaction assays.[26,27] Although the biochemical definition of these complexes, through their purification from cell extracts, has revealed a large variety of them,[28] they can be categorized in either of two major groups: Polycomb repressive complexes of type 1 (PRC1) and Polycomb repressive complexes of type 2 (PRC2). These are compositionally defined by a core of mutually exclusive, rather invariant, PcG-encoded subunits, which in turn confer them specific histone-modifying activites: PRC1 complexes contain a histone H2A E3 monoubiquitin ligase activity, whereas PRC2 complexes have a histone methyltransferase that methylates the lysine 27 of histone H3. In addition, a variety of complexes that do not contain complete sets of these core subunits, but only some PcG subunits, has also been identified.

8.3.1 POLYCOMB REPRESSIVE COMPLEX OF TYPE 2: MARKING CHROMATIN FOR REPRESSION

The histone methyltransferase activity of PcG complexes is associated with Enhancer of zeste [E(z) in flies, EZH2 in vertebrates], a protein containing a SET domain similar to those identified in other histone methyltransferases. E(z) or EZH2 complexes have been isolated either using an epitope tag approach[29,30] or by unbiased isolation following the enzymatic activity.[30–32] The complexes thus isolated differed slightly in their subunit composition, possibly as a consequence of the distinct purification schemes. However, they all share subunits that define the core of PcG complexes with histone H3K27 methyltransferase activity. These include, in addition to E(z)/EZH2, the following PcG gene products: extra sex combs (Esc in flies, EED in vertebrates), Suppressor of zeste 12 (SUZ12), and the histone-binding proteins RbAp46/RBBP7 or Rbbp48/Nurf55/RBBP4. Biochemical analysis of the isolated complexes, or of their reconstituted counterparts using recombinant subunits, shows that their histone methyltransferase activity is specific for the lysine K27 of H3 in either histone octamers or oligonucleosomes (reviewed by Cao and Zhang, 2004).[33] Enhancer of zeste complexes are also found in plants and in worms and seem to correspond to the earliest PcG genes that appeared during evolution (reviewed by Whitcomb et al., 2007).[3]

The relevance of H3K27 methylation for gene repression was first observed in rescue experiments in the fly. Thus, enzymatically inactive E(z) proteins, but not the wild-type form, failed to silence Ubx,[29] a Hox gene regulated by the PcG system. Also, H3K27me3 marks, detected by chromatin immunoprecipitation experiments, were abolished in E(z) mutant embryos with concurrent upregulation of the Ubx gene.[34] Also important, the inactivation of E(z) was found to affect the occupancy of the Ubx regulatory region by PC, a subunit of the other large group of Polycomb complexes, PRC1, thus linking the activity of the two complexes to gene repression.[31] This led to the proposition that PRC2 complexes are involved in the establishment of a transcriptionally repressed state (Polycomb initiation complex) by marking regulatory regions for their eventual repression.

Currently, the notion that H3K27me3 marks associate to gene repression is firmly established from genome-wide expression analysis of ES cells deficient in core subunits of PRC2 complexes. In these mutant cells, transcription factors and other developmental regulators account for a large fraction of the genes derepressed in a H3K27me3 dependent.[35-37] The activity of PRC2 complexes is also relevant to regulatory processes involved in stable gene silencing such as X chromosome inactivation[34,38,39] and genomic imprinting.[40] In agreement with this broad regulatory role, mutant alleles of the mouse *Ezh2* gene are embryonic lethal[41,42] at preimplantation stages of development, and no *Ezh2−/−* ES cells can be established.[41]

8.3.1.1 The Activity of PRC2 Complexes Depends on Subunits Without Histone Methyltransferase Activity

Although EZH2 is the enzymatically active subunit of PRC2 complexes, the other core subunits are also essential to the activity of the complex, as shown by the inactivation of core subunits Eed[43] or Suz12.[37,44] In part this is due to their contribution to the stability of the complex since, in the absence of Eed or Suz12, no PRC2 core subunits are detected despite the fact that their stationary mRNA levels are not affected. In addition, Suz12 is needed for the recruitment of RbAp48/Rbbp4 to the complex,[37] a subunit that mediates nucleosome binding.[37,45] Analysis of global methylation of H3K27 in Suz12- or Eed-deficient mouse ES cells shows no tri- and dimethylated forms of H3K27, suggesting that the activity of PRC2 is specific. A difference between these cells, however, is that whereas H3K27me1 is readily detected in Suz12−/− cells, it cannot be detected in Eed−/− cells.[43] Whether this relates to an as yet uncharacterized histone H3K27 methyltransferase is not known.

Two more EZH2-SUZ12-RBBP4/7-containing complexes have been isolated that differ in the EED form they contain,[46,47] resulting from the use of alternate translation initiation sites on *EED* mRNA. Thus, in human cells, PRC2 contains a mixture of EED (amino acids 35–535) and full-length EED (EED1, amino acids 1–535) and shows no H3K27 methyltransferase activity on nucleosomal templates. In contrast, a so-called PRC3 complex, which has only the EED (amino acids 95–535) form, is active on nucleosomal substrates. Finally, a PRC4 complex contains the EED2 (amino acids 35–535) form.[46] PRC3 and PRC4 complexes also show the ability to methylate in vitro the lysine 26 of linker histone H1, an activity that depends on the presence of SUZ12.[48] The substrate preferences and specificity associated with the various EED forms, however, are controversial,[49] and in vivo evidence is needed to evaluate the putative regulatory differences among PRC2, -3, and -4 complexes.

PRC4 was identified in tissue culture cells that overexpress EZH2, and contains SIRT1, a NAD+-dependent histone deacetylase that is not present in the other EZH2-containing complexes.[46] Eed2, the Eed form of PRC4 complexes, is expressed predominantly in undifferentiated ES cells and in cancer cells, but not in normal cells or differentiating ES cells. This is consistent with the formation of PRC4 complexes in tumor cells since usually cancer progression is accompanied by increases in the levels of EZH2.[50] However, the functional relevance of the proposed changes in EZH2-containing complexes is yet to be determined.

Another example of the dependence of PRC2 activity on their subunits is that of a newly identified complex, Pcl-PRC2, isolated from *Drosophila* embryos, which contains the product of the *polycomblike (Pcl)* gene product. This complex, for which no mammalian counterpart has yet been found, seems responsible for the high density of H3K27me3 marks at PcG target genes. Inactivation of *Pcl* has no effect on PRC2 recruiting, and the associated gene derepression occurs with a decrease (but not elimination) of H3K27me3, leaving H3K27me2 and H3K27me1 marks intact.[51]

8.3.2 POLYCOMB REPRESSIVE COMPLEX I: MAINTAINING THE REPRESSED STATE

Most of the PcG products not associated with PRC2 complexes belong in the other major Polycomb multiprotein complex, PRC1, also named as the maintenance complex because it acts to repress genes marked in the first place by the PRC2 (initiating) complex. Although this is a simplified description of the function of both types of complexes, it is useful in delineating fundamental differences between the activities of PcG complexes as transcriptional repressors.

Using an epitope tag approach with early *D. melanogaster* embryos that expressed the FLAG-tagged PcG products Polyhomeotic or Posterior sex combs, researchers isolated a complex that also contained two other PcG products: Polycomb and the homolog of vertebrate Ring1 proteins.[52,131] Proteomic analysis of the complex identified a relatively large set of non-PcG products that included some components of the transcription initiation machinery.[53] This complex was named PRC1. A related complex, isolated in a similar fashion from human tissue culture cells, contained a much simpler (fewer subunits) complex, including the homologs of Polycomb (M33/CBX2 and PC3/CBX8), Posterior sex combs (BMI1/PCGF4), Polyhomeotic (HPH1/PHC1, HPH2/PHC2 and HPH3/PHC3), and Sex combs extra (RING1A/RING1 and RING1B/RNF2), which are the subunits that define the core of PRC1 complexes. The complex also contained substoichiometric amounts of another PcG product, the homolog of Sex combs on midleg (SCMH1), and a few non-PcG subunits.[54] Similar PRC1 complexes differing in the homologs that compose the core subunits, possibly reflecting their relative abundance, have also been isolated or identified.[55–57]

The RING1B and RING1A subunits of PRC1 complexes are E3 ligases that monoubiquitylate the lysine 119 of histone H2A (H2AK119Ub1). They were found in a purification scheme using an in vitro ubiquitylation assay with a nucleosomal substrate in the presence of added ubiquitin and E1 (ubiquitin-activating) and E2 (activated ubiquitin transfer) ligases.[58] The complex, named PRC1l, contained the PRC1 core subunits RING1A/RING1, RING1B/RNF2, BMI1/PCGF4, PH1/PHC1, and two unidentified components. RING1B/RNF2 is the essential component that interacts with the UbcH5c E2 ligase, but maximal activity of the complex depends on BMI1/PCGF4, which acts as a coactivator.[59–61] The E3 ligase activity of the paralog RING1A/RING1 is much weaker than that of RING1B/RNF2, at least in vitro.[59,61]

Inactivation of *Ring1B/Rnf2* or *Bmi1/Pcgf4* results in *Hox* gene derepression, together with a decrease in the levels of the H2AK119Ub1 marks at regulatory regions of these genes.[60,61]

The integrity of the PRC1 complex and of its subunits depends on Ring1B/Rnf2, in a parallel situation to that of PRC2 complex. *Ring1B/Rnf2⁻/⁻* ES cells are almost depleted of PRC1 proteins, even though the stationary levels of the corresponding mRNAs are unaffected.[63,64] Therefore, the wide gene derepression that accompanies *Ring1B/Rnf2* inactivation, which comprises a set of genes largely overlapping with that of PRC2 targets,[63,64] cannot be associated only with the decrease of H2K119Ub1 marks observed both globally and at specific targets.

8.3.3 OTHER PcG COMPLEXES

Several multiprotein complexes that contain PcG subunits have been identified. Two of particular interest contain the only DNA-binding protein encoded by a PcG gene the product of the only PcG gene: Pleiohomeotic (Pho) and the fly homolog of vertebrate YY1 proteins (YY1, YY2, and Rex-1),[65] which have a conserved four-zinc-finger domain with the ability to bind DNA. The complexes, however, have been identified only in extracts from *D. melanogaster* embryos.[66,132] In one, Pho is associated with dINO80, a component of nucleosomal remodeling activities. In the other, a so-called Pho repressive complex (PhoRC), Pho is associated with dSfmbt, a protein with the ability to bind mono- and dimethylated forms of histone H3K9 and H4K20 through MBT domains. PhoRC acts as a *Hox* gene repressor and involves histone marks other than H3K27 and H2AK119Ub1 in PcG-mediated gene repression. However, no mammalian counterpart of this complex has been found yet. As for other protein complexes containing PcG subunits, those containing Ring1A/Ring1 and Ring1B/Rnf2 are predominant.

8.3.3.1 Complexes Containing the PRC1 Subunits Ring1A/Ring1B

During the biochemical characterization of a number of transcriptional regulators several complexes have been isolated that contain the E3 ligases initially found in PRC1 complexes. All of them were identified using epitope tagged forms of non-PcG regulators. However, it is likely that they coexist, together with PRC1, in most cell types, as suggested by the proteomic profiling of Ring1B/Rnf2-containing complexes in a murine hematopoietic cell line.[57]

One of these complexes, E2F6.com-1, was isolated from human Hela cells that expressed as a doubly FLAG, HA-tagged E2F6 cDNA.[67] E2F6 and DP1, a heteromeric partner of the E2F family of transcription factors, form a DNA-binding module that targets the complex to E2F targets; MAX and MGA, another heteromeric transcription factor that recognizes E2 DNA-binding motifs, also copurified with this complex. MGA, in addition, has a motif found in the T-box gene family of DNA-binding proteins. Additional subunits associated with this purification are the heterodimeric GLP/EHMT1-G9a/EHMT2 histone methyltransferase, specific for the lysine 9 of histone H3; the PcG-related proteins RING1A, RING1B, and MBLR/PCGF6; and YAF2, a paralog of the Ring1 and YY1-binding protein RYBP.[133,134] Subunits that recognize methylated histone residues, such as HP1γ, and L3MBTL2, were also identified in this purification. L3MBTL2 is one of several related proteins that contain Mbt repeats, including dSfmbt of the PhoRC complex of *Drosophila*.[66] It is of note that the related L3MBTL1 protein has the ability to compact nucleosomal arrays in a histone modification-specific manner (methylated H4K20 and H1bK26).[68] Such a chromatin compaction activity depends on the SAM domain of L3MBTL1, a protein motif found in the polyhomeotic subunits of PRC1, PHC1, PHC2, and PHC3, which is known for its ability to participate in large homo- and heterodimeric SAM polymers.[69,70] It is not clear, however, whether all these subunits are a single biochemical entity or whether, more likely, they represent a mix of two or more complexes isolated by their content in E2F6.

A related complex, containing most of the subunits identified in the isolation of E2F6.com-1, except for MBLR/PCGF6, was purified from nuclear extracts from Hela cells expressing a FLAG and HA-tagged form of the histone H3K4me3 demethylase JARID1C/SMX.[71] Among the co-purifying subunits were histone deacetylases 1 and 2 (HDAC1, 2), the corepressor NCOR1, and the DNA-binding repressor REST.

A rather different complex, identified during a purification scheme to isolate complexes that contained a BCL6-associated corepressor, BCOR,[72] contained the PRC1 subunits RING1A, RING1B, their interactor RYBP and the BMI1 paralog NSPC1/PCGF1, and subunits with a distinct regulatory potential. These are SKP1 and FBXL10/JHDM1b, which may play roles in ubiquitylation processes; additionally, a histone H3K36 demethylase activity has been associated with FBXL10.[73] This BCOR/FBXL10 complex was shown to associate and modify histone H2A in a BCL6-dependent manner.[72]

Most of these subunits have also been identified during the identification of components of Ring1B/Rnf2-containing complexes.[57] In this study, several other proteins previously not known to be associated with Ring1B/Rnf2 were identified, including histone demethylases (LSD1) and kinases (Ck2a,b), expanding the regulatory potential of complexes containing the PRC1 subunit Ring1B/Rnf2.

A CTBP repressor complex,[74] containing HDACs and the GLP/EHMT1-G9a/EHMT2 histone methyltransferase, among other subunits, was found to contain PC2/CBX4, which is the only Polycomb paralog that has an associated E3 SUMO ligase activity.[75] Several substrates have been found to be sumoylated in a PC2/CBX4-dependent manner, such as Smad-interacting protein 1,[76] homeodomain-interacting protein kinase 2,[77] and Dnmt3a,[78] but it is not known whether they are associated with PRC1 complexes or are part of other regulatory ensembles.

Whereas all these Ring1A/Ring1B-containing complexes contain subunits that share transcriptional repression functions, there is at least one example of a complex involved in activating function. This complex, isolated by affinity purification of FLAG-tagged WDR5, which contains the H3K4 histone methyltransferase MLL1 and the H4K16 histone acetyltransferase MOF, among many other subunits, also has RING1B/RNF2.[79] Again, the nature of the purification scheme makes it likely that these subunits belong to more than one WDR5-containing complex.

The function of PcG subunits, mostly RING1A and RING1B, in such a heterogeneous mix of complexes is unknown. They may act as histone H2A E3 monoubiquitin ligases, but they could also play a scaffold/structural role, or even act as ubiquitin ligases modifying some of their interacting partners or nearby chromatin proteins to which they are recruited. It is also possible that some of the subunits in these complexes mediate or contribute to H3K27me3-independent recruitment of Ring1B.[80]

8.4   REVERSAL OF PCG-DEPENDENT EPIGENETIC MARKS

Resolution of bivalent marks during differentiation of ES cells involves the loss of H3K27 marks. This means that histone modifications are removed or the nucleosomes that contain the modified histones are replaced by new ones assembled from histones marked in some other manner. Whereas global changes in the histone marks could result from variations in the expression levels of PcG subunits, it is likely that loci-specific processes are responsible for the cell type changes occurring during differentiation, such as the resolution of bivalent marks. Recent reports have identified both H3 lysine 27 demethylases and H2A deubiquitinases involved in control of PcG targets.

8.4.1   HISTONE H3K27 DEMETHYLASES

A large family of histone demethylases share a Jumonji C (JmjC) motif as the catalytically active domain in a hydroxylation reaction that uses Fe(II) and α-ketoglutarate as cofactors (reviewed by Klose et al., 2006, and by Shi and Whetstine, 2007).[81,82] Following a candidate approach, researchers identified a phylogenetic subcluster of JmjC-domain proteins of unknown function (UTX, UTY, and JMJD3) in silico. In vitro work with recombinant forms of at least two of them, UTX and JMJD3, showed they demethylate both H3K27me3 and H3K27me2, but not H3K27me1.[83,84] The reaction was specific, and peptides with trimethylated H3K4, H3K9, or H4K20 were not modified. Likewise, their overexpression in tissue culture cells resulted in decreased immunofluorescent signals for H3K27me3 and H3K27me2.[84] Jmjd3 was also independently identified among strongly induced genes in macrophages subjected to inflammatory stimuli.[85]

UTX and JMJD3 bind to active Hox genes, both in tissue culture cells and in bone marrow cells, becoming upregulated upon inactivation of the H3K27 demethylases. In the pluripotent NT2/D1 embryonal carcinoma cell line, downregulation of UTX results in enhanced occupancy of the promoter region of two Hox genes by PRC1 products BMI1 and RING1A and enrichment in H2A119Ub1 marks.[84] These results are consistent with the proposed role of PRC1 complexes maintaining transcriptional repression of genes marked by H3K27me3-marked nucleosomes, which act as a docking surface for the chromodomain-containing subunits of PRC1 (for instance, M33/Cbx2, Pc2/Cbx4, or Pc3/Cbx8) that specifically recognize H3 tails trimethylated at K27.[86–88]

UTX binding to Hox genes in primary fibroblasts shows a distinctive pattern in very discrete regions of about 500 bp downstream of the transcriptional start site, contrasting with the broad binding domains of SUZ12 or the histone marks H3K27me3 and H3K4me2.[83] Chromatin immunoprecipitation (ChIP) analysis using ES cells showed no binding to the Hox clusters, suggesting an important regulatory role for the recruiting of these histone H3K27 demethylases. In this regard, in a model of cell differentiation induced by retinoic acid, Hox gene upregulation, enhanced UTX association to Hox targets, and decrease of nucleosomal H3K27me3 marks were all preceded by association of Ash2L, a subunit needed for maximal activity of complexes that trimethylate the lysine 4 of histone H3.[89] In fact, UTX is a subunit of one of these complexes, containing, in addition to ASHL2, RBBP5,

WDR5, and the H3K4, methyltransferase MLL2.[84] These results support a stepwise recruiting model in which a H3K4 methyltransferase core complex devoid of H3K27 demethylase activity is recruited first and subsequently exchanged by UTX/JMJD3-containing Mll complexes.

8.4.2 Histone H2A Deubiquitinases

Using an in vitro deubiquitination assay, investigators determined the major H2A deubiquitinase activity of Hela cells were associated with a homotetramer of Ubp-M/USP16, a previously known ubiquitin-processing protease.[90,91] Ubp-M/USP16 contains a ubiquitin carboxyl-terminal hydrolase-like zinc finger domain also present in a small subfamily of ubiquitinases. Ubp-M/USP16 specifically deubiquitinates H2AK119Ubl-containing nucleosomes. Its downregulation in tissue culture cells, by siRNA, results in an increase of global ubiquitylated H2A and defective progression through the M phase of the cell cycle, a finding consistent with impaired recruitment of Aurora B kinase to nucleosomes and concomitant phosphorylation of serine 10 of histone H3.[91] Loss-of-function experiments also showed that *Hox* gene regulation and body patterning in *Xenopus laevis* were affected.

It is not clear whether Ubp-M/USP16 works mostly as a cell cycle progression deubiquitinase and also as a regulator of gene expression because its purifications produced no subunits that may help to distinguish between these two possibilities. A different situation, however, is that of MYSM1 (myb-like, SWIRM, and MPN domains 1, also know as 2A-DUB), another H2A deubiquitinase identified in a screen for positive coregulators of the androgen receptor.[92] MYSM1/2A-DUB contains a JAMM/MPN+ domain similar to others found in metalloproteases that hydrolyze the isopeptidase bond of nedd8 and/or ubiquitin chains.[93] MYSM1/2A-DUB also contains a SANT domain known to be involved in binding to histones (reviewed by Boyer et al., 2004)[94] and a SWIRM domain that recognizes nucleosomal linker DNA sequences and/or N-terminal tails of histone H3.[95,96] In vitro deubiquitination assays with recombinant MYSM1/2A-DUB, together with overexpression/knockdown experiments, showed that MYSM1/2A-DUB is a H2AK119Ubl deubiquitinase.[92] Whereas androgen and estrogen receptor-dependent targets were affected by MYSM1/2A-DUB knockdown, genes controlled by retinoic acid receptor and thyroid hormone receptor were not, suggesting that other uncharacterized H2A deubiquitinases contribute to the regulation of specific subsets of H2AK119Ubl-dependent targets.

MYSM1/2A-DUB is found in complexes that include the histone acetyl transferase (HAT) PCAF (p300/CBP-associated factor).[92] Treatment of tissue culture cells with HAT inhibitors (which results in hyperacetylated nucleosomes) enhances the deubiquitinating activity of MYSM1/2A-DUB. Also, deubiquitination of H2A correlates with phosphorylation of linker histone H1 and its release from nucleosomes.[92] Altogether, the results suggest that the removal of the repressive mark, H2AK119Ubl, is coordinated with histone acetylation and histone H1 dissociation from nucleosomes. Since previously isolated HAT-containing complexes do not contain MYSM1/2A-DUB, it is likely that both types of HAT complexes, with and without deubiquitinase subunits, are used in a stepwise manner to achieve full transcriptional activation. It seems that the reversion of PcG-dependent histone marks followed parallel strategies in which facilitating complexes (H3K4 methyltransferases, HATs), without demethylation/deubiquitination activities, act first, facilitating the subsequent reversal of histone marks through recruitment of similar complexes, now containing demethylase/deubiquitinase subunits.

Histone H2A ubiquitylation/deubiquitylation appears to play important roles in transcriptional initiation and elongation, as suggested by studies on promoter occupancy by RNApolII forms specific to transcriptional state. These forms are defined by specific phosphorylation of residues in the C-terminal repeat of the RNApolII, which determines the regulators to which it binds at any given point in the transcription cycle (reviewed by Phatnani and Greenleaf, 2006).[97] Phosphorylation occurs mostly at the Ser2 and Ser5 in the repeats. Normally, the Ser5 phosphorylated (Ser5P) form is associated with the beginning of genes and is thought to be involved in transcription initiation. In contrast, the form phosphorylated at Ser2 (Ser2P) is found in exons and intervening sequences and is thought to be the predominant form in transcription elongation. Using as a model the hormone-inducible PSA gene, researchers found that promoter occupancy by the Ser5P form of RNApolII decreases with the knockdown of MYSM1/2A-DUB,[92] suggesting that dismissal of the deubiquitinase results in inefficient initiation; moreover, occupancy of coding sequences by the elongation-specific marks, RNApolII (Ser2P) and H3K36me3, also decreased upon MYSM1/2A-DUB inactivation. Similarly, studies that use a conditional model of inactivation of the E3 ligases Ring1A/Ring1 and Ring1B/Rnf2 in ES cells show concurrent loss of H2AK119Ub1 marks at promoters in bivalent domains, conformational changes in RNApolII conformation, and transcriptional activation.[98] Since Ring1B inactivation has no effect on H3K27me3 marks,[63,64,98] the results imply a role for PRC1-dependent regulation of transcription initiation, and possibly elongation, involving histone H2A ubiquitylation, in the maintenance of the repressed state of bivalent domains.[98]

8.5 INTERSECTION OF PcG AND DNA METHYLATION PATHWAYS

DNA methylation patterns undergo dynamic changes during the early stages of development (reviewed by Li, 2002).[99] Following fertilization, a generalized demethylation takes place, followed by de novo methylation. DNA marks corresponding to imprinted genes are not affected and are erased only in the primordial germ cells prior to reestablishment in a sex-dependent manner. De novo DNA methylation depends on the methyltransferases encoded by the *Dnmt3a* and *Dnmt3b* genes, while maintenance of preexistent patterns is carried out by Dnmt1 (reviewed by Goll and Bestor, 2005).[20] De novo methylation during gametogenesis also uses Dnmt3L, an enzymatically inactive paralog of Dnmt3a/b, which interacts with the tail of histone H3 whose lysine 4 is unmethylated and activates Dnmt3a2.[100,101] DNA methylation can have either direct or indirect effects on gene activity, by altering binding of transcription factor or through methyl-CpG-binding proteins (MBD).[102]

Abnormal DNA methylation patterns, in which sequences at some loci hypomethylated whereas at others, including tumor suppressor genes, they hypermethylated and were transcriptionally inactivated, are a feature of tumor cells and neoplasms (reviewed by Jones and Baylin, 2002).[103]

Links between the PcG and the DNA methylation regulatory systems had been found in the biallelic expression of a subset of imprinted loci and concurrent DNA methylation in mice deficient in the PRC2 subunit eed.[103] The closest link, however, is the direct associacion of the PRC2 subunit EZH2 with the DNA methyltransferases DNMT1, DNMT3a, and DNMT3b. Thus, in tissue culture cells, DNA methyltransferases occupancy of a subset of promoters, and the extent of their DNA methylation is found to depend on EZH2.[104] Moreover, methylation of promoter targets bound by PML-RARa, an oncogenic transcription factor responsible for almost all cases of acute promyeolocytic leukemia, was also dependent on the recruitment of PRC2-type complexes and H3K27 hypermethylation.[105] In the same line, another study used a stable cell line expressing a histone H3R27 variant that cannot be methylated by EZH2 and reported a generalized reduction of H3K27me3 and DNA methylation levels, together with an enhanced expression of tumor suppressor genes.[106]

The possible role of PRC2 products targeting DNA methylation is relevant considering that the abnormal DNA methylation patterns of tumor cells do not correlate with differences in the expression of DNA methyltransferases, and therefore it is likely to result from their mislocalization. However, the interplay between the activities of the PcG and DNA methylation systems is still poorly characterized. For instance, there are reports of transcriptional derepression subsequent to the abrogation of the H3K27me3 activity, by the use of Ezh2 shRNA[107] or by pharmacological disruption of the PRC2 complex[108] that is not accompanied by DNA demethylation. In this regard, it is of note that inhibition of SIRT1, the class III histone deacetylase associated with cancer cell-enriched EZH2-containing PRC4 complexes, reactivates silenced promoters in tumor cells while their DNA hypermethylation status remains unchanged.[109]

The intersection between DNA methylation and PcG-dependent repression extends also to PRC1 complexes, as shown by the direct association between PRC1 subunits and cofactors/effectors of DNA methylation. For instance, BMI1/PCGF4 interacts directly with DMAP1, the DNMT1-associated protein 1,[110] and both Bmi1/Pcgf4 and M33/Cbx2 are found in complexes with DNA methyltransferase activity that contain Dnmt3a/b.[111] On the other hand, RING1B/RNF2 and PH2/PHC2 have been identified as direct interactors of MBD1,[112] a transcriptional repressor known to bind methylated DNA. Lsh, a member of the SNF2 family of remodelers with a role in de novo methylation, also associates with Bmi1/Pcgf4 and M33/Cbx2.[111]

These associations seem functionally relevant since the inactivation of these proteins, by iRNA (MBD1, DMAP1) or by gene deletion (Lsh), correlates with derepression of Hox genes. Moreover, concurrent gene activation of PcG targets and changes in DNA hypomethylation take place in cells lacking Lsh or Bmi1.[110,111] Changes associated with loss of function of Bmi1/Pcgf4, however, may be related to their association with DNA methyltransferases rather than with Dmap1, since the Dnmt1 domain that binds Dmap1 is dispensable for in vivo function.[113] Both occupancy of Hox gene promoters by Dnmt3b and PRC1 and H2AK119Ub1 marks decrease in Lsh-deficient cells,[111] making it difficult to determine whether they are mechanistically independent events or whether they affect each other. For instance, alteration of DNA methylation patterns may have an effect on PRC1 recruiting, because it seems to suggest the disruption of PcG bodies, large intranuclear structures that form on pericentromeric domains,[114] resulting from the treatment of tissue culture cells with the DNA demethylating agent 5-aza-deoxycytidine.[115] However, similar experiments with a different cell line failed to produce any effect on PcG bodies while modifying the localization of MBD1.[112]

When the activity of PcG-antagonizing complexes is altered—for instance, by inactivation of histone H3K4 methyltransferase complexes—Hox gene downregulation is found associated with an increased degree of DNA methylation.[116,117] This suggests that, at least at some loci, PcG repressive mechanisms and DNA methylation act as normally intertwined regulatory pathways, thus explaining that, when disturbed, the probability of PcG targets becoming DNA methylated increases in neoplastic cells and tissues.[23-25,118]

8.5 RECRUITING OF PCG COMPLEXES AND MECHANISMS OF REPRESSION

PcG complex(es) recruiting to targets seems to differ between insects and vertebrates. In flies, genetic analysis has identified a collection of DNA sequences functionally relevant to PcG function (reviewed by Ringrose and Paro, 2007;[119] Schwartz et al., 2006[120]). No such sequences, also known as Polycomb response elements (PREs) have been identified yet in mammalian cells. There is, however, evidence for a role in YY1 recruiting PRC2 complexes to targets.[121] Genome-wide ChIP-on-ChIP experiments analyzing binding to about 10 kb of sequences around the transcription start site show that mammalian PcG products associate with a rather restricted area around promoters.[36,121,122] Broader areas, however, are occupied by PcG products when the arrays include entire loci (for instance, Hox clusters or the *INK4/ARF* locus),[122,123] in agreement with broader H3K27me3 domains seen for some loci when the ChIP analysis is based on sequencing of the immunoprecipitated DNA.[12,14] Recently, it has been shown that PRC2 occupancy and H3K27 trimethylation of nucleosomes at the *HOXA* and *HOXD* clusters are dependent on noncoding RNAs.[124,125]

Just as PcG recruiting is poorly understood, the mechanisms that mediate PcG transcriptional repression remain little known. It is likely that several mechanisms are used. Some evidence exists for PcG targeting the transcriptional initiation machinery at the promoter.[92,98,126] Also, evidence indicates that PRC1 complexes inhibit SWI/SNF-nucleosome remodeling and promote formation of high-order chromatin structures,[52,127] but evidence for their relevance in PcG function in vivo is lacking. It is also worth noting that silencing of *Hox* gene clusters in *Drosophila* involves the formation of high-order chromatin structures, which depends on the activity of PcG products.[128,129] Although direct evidence for such a function for mammalian PcG complexes is still lacking, a role similar to that seen in *Drosophila* could be anticipated, at least for the mammalian Hox clusters, which are also regulated by high-order chromatin structures.[130]

8.6 PERSPECTIVES

The association of PcG regulation to developmental processes reflects their suitability as a flexible, dynamic system for transcriptional regulation of a subset of genes with roles in the determination of cell states and transitions between them.

Newly developed techniques for the analysis of chromatin-immunoprecipitated DNA (by massive sequencing of DNA fragments, ChIP-seq) will facilitate the generation of large collections of chromatin states of many cell types and contribute to the description of the dynamic changes in PcG targeting activity that accompany the transitions between cell types along differentiating pathways. Likewise, further development of proteomic methods will allow greater refinement in the description of PcG complexes. However, their recruitment to targets is one of the main areas in need of progress. Also, although not specific to PcG regulation, it will be important to determine the contribution of histone modifications to the heritability of epigenetic states given the absence of a histone-based mechanism for maintenance of chromatin modifications during cell division. Finally, gaining insight into PcG-related transcriptional repression mechanisms, be it the initiation/elongation steps or the formation of higher order chromatin structures and nuclear localization, is essential for understanding PcG function.

ACKNOWLEDGMENTS

Work in the author's laboratory is funded by a grant from the Education Ministry (SAF2007-06952-CO2-01) and the OncoCycle program from the Comunidad de Madrid.

REFERENCES

1. Kennison, J. A., The Polycomb and trithorax group proteins of Drosophila: trans-regulators of homeotic gene function, *Annu Rev Genet* 29, 289–303, 1995.
2. Schuettengruber, B., Chourrout D., Vervoort M., Leblanc B. and Cavalli G., Genome regulation by polycomb and trithorax proteins, *Cell* 128 (4), 735–745, 2007.
3. Whitcomb, S. J., Basu A., Allis C. D. and Bernstein E., Polycomb Group proteins: an evolutionary perspective, *Trends Genet* 2007.
4. Ptashne, M., On speaking, writing and inspiration, *Curr Biol* 17 (10), R348–49, 2007.
5. Li, B., Carey M. and Workman J. L., The role of chromatin during transcription, *Cell* 128 (4), 707–719, 2007.
6. Berger, S. L., The complex language of chromatin regulation during transcription, *Nature* 447 (7143), 407–412, 2007.
7. Kouzarides, T., Chromatin modifications and their function, *Cell* 128 (4), 693–705, 2007.
8. Sparmann, A. and van Lohuizen M., Polycomb silencers control cell fate, development and cancer, *Nat Rev Cancer* 6 (11), 846–856, 2006.
9. Martinez, A. M. and Cavalli G., The role of Polycomb Group proteins in cell cycle regulation during development, *Cell Cycle* 5 (11), 1189–1197, 2006.
10. Spivakov, M. and Fisher A. G., Epigenetic signatures of stem-cell identity, *Nat Rev Genet* 8 (4), 263–271, 2007.
11. Boyer, L. A., Mathur D. and Jaenisch R., Molecular control of pluripotency, *Curr Opin Genet Dev* 16 (5), 455–462, 2006.
12. Barski, A., Cuddapah S., Cui K., Roh T. Y., Schones D. E., Wang Z., Wei G., Chepelev I. and Zhao K., High-resolution profiling of histone methylations in the human genome, *Cell* 129 (4), 823–837, 2007.
13. Guenther, M. G., Levine S. S., Boyer L. A., Jaenisch R. and Young R. A., A chromatin landmark and transcription initiation at most promoters in human cells, *Cell* 130 (1), 77–88, 2007.

14. Mikkelsen, T. S., Ku M., Jaffe D. B., Issac B., Lieberman E., Giannoukos G., Alvarez P., Brockman W., Kim T. K., Koche R. P., Lee W., Mendenhall E., O'donovan A., Presser A., Russ C., Xie X., Meissner A., Wernig M., Jaenisch R., Nusbaum C., Lander E. S. and Bernstein B. E., Genome-wide maps of chromatin state in pluripotent and lineage-committed cells, *Nature* 448, 553–560, 2007.

15. Weber, M., Hellmann I., Stadler M. B., Ramos L., Pääbo S., Rebhan M. and Schübeler D., Distribution, silencing potential and evolutionary impact of promoter DNA methylation in the human genome, *Nat Genet* 39 (4), 457–466, 2007.

16. Guccione, E., Bassi C., Casadio F., Martinato F., Cesaroni M., Schuchlautz H., Lüscher B. and Amati B., Methylation of histone H3R2 by PRMT6 and H3K4 by an MLL complex are mutually exclusive, *Nature* 2007.

17. Bannister, A. J., Schneider R., Myers F. A., Thorne A. W., Crane-Robinson C. and Kouzarides T., Spatial distribution of di-and tri-methyl lysine 36 of histone H3 at active genes, *J Biol Chem* 280 (18), 17732–17736, 2005.

18. Bernstein, B. E., Mikkelsen T. S., Xie X., Kamal M., Huebert D. J., Cuff J., Fry B., Meissner A., Wernig M., Plath K., Jaenisch R., Wagschal A., Feil R., Schreiber S. L. and Lander E. S., A bivalent chromatin structure marks key developmental genes in embryonic stem cells, *Cell* 125 (2), 315–326, 2006.

19. Azuara, V., Perry P., Sauer S., Spivakov M., Jørgensen H. F., John R. M., Gouti M., Casanova M., Warnes G., Merkenschlager M. and Fisher A. G., Chromatin signatures of pluripotent cell lines, *Nat Cell Biol* 8 (5), 532–538, 2006.

20. Goll, M. G. and Bestor T. H., Eukaryotic cytosine methyltransferases, *Annu Rev Biochem* 74, 481–514, 2005.

21. Eden, A., Gaudet F., Waghmare A. and Jaenisch R., Chromosomal instability and tumors promoted by DNA hypomethylation, *Science* 300 (5618), 455, 2003.

22. Tanay, A., O'donnell A. H., Damelin M. and Bestor T. H., Hyperconserved CpG domains underlie Polycomb-binding sites, *Proc Natl Acad Sci USA* 104 (13), 5521–5526, 2007.

23. Schlesinger, Y., Straussman R., Keshet I., Farkash S., Hecht M., Zimmerman J., Eden E., Yakhini Z., Ben-Shushan E., Reubinoff B. E., Bergman Y., Simon I. and Cedar H., Polycomb-mediated methylation on Lys27 of histone H3 pre-marks genes for de novo methylation in cancer, *Nat Genet* 32 (2), 232–236, 2006.

24. Widschwendter, M., Fiegl H., Egle D., Mueller-Holzner E., Spizzo G., Marth C., Weisenberger D. J., Campan M., Young J., Jacobs I. and Laird P. W., Epigenetic stem cell signature in cancer, *Nat Genet* 39 (2), 157–158, 2006.

25. Ohm, J., McGarvey K., Yu X., Cheng L., Schuebel K., Cope L., Mohammad H., Chen W., Daniel V., Yu W., Berman D., Jenuwein T., Pruitt K., Sharkis S., Watkins D. N., Herman J. and Baylin S., A stem cell-like chromatin pattern may predispose tumor suppressor genes to DNA hypermethylation and heritable silencing, *Nat Genet* 39 (2), 237–242, 2007.

26. van Lohuizen, M., Tijms M., Voncken J. W., Schumacher A., Magnuson T. and Wientjens E., Interaction of mouse polycomb-group (Pc-G) proteins Enx1 and Enx2 with Eed: indication for separate Pc-G complexes, *Mol Cell Biol* 18 (6), 3572–3579, 1998.

27. Sewalt, R. G., van der Vlag J., Gunster M. J., Hamer K. M., den Blaauwen J. L., Satijn D. P., Hendrix T., van Driel R. and Otte A. P., Characterization of interactions between the mammalian polycomb-group proteins Enx1/EZH2 and EED suggests the existence of different mammalian polycomb-group protein complexes, *Mol Cell Biol* 18 (6), 3586–3595, 1998.

28. Otte, A. P. and Kwaks T. H., Gene repression by Polycomb group protein complexes: a distinct complex for every occasion? *Curr Opin Genet Dev* 13 (5), 448–454, 2003.

29. Müller, J., Hart C. M., Francis N. J., Vargas M. L., Sengupta A., Wild B., Miller E. L., O'Connor M. B., Kingston R. E. and Simon J. A., Histone methyltransferase activity of a Drosophila Polycomb group repressor complex, *Cell* 111 (2), 197–208, 2002.

30. Kuzmichev, A., Nishioka K., Erdjument-Bromage H., Tempst P. and Reinberg D., Histone methyltransferase activity associated with a human multiprotein complex containing the Enhancer of Zeste protein, *Genes Dev* 16 (22), 2893–2905, 2002.

31. Cao, R., Wang L., Wang H., Xia L., Erdjument-Bromage H., Tempst P., Jones R. S. and Zhang Y., Role of histone H3 lysine 27 methylation in Polycomb-group silencing, *Science* 298 (5595), 1039–1043, 2002.

32. Czermin, B., Melfi R., McCabe D., Seitz V., Imhof A. and Pirrotta V., Drosophila enhancer of Zeste/ESC complexes have a histone H3 methyltransferase activity that marks chromosomal Polycomb sites, *Cell* 111 (2), 185–196, 2002.

33. Cao, R. and Zhang Y., The functions of E(Z)/EZH2-mediated methylation of lysine 27 in histone H3, *Curr Opin Genet Dev* 14 (2), 155–164, 2004.

34. Plath, K., Fang J., Mlynarczyk-Evans S. K., Cao R., Worringer K. A., Wang H., de la Cruz C. C., Otte A. P., Panning B. and Zhang Y., Role of histone H3 lysine 27 methylation in X inactivation, *Science* 300 (5616), 131–135, 2003.

35. Boyer, L. A., Plath K., Zeitlinger J., Brambrink T., Medeiros L. A., Lee T. I., Levine S. S., Wernig M., Tajonar A., Ray M. K., Bell G. W., Otte A. P., Vidal M., Gifford D. K., Young R. A. and Jaenisch R., Polycomb complexes repress developmental regulators in murine embryonic stem cells, *Nature* 441 (7091), 349–353, 2006.

36. Lee, T. I., Jenner R. G., Boyer L. A., Guenther M. G., Levine S. S., Kumar R. M., Chevalier B., Johnstone S. E., Cole M. F., Isono K., Koseki H., Fuchikami T., Abe K., Murray H. L., Zucker J. P., Yuan B., Bell G. W., Herbolsheimer E., Hannett N. M., Sun K., Odom D. T., Otte A. P., Volkert T. L., Bartel D. P., Melton D. A., Gifford D. K., Jaenisch R. and Young R. A., Control of developmental regulators by polycomb in human embryonic stem cells, *Cell* 125 (2), 301–313, 2006.

37. Pasini, D., Bracken A. P., Hansen J. B., Capillo M. and Helin K., The Polycomb Group protein Suz12 is required for Embryonic Stem Cell differentiation, *Mol Cell Biol* 27 (10), 3769–3779, 2007.

38. Silva, J., Mak W., Zvetkova I., Appanah R., Nesterova T. B., Webster Z., Peters A. H., Jenuwein T., Otte A. P. and Brockdorff N., Establishment of histone h3 methylation on the inactive X chromosome requires transient recruitment of Eed-Enx1 polycomb group complexes, *Dev Cell* 4 (4), 481–495, 2003.

39. Mak, W., Baxter J., Silva J., Newall A. E., Otte A. P. and Brockdorff N., Mitotically stable association of polycomb group proteins eed and enx1 with the inactive x chromosome in trophoblast stem cells, *Curr Biol* 12 (12), 1016–1020, 2002.

40. Mager, J., Montgomery N. D., de Villena F. P. and Magnuson T., Genome imprinting regulated by the mouse Polycomb group protein Eed, *Nat Genet* 33 (4), 502–507, 2003.

41. O'Carroll, D., Erhardt S., Pagani M., Barton S. C., Surani M. A. and Jenuwein T., The polycomb-group gene Ezh2 is required for early mouse development, *Mol Cell Biol* 21 (13), 4330–4336, 2001.

42. Erhardt, S., Su I. H., Schneider R., Barton S., Bannister A. J., Perez-Burgos L., Jenuwein T., Kouzarides T., Tarakhovsky A. and Surani M. A., Consequences of the depletion of zygotic and embryonic enhancer of zeste 2 during preimplantation mouse development, *Development* 130 (18), 4235–4248, 2003.

43. Montgomery, N. D., Yee D., Chen A., Kalantry S., Chamberlain S. J., Otte A. P. and Magnuson T., The Murine Polycomb Group Protein Eed Is Required for Global Histone H3 Lysine-27 Methylation, *Curr Biol* 15 (10), 942–947, 2005.

44. Cao, R. and Zhang Y., SUZ12 is required for both the histone methyltransferase activity and the silencing function of the EED-EZH2 complex, *Mol Cell* 15 (1), 57–67, 2004.

45. Nekrasov, M., Wild B. and Müller J., Nucleosome binding and histone methyltransferase activity of Drosophila PRC2, *EMBO Rep* 6 (4), 348–353, 2005.

46. Kuzmichev, A., Margueron R., Vaquero A., Preissner T. S., Scher M., Kirmizis A., Ouyang X., Brockdorff N., Abate-Shen C., Farnham P. and Reinberg D., Composition and histone substrates of polycomb repressive group complexes change during cellular differentiation, *Proc Natl Acad Sci USA* 102 (6), 1859–1864, 2005.

47. Kirmizis, A., Bartley S. M., Kuzmichev A., Margueron R., Reinberg D., Green R. and Farnham P. J., Silencing of human polycomb target genes is associated with methylation of histone H3 Lys 27, *Genes Dev* 18 (13), 1592–1605, 2004.

48. Pasini, D., Bracken A. P., Jensen M. R., Denchi E. L. and Helin K., Suz12 is essential for mouse development and for EZH2 histone methyltransferase activity, *EMBO J* 23 (20), 4061–4071, 2004.

49. Martin, C., Cao R. and Zhang Y., Substrate preferences of the EZH2 histone methyltransferase complex, *J Biol Chem* 281 (13), 8365–8370, 2006.

50. Varambally, S., Dhanasekaran S. M., Zhou M., Barrette T. R., Kumar-Sinha C., Sanda M. G., Ghosh D., Pienta K. J., Sewalt R. G., Otte A. P., Rubin M. A. and Chinnaiyan A. M., The polycomb group protein EZH2 is involved in progression of prostate cancer, *Nature* 419 (6907), 624–629, 2002.

51. Nekrasov, M., Klymenko T., Fraterman S., Papp B., Oktaba K., Köcher T., Cohen A., Stunnenberg H. G., Wilm M. and Müller J., Pcl-PRC2 is needed to generate high levels of H3-K27 trimethylation at Polycomb target genes, *EMBO J* 26 (18), 4078–4088, 2007.

52. Shao, Z., Raible F., Mollaaghababa R., Guyon J. R., Wu C. T., Bender W. and Kingston R. E., Stabilization of chromatin structure by PRC1, a Polycomb complex, *Cell* 98 (1), 37–46, 1999.

53. Saurin, A. J., Shao Z., Erdjument-Bromage H., Tempst P. and Kingston R. E., A Drosophila Polycomb group complex includes Zeste and dTAFII proteins, *Nature* 412 (6847), 655–660, 2001.

54. Levine, S. S., Weiss A., Erdjument-Bromage H., Shao Z., Tempst P. and Kingston R. E., The core of the polycomb repressive complex is compositionally and functionally conserved in flies and humans, *Mol Cell Biol* 22 (17), 6070–6078, 2002.

55. Wiederschain, D., Chen L., Johnson B., Bettano K., Jackson D., Taraszka J., Wang Y. K., Jones M. D., Morrissey M., Deeds J., Mosher R., Fordjour P., Lengauer C. and Benson J. D., Contribution of polycomb homologues Bmi-1 and Mel-18 to medulloblastoma pathogenesis, *Mol Cell Biol* 27 (13), 4968–4979, 2007.

56. Dietrich, N., Bracken A. P., Trinh E., Schjerling C. K., Koseki H., Rappsilber J., Helin K. and Hansen K. H., Bypass of senescence by the polycomb group protein CBX8 through direct binding to the INK4A-ARF locus, *EMBO J* 26 (6), 1637–1648, 2007.

57. Sánchez, C., Sánchez I., Demmers J. A., Rodriguez P., Strouboulis J. and Vidal M., Proteomic analysis of Ring1B/Rnf2 interactors identifies a novel complex with the Fbxl10/Jmjd1B histone demethylase and the BcoR corepressor, *Mol Cell Proteomics* 6 (5), 820–834, 2007.

58. Wang, H., Wang L., Erdjument-Bromage H., Vidal M., Tempst P., Jones R. S. and Zhang Y., Role of histone H2A ubiquitination in Polycomb silencing, *Nature* 431 (7010), 873–878, 2004.

59. Buchwald, G., van der Stoop P., Weichenrieder O., Perrakis A., van Lohuizen M. and Sixma T. K., Structure and E3-ligase activity of the Ring-Ring complex of Polycomb proteins Bmi1 and Ring1b, *EMBO J* 25 (11), 2465–2474, 2006.

60. Cao, R., Tsukada Y. I. and Zhang Y., Role of Bmi-1 and Ring1A in H2A Ubiquitylation and Hox Gene Silencing, *Mol Cell* 20 (6), 845–854, 2005.

61. Wei, J., Zhai L., Xu J. and Wang H., Role of Bmi1 in H2A ubiquitylation and hox gene silencing, *J Biol Chem* 281 (32), 22537–22544, 2006.

62. de Napoles, M., Mermoud J. E., Wakao R., Tang Y. A., Endoh M., Appanah R., Nesterova T. B., Silva J., Otte A. P., Vidal M., Koseki H. and Brockdorff N., Polycomb Group Proteins Ring1A/B Link Ubiquitylation of Histone H2A to Heritable Gene Silencing and X Inactivation, *Dev Cell* 7 (5), 663–676, 2004.

63. Endoh, Endo A., Endoh, Fujimura, Ohara, Toyada, Brockdorff, Okano, Niwa, Vidal and Koseki, Polycomb group proteins Ring1A/B are functionally linked to the core transcriptional regulatory circuitry to maintain ES cell identity, *Development.*

64. Leeb, M. and Wutz A., Ring1B is crucial for the regulation of developmental control genes and PRC1 proteins but not X inactivation in embryonic cells, *J Cell Biol* 178 (2), 219–229, 2007.

65. Brown, J. L., Mucci, D., Whiteley, M., Dirksen, M. L. and Kassis, J. A. The Drosophila Polycomb group gene pleiohomeotic encodes a DNA binding protein with homology to the transcription factor YY1, *Mol Cell* 1 (7), 1057–1064, 1998.

66. Klymenko, T., Papp B., Fischle W., Köcher T., Schelder M., Fritsch C., Wild B., Wilm M. and Müller J., A Polycomb group protein complex with sequence-specific DNA-binding and selective methyl-lysine-binding activities, *Genes Dev* 20 (9), 1110–1122, 2006.

67. Ogawa, H., Ishiguro K., Gaubatz S., Livingston D. M. and Nakatani Y., A complex with chromatin modifiers that occupies E2F- and Myc-responsive genes in G0 cells, *Science* 296 (5570), 1132–1136, 2002.

68. Trojer, P., Li G., Sims R. J., Vaquero A., Kalakonda N., Boccuni P., Lee D., Erdjument-Bromage H., Tempst P., Nimer S. D., Wang Y. H. and Reinberg D., L3MBTL1, a histone-methylation-dependent chromatin lock, *Cell* 129 (5), 915–928, 2007.

69. Kim, C. A., Gingery M., Pilpa R. M. and Bowie J. U., The SAM domain of polyhomeotic forms a helical polymer, *Nat Struct Biol* 9 (6), 453–457, 2002.

70. Kim, C. A., Sawaya M. R., Cascio D., Kim W. and Bowie J. U., Structural organization of a sex-comb-on-midleg/polyhomeotic copolymer, *J Biol Chem* 280 (30), 27769–27775, 2005.

71. Tahiliani, M., Mei P., Fang R., Leonor T., Rutenberg M., Shimizu F., Li J., Rao A. and Shi Y., The histone H3K4 demethylase SMCX links REST target genes to X-linked mental retardation, *Nature* 447 (7144), 601–605, 2007.

72. Gearhart, M. D., Corcoran C. M., Wamstad J. A. and Bardwell V. J., Polycomb Group and SCF ubiquitin ligases are found in a novel BCOR complex that is recruited to BCL6 targets, *Mol Cell Biol* 26 (18), 6880–6889, 2006.

73. Tsukada, Y. I., Fang J., Erdjument-Bromage H., Warren M. E., Borchers C. H., Tempst P. and Zhang Y., Histone demethylation by a family of JmjC domain-containing proteins, *Nature* 439 (7078), 811–816, 2005.

74. Shi, Y., Sawada J., Sui G., Affar E. B., Whetstine J. R., Lan F., Ogawa H., Luke M. P., Nakatani Y. and Shi Y., Coordinated histone modifications mediated by a CtBP corepressor complex, *Nature* 422 (6933), 735–738, 2003.

75. Kagey, M. H., Melhuish T. A. and Wotton D., The polycomb protein Pc2 is a SUMO E3, *Cell* 113 (1), 127–137, 2003.

76. Long, J., Zuo D. and Park M., Pc2-mediated SUMOylation of Smad-interacting protein 1 attenuates transcriptional repression of E-cadherin, *J Biol Chem* 280 (42), 35477–35489, 2005.

77. Roscic, A., Möller A., Calzado M. A., Renner F., Wimmer V. C., Gresko E., Lüdi K. S. and Schmitz M. L., Phosphorylation-Dependent Control of Pc2 SUMO E3 Ligase Activity by Its Substrate Protein HIPK2, *Mol Cell* 24 (1), 77–89, 2006.

78. Li, B., Zhou J., Liu P., Hu J., Jin H., Shimono Y., Takahashi M. and Xu G., Polycomb protein Cbx4 promotes SUMO modification of de novo DNA methyltransferase Dnmt3a, *Biochem J* 405 (2), 369–378, 2007.

79. Dou, Y., Milne T. A., Tackett A. J., Smith E. R., Fukuda A., Wysocka J., Allis C. D., Chait B. T., Hess J. L. and Roeder R. G., Physical Association and Coordinate Function of the H3 K4 Methyltransferase MLL1 and the H4 K16 Acetyltransferase MOF, *Cell* 121 (6), 873–885, 2005.

80. Schoeftner, S., Sengupta A. K., Kubicek S., Mechtler K., Spahn L., Koseki H., Jenuwein T. and Wutz A., Recruitment of PRC1 function at the initiation of X inactivation independent of PRC2 and silencing, *EMBO J* 25 (13), 3110–3122, 2006.

81. Klose, R. J., Kallin E. M. and Zhang Y., JmjC-domain-containing proteins and histone demethylation, *Nat Rev Genet* 7 (9), 715–727, 2006.

82. Shi, Y. and Whetstine J. R., Dynamic regulation of histone lysine methylation by demethylases, *Mol Cell* 25 (1), 1–14, 2007.

83. Lan, F., Bayliss P. E., Rinn J. L., Whetstine J. R., Wang J. K., Chen S., Iwase S., Alpatov R., Issaeva I., Canaani E., Roberts T. M., Chang H. Y. and Shi Y., A histone H3 lysine 27 demethylase regulates animal posterior development, *Nature* 2007.

84. Lee, M. G., Villa R., Trojer P., Norman J., Yan K. P., Reinberg D., Di Croce L. and Shiekhattar R., Demethylation of H3K27 regulates Polycomb recruitment and H2A ubiquitination, *Science* 2007.

85. De Santa, F., Totaro M. G., Prosperini E., Notarbartolo S., Testa G. and Natoli G., The histone H3 lysine-27 demethylase Jmjd3 links inflammation to inhibition of Polycomb-mediated gene silencing, *Cell* 130 (6), 1083–1094, 2007.

86. Min, J., Zhang Y. and Xu R. M., Structural basis for specific binding of Polycomb chromo-domain to histone H3 methylated at Lys 27, *Genes Dev* 17 (15), 182318–182328, 2003.

87. Fischle, W., Wang Y., Jacobs S. A., Kim Y., Allis C. D. and Khorasanizadeh S., Molecular basis for the discrimination of repressive methyl-lysine marks in histone H3 by Polycomb and HP1 chromodomains, *Genes Dev* 17 (15), 18701–18881, 2003.

88. Bernstein, E., Duncan E. M., Masui O., Gil J., Heard E. and Allis C. D., Mouse polycomb proteins bind differentially to methylated histone H3 and RNA and are enriched in facultative heterochromatin, *Mol Cell Biol* 26 (7), 2560–2569, 2006.

89. Dou, Y., Milne T. A., Ruthenburg A. J., Lee S., Lee J. W., Verdine G. L., Allis C. D. and Roeder R. G., Regulation of MLL1 H3K4 methyltransferase activity by its core components, *Nat Struct Mol Biol* 13 (8), 713–719, 2006.

90. Cai, S. Y., Babbitt R. W. and Marchesi V. T., A mutant deubiquitinating enzyme (Ubp-M) associates with mitotic chromosomes and blocks cell division, *Proc Natl Acad Sci USA* 96 (6), 2828–2833, 1999.

91. Joo, H. Y., Zhai L., Yang C., Nie S., Erdjument-Bromage H., Tempst P., Chang C. and Wang H., Regulation of cell cycle progression and gene expression by H2A deubiquitination, *Nature* 2007.

92. Zhu, P., Zhou W., Wang J., Puc J., Ohgi K. A., Erdjument-Bromage H., Tempst P., Glass C. K. and Rosenfeld M. G., A Histone H2A Deubiquitinase Complex Coordinating Histone Acetylation and H1 Dissociation in Transcriptional Regulation, *Mol Cell* 27 (4), 609–621, 2007.

93. Ambroggio, X. I., Rees D. C. and Deshaies R. J., JAMM: a metalloprotease-like zinc site in the proteasome and signalosome, *PLoS Biol* 2 (1), E2, 2004.

94. Boyer, L. A., Latek R. R. and Peterson C. L., The SANT domain: a unique histone-tail-binding module? *Nat Rev Mol Cell Biol* 5 (2), 158–163, 2004.

95. Tochio, N., Umehara T., Koshiba S., Inoue M., Yabuki T., Aoki M., Seki E., Watanabe S., Tomo Y., Hanada M., Ikari M., Sato M., Terada T., Nagase T., Ohara O., Shirouzu M., Tanaka A., Kigawa T. and Yokoyama S., Solution structure of the SWIRM domain of human histone demethylase LSD1, *Structure* 14 (3), 457–468, 2006.

96. Yoneyama, M., Tochio N., Umehara T., Koshiba S., Inoue M., Yabuki T., Aoki M., Seki E., Matsuda T., Watanabe S., Tomo Y., Nishimura Y., Harada T., Terada T., Shirouzu M., Hayashizaki Y., Ohara O., Tanaka A., Kigawa T. and Yokoyama S., Structural and functional differences of SWIRM domain subtypes, *J Mol Biol* 369 (1), 222–238, 2007.

97. Phatnani, H. P. and Greenleaf A. L., Phosphorylation and functions of the RNA polymerase II CTD, *Genes Dev* 20 (21), 2922–2936, 2006.

98. Stock, K., Giadrossi, Casanova, Brookes, Vidal, Koseki, Brockdorff, Fisher G. and Pombo, Ring1-mediated ubiquitination of H2A restrains poised RNA polymerase II at bivalent genes in ES cells, *Nat. Cell Biol.*

99. Li, E., Chromatin modification and epigenetic reprogramming in mammalian development, *Nat Rev Genet* 3 (9), 662–673, 2002.

100. Jia, D., Jurkowska R. Z., Zhang X., Jeltsch A. and Cheng X., Structure of Dnmt3a bound to Dnmt3L suggests a model for de novo DNA methylation, *Nature* 449 (7159), 248–251, 2007.

101. Ooi, S. K., Qiu C., Bernstein E., Li K., Jia D., Yang Z., Erdjument-Bromage H., Tempst P., Lin S. P., Allis C. D., Cheng X. and Bestor T. H., DNMT3L connects unmethylated lysine 4 of histone H3 to de novo methylation of DNA, *Nature* 448 (7154), 714–717, 2007.

102. Klose, R. J. and Bird A. P., Genomic DNA methylation: the mark and its mediators, *Trends Biochem Sci* 31 (2), 89–97, 2006.

103. Jones, P. A. and Baylin S. B., The fundamental role of epigenetic events in cancer, *Nat Rev Genet* 3 (6), 415–428, 2002.

104. Viré, E., Brenner C., Deplus R., Blanchon L., Fraga M., Didelot C., Morey L., Van Eynde A., Bernard D., Vanderwinden J. M., Bollen M., Esteller M., Di Croce L., de Launoit Y. and Fuks F., The Polycomb group protein EZH2 directly controls DNA methylation, *Nature* 439 (7078), 871–874, 2005.

105. Villa, R., Pasini D., Gutierrez A., Morey L., Occhionorelli M., Viré E., Nomdedeu J. F., Jenuwein T., Pelicci P. G., Minucci S., Fuks F., Helin K. and Di Croce L., Role of the polycomb repressive complex 2 in acute promyelocytic leukemia, *Cancer Cell* 11 (6), 513–525, 2007.

106. Abbosh, P. H., Montgomery J. S., Starkey J. A., Novotny M., Zuhowski E. G., Egorin M. J., Moseman A. P., Golas A., Brannon K. M., Balch C., Huang T. H. and Nephew K. P., Dominant-negative histone h3 lysine 27 mutant derepresses silenced tumor suppressor genes and reverses the drug-resistant phenotype in cancer cells, *Cancer Res* 66 (11), 5582–5591, 2006.

107. McGarvey, K. M., Greene E., Fahrner J. A., Jenuwein T. and Baylin S. B., DNA Methylation and Complete Transcriptional Silencing of Cancer Genes Persist after Depletion of EZH2, *Cancer Res* 67 (11), 5097–5102, 2007.

108. Tan, J., Yang X., Zhuang L., Jiang X., Chen W., Lee P. L., Karuturi R. K., Tan P. B., Liu E. T. and Yu Q., Pharmacologic disruption of Polycomb-repressive complex 2-mediated gene repression selectively induces apoptosis in cancer cells, *Genes Dev* 21 (9), 1050–1063, 2007.

109. Pruitt, K., Zinn R. L., Ohm J. E., McGarvey K. M., Kang S. H., Watkins D. N., Herman J. G. and Baylin S. B., Inhibition of SIRT1 reactivates silenced cancer genes without loss of promoter DNA hypermethylation, *PLoS Genet* 2 (3), e40, 2006.

110. Negishi, M., Saraya A., Miyagi S., Nagao K., Inagaki Y., Nishikawa M., Tajima S., Koseki H., Tsuda H., Takasaki Y., Nakauchi H. and Iwama A., Bmi1 cooperates with Dnmt1-associated protein 1 in gene silencing, *Biochem Biophys Res Commun* 353 (4), 992–998, 2007.

111. Xi, S., Zhu H., Xu H., Schmidtmann A., Geiman T. M. and Muegge K., Lsh controls Hox gene silencing during development, *Proc Natl Acad Sci USA* 104 (36), 14366–14371, 2007.

112. Sakamoto, Y., Watanabe S., Ichimura T., Kawasuji M., Koseki H., Baba H. and Nakao M., Overlapping roles of the methylated DNA binding protein MBD1 and polycomb group proteins in transcriptional repression of HOXA genes and heterochromatin foci formation, *J Biol Chem* 282 (22), 16391–16400, 2007.

113. Ding, F. and Chaillet J. R., In vivo stabilization of the Dnmt1 (cytosine-5)- methyltransferase protein, *Proc Natl Acad Sci USA* 99 (23), 14861–14866, 2002.

114. Saurin, A. J., Shiels C., Williamson J., Satijn D. P., Otte A. P., Sheer D. and Freemont P. S., The human polycomb group complex associates with pericentromeric heterochromatin to form a novel nuclear domain, *J Cell Biol* 142 (4), 887–898, 1998.

115. Hernández-Muñoz, I., Taghavi P., Kuijl C., Neefjes J. and van Lohuizen M., Association of BMI1 with polycomb bodies is dynamic and requires PRC2/EZH2 and the maintenance DNA methyltransferase DNMT1, *Mol Cell Biol* 25 (24), 11047–11058, 2005.

116. Milne, T. A., Briggs S. D., Brock H. W., Martin M. E., Gibbs D., Allis C. D. and Hess J. L., MLL targets SET domain methyltransferase activity to Hox gene promoters, *Mol Cell* 10 (5), 1107–1117, 2002.

117. Terranova, R., Agherbi H., Boned A., Meresse S. and Djabali M., Histone and DNA methylation defects at Hox genes in mice expressing a SET domain-truncated form of Mll, *Proc Natl Acad Sci USA* 103 (17), 6629–6634, 2006.

118. Rauch, T., Wang Z., Zhang X., Zhong X., Wu X., Lau S. K., Kernstine K. H., Riggs A. D. and Pfeifer G. P., Homeobox gene methylation in lung cancer studied by genome-wide analysis with a microarray-based methylated CpG island recovery assay, *Proc Natl Acad Sci USA* 104 (13), 5527–5532, 2007.

119. Ringrose, L. and Paro R., Polycomb/Trithorax response elements and epigenetic memory of cell identity, *Development* 134 (2), 223–232, 2007.

120. Schwartz, Y. B., Kahn T. G., Nix D. A., Li X. Y., Bourgon R., Biggin M. and Pirrotta V., Genome-wide analysis of Polycomb targets in Drosophila melanogaster, *Nat Genet* 38 (6), 700–705, 2006.

121. Caretti, G., Di Padova M., Micales B., Lyons G. E. and Sartorelli V., The Polycomb Ezh2 methyltransferase regulates muscle gene expression and skeletal muscle differentiation, *Genes Dev* 18 (21), 2627–2638, 2004.

122. Bracken, A. P., Dietrich N., Pasini D., Hansen K. H. and Helin K., Genome-wide mapping of Polycomb target genes unravels their roles in cell fate transitions, *Genes Dev* 20 (9), 1123–1136, 2006.

123. Bracken, A. P., Kleine-Kohlbrecher D., Dietrich N., Pasini D., Gargiulo G., Beekman C., Theilgaard-Mönch K., Minucci S., Porse B. T., Marine J. C., Hansen K. H. and Helin K., The Polycomb group proteins bind throughout the INK4A-ARF locus and are disassociated in senescent cells, *Genes Dev* 21 (5), 525–530, 2007.

124. Sessa, L., Breiling A., Lavorgna G., Silvestri L., Casari G. and Orlando V., Noncoding RNA synthesis and loss of Polycomb group repression accompanies the colinear activation of the human HOXA cluster, *RNA* 13 (2), 223–239, 2007.

125. Rinn, J. L., Kertesz M., Wang J. K., Squazzo S. L., Xu X., Brugmann S. A., Goodnough L. H., Helms J. A., Farnham P. J., Segal E. and Chang H. Y., Functional demarcation of active and silent chromatin domains in human HOX loci by noncoding RNAs, *Cell* 129 (7), 1311–1323, 2007.

126. Dellino, G. I., Schwartz Y. B., Farkas G., McCabe D., Elgin S. C. and Pirrotta V., Polycomb silencing blocks transcription initiation, *Mol Cell* 13 (6), 887–893, 2004.

127. Francis, N. J., Kingston R. E. and Woodcock C. L., Chromatin compaction by a polycomb group protein complex, *Science* 306 (5701), 1574–1577, 2004.

128. Lanzuolo, C., Roure V., Dekker J., Bantignies F. and Orlando V., Polycomb response elements mediate the formation of chromosome higher-order structures in the bithorax complex, *Nat Cell Biol* 2007.

129. Grimaud, C., Bantignies F., Pal-Bhadra M., Ghana P., Bhadra U. and Cavalli G., RNAi components are required for nuclear clustering of Polycomb group response elements, *Cell* 124 (5), 957–971, 2006.

130. Chambeyron, S. and Bickmore W. A., Chromatin decondensation and nuclear reorganization of the HoxB locus upon induction of transcription, *Genes Dev* 18 (10), 1119–1130, 2004.

131. Schoorlemmer, J., Marcos-Gutiérrez C., Were F., Martínez R., García E., Satijn D. P., Otte A.P. and Vidal M., Ring1A is a transcriptional repressor that interacts with the Polycomb-M33 protein and is expressed at rhombomere boundaries in the mouse hindbrain, *EMBO J* 16 (19), 5930–5942, 1997.

132. Wu, S., Mulligan P., Gay F., Landry J., Liu H., Lu J., Qi H. Q., Wang W., Nickoloff J. A., Wu C. D. A. and Shi Y. A., YY1-INO80 complex regulates genomic stability through homologous recombination-based repair, *Nat Struct Mol Biol* 14 (12), 1165–1171, 2007.

133. Kalenik, J. L., Chen D., Bradley M. E., Chen S. J., Lee T. C., Yeast two-hybrid cloning of a novel zinc finger protein that interacts with the multifunctional transcription factor YY1, *Nucleic Acid Res* 1997, 25 (4), 843–849, 1997.

134. García, E., Marcos-Gutiérrez C., del Mar Lorente M., Moreno J. C. and Vidal M., RYBP, a new repressor protein that interacts with components of the mammalian Polycomb complex, and with the transcription factor YY1, *EMBO J* 18 (12), 3404–3418, 1999.

9 Epigenetics and its Genetic Syndromes

Richard J. Gibbons

CONTENTS

9.1 Introduction 155
9.2 Chromatin Remodeling 156
    9.2.1 X-Linked α Thalassemia Mental Retardation (ATR-X)
        Syndrome 156
    9.2.2 CHARGE Syndrome... 158
    9.2.3 Cockayne Syndrome B (CSB)... 158
    9.2.4 Schimke Immuno-Osseous Dysplasia (SIOD)................................... 159
9.3 Establishing DNA Methylation 159
    9.3.1 Immunodeficiency, Centromeric Instability, and Facial
        Anomalies (ICF) Syndrome.. 159
9.4 Reading DNA Methylation 160
    9.4.1 Rett syndrome 160
9.5 Modifying Histones 163
    9.5.1 Coffin–Lowry Syndrome (CLS) 163
    9.5.2 Rubenstein–Taybi Syndrome (RSTS) 164
    9.5.3 Sotos Syndrome 164
9.6 Silencing by Position Effect 165
    9.6.1 Facioscapulohumeral Dystrophy (FSHD)... 165
9.7 Concluding Remarks.. 167
References... 168

9.1 INTRODUCTION

The study of human genetic conditions has had a major impact not only by providing a better understanding of the pathogenesis of disease but also by helping elucidate the function of proteins. This largely observational exercise complements the experimental approach, which, by its nature, usually only discloses what one is looking for. Through observation of the disease phenotype, one may reveal unknown properties of a protein or its participation in a pathway that had not been foreseen.

In this chapter, I explore the contribution of a growing number of genetic syndromes in defining the function of the components of chromatin and of the enzymes that modify chromatin structure. The perturbation of gene expression is a common theme in these "diseases of chromatin" and they are of great interest in attempts to understand how the epigenome is established and regulated. In most cases these conditions are caused by mutations in trans-acting factors, and here they are classified as being involved in chromatin remodeling, establishing or reading DNA methylation, and modifying histones. Imprinting disorders are not discussed as these are dealt with elsewhere (Chapter 11). Finally, there is a discussion on the chromosomal deletion seen in facioscapulohumeral dystrophy, which affects the expression of genes in *cis* in a manner reminiscent of position-effect variegation.

9.2   CHROMATIN REMODELING

9.2.1   X-Linked α Thalassemia Mental Retardation (ATR-X) Syndrome

Mutations in *ATRX* give rise to an X-linked syndrome (OMIM 301040) associated with severe mental retardation, microcephaly, characteristic facies, genital abnormalities, and α thalassemia.[1] The distinctive facial traits are most readily recognized in early childhood and the gestalt is probably secondary to facial hypotonia. The frontal hair is often upswept; the eyes are widely spaced; there are epicanthic folds, a flat nasal bridge, and midface hypoplasia; the nose is small triangular, and upturned; the upper lip is tented; and the lower lip full and everted. The frontal incisors are frequently widely spaced, the tongue protrudes, and there is prodigious dribbling. Genital abnormalities are seen in 80% of children, and these may be very mild (e.g., undescended testes or deficient prepuce), but the spectrum of abnormality extends through hypospadias and micropenis to ambiguous female external genitalia. The most severely affected children, who are clinically defined as male pseudohermaphrodites, are usually raised as females. In such cases there are no Mullerian structures present and intra-abdominal, dysgenetic testes (streak gonads) may be found (reviewed by Gibbons and Higgs, 2000).[2] α Thalassemia is a form of anemia that is caused by a defect in the production of α globin. ATR-X syndrome has been shown to be due to a defect in transcription, distinguishing it from most other forms of this anemia, which commonly result from deletions of the α globin genes. Although α thalassemia was initially one of the defining elements of the syndrome, it is clear that there is considerable variation in the hematological manifestations associated with *ATRX* mutations, and up to 15% of cases of ATR-X syndrome show no signs of α thalassemia. At the other extreme, α globin expression is virtually extinguished in individuals with acquired mutations in *ATRX,* which occur as a rare complication in the hematological disorder myelodysplasia.[3]

Female carriers are usually physically and intellectually normal, and this is associated with a highly skewed pattern of X chromosome inactivation, with the normal allele preferentially on the active chromosome.[4]

Missense mutations cluster in two regions of the gene. One is an N-terminal zinc finger domain (called the ADD domain), which is related to sequences seen in the DNA methyltransferase 3 (DNMT3) family of de novo DNA methyltransferases.[5] Almost one half of the disease-causing missense mutations lie in this region, which accounts for only 4% of the coding sequence. NMR-based structural analysis of the ADD domain of ATRX shows that it has an N-terminal GATA-like zinc finger, a PHD finger, and a long C-terminal α helix packed together to form a single globular domain (Argentaro et al., submitted). The α helix of the GATA-like finger is exposed and highly basic, suggesting a DNA-binding function for ATRX, and this is consistent with DNA-binding studies by Cardoso and colleagues.[6] The disease-causing mutations affecting the ADD domain fall into two groups: the majority affect buried residues and hence affect the structural integrity of the ADD domain; another group affects a cluster of surface residues, and these are likely to perturb a potential protein interaction site (Argentaro et al., submitted). The second cluster of missense mutations is seen in the C-terminal half of ATRX in a conserved domain that defines ATRX as a member of the SNF2 family of helicase/ATPases. These proteins are thought to be molecular motors that are able to remodel chromatin and are powered by the hydrolysis of ATP. Although ATRX appears to have only weak chromatin remodeling activity, it is a translocase that can actively move along double-stranded DNA. Members of this family are involved in a wide variety of cellular functions, including the regulation of transcription (SNF2, MOT1, and brahma), control of the cell cycle (NPS1), DNA repair (RAD16, RAD54, and ERCC6), and mitotic chromosome segregation (lodestar). Given that mutations in ATRX lead to downregulation of α-globin gene expression, it is thought that ATRX may play a role in gene expression.

Interestingly, even the most deleterious mutations in *ATRX* are associated with detectable protein, and this even extends to nonsense mutations, where use of an alternate translational start site or mRNA splicing leads to a degree of phenotypic rescue (unpublished; Howard et al., 2004).[7] In mice, *ATRX* knockout leads to early embryonic lethality,[8] so it seems possible that none of the human constitutional mutations are true nulls.

Protein studies have shown that ATRX is a nuclear protein with a punctate staining pattern. In mouse cells, and to a lesser extent in human cells, the majority of the protein is associated with DAPI-bright regions of the nucleus, which are known to represent pericentromeric heterochromatin.[9] In most cell types, it appears likely that the recruitment of ATRX to these sites is via the heterochromatic proteins HP1α and HP1β, which are bound via H3K9me3.[10] In ES cells knocked out for the H3K9 histone methyltransferases Suv39H1/2, both HP1 and ATRX are delocalized.[11] In mature neurones, however, ATRX appears to be recruited to pericentromeric heterochromatin by MECP$_2$.[12] ATRX is also found in promyelocytic leukemia nuclear bodies, where it interacts with the transcription cofactor Daxx.[13] One additional striking finding in human metaphase preparations is that anti-ATRX antibodies consistently localize to the short arms of acrocentric chromosomes and colocalize with a transcription factor (upstream binding factor) that is known to bind the ribosomal DNA (*rDNA*) arrays in nucleolar organizer regions.[14]

Mutations in *ATRX* are associated with changes in DNA methylation; although there is no global reduction in methylcytosine, both hypomethylation (e.g., at *rDNA*) and hypermethylation (e.g., at the repeat *DYZ2*) are observed and suggest ATRX plays a role in establishing or maintaining the *pattern* of DNA methylation.[14] To date, no change in the pattern of methylation has been detected in the α-globin gene cluster that might explain the reduced expression of the α-globin genes. In fact, at present it is not known whether these genes are direct or indirect targets. However, it does appear that a determinant of the severity of the α thalassemia in this condition does map to a block of linkage disequilibrium that includes the α-globin gene loci (unpublished). Intriguingly, in the conditional *ATRX* knockout in mouse, α-globin expression is not perturbed, raising the possibility that the targets in these different species may be different (unpublished).

9.2.2 CHARGE SYNDROME

CHARGE syndrome (OMIM 214800) is a nonrandom pattern of congenital anomalies. CHARGE is an acronym for the features commonly seen in this condition: coloboma, heart defect, atresia choanae, retarded growth and development, genital hypoplasia, and ear anomalies/deafness (reviewed in Blake et al., 2006).[15] The birth incidence is approximately 1:10,000 and cases are usually sporadic. It arises due to deletions of, or point mutations within, *CHD7* (8q12).[16] *CHD7* encodes a chromodomain helicase DNA (CHD)-binding protein, which is an SNF2-like ATPase and by association is thought to play a role in chromatin structure and gene expression.

All the malformations in CHARGE occur during the first 3 months of pregnancy. In situ hybridization analysis of the *CHD7* gene during early human development shows a good correlation between CHD7 expression patterns and the developmental anomalies observed in the syndrome.[17] Most mutations are stop or frameshift, leading to protein truncation, and this suggests that the phenotype of CHARGE syndrome arises due to haploinsufficiency.

Although at present there is little published literature on the function of CHD7, analysis of other members of the CHD subfamily indicates some of the likely characteristics. CHD proteins are characterized by the presence of tandem chromodomains, which are highly conserved sequence motifs involved in chromatin remodeling and the regulation of gene expression in eukaryotes during development (reviewed in Jones et al., 2000).[18] Functional analyses have demonstrated that the chromodomain serves as a module to mediate chromatin interactions by binding directly to DNA, RNA, and/or methylated histone H3.[19–21] Two members of the subfamily, Chd3 (Mi-2α) and Chd4 (Mi-2β), are well-characterized biochemically as components of a chromatin remodeling and histone-deacetylating enzyme complex (NURD) that contributes to appropriate gene regulation.[22]

9.2.3 COCKAYNE SYNDROME B (CSB)

Cockayne syndrome (CS) is a DNA repair disorder and is characterized by stunted growth that is apparent in the first few years of life. The affected individuals exhibit skin photosensitivity, thin dry hair, and a progeroid (prematurely aged) appearance; they also develop a progressive pigmentary retinopathy and sensorineural deafness. In marked contrast with the other DNA repair disorder xeroderma pigmentosa, CS is not associated with a significant increase in skin cancer or a predisposition to infection. Affected individuals become progressively cachectic, and death commonly occurs within the first two decades.[23]

Fibroblasts from individuals with CS exhibit increased sensitivity to UV irradiation, and particularly marked is the failure of RNA synthesis to recover to normal rates after UV exposure.[24,25] Cell fusion was used to demonstrate two major complementation groups, A and B.[26] CSA (OMIM 216400) is caused by mutations in the group 8 excision repair cross-complementing protein (*ERCC8*)[27] and CSB (OMIM 133540) is caused by mutations in *ERCC6*.[28] Both proteins are part of a nucleotide excision repair (NER) pathway that removes DNA damage arising from UV irradiation. This particular NER pathway preferentially repairs lesions on the transcribed strand of active genes, a process that normally proceeds at a faster rate than the repairs on nontranscribed strands.

ERCC6 encodes a Snf2-related DNA-dependent ATPase. It actively wraps around DNA[29] and has ATP-dependent nucleosome remodeling activity.[30] When cells from individuals with CSB were rescued with wild-type ERCC6, a group of genes with significant changes in gene expression were identified. These included many genes that were also regulated by inhibitors of HDAC and DNA methylation. This finding is consistent with a possible role for ERCC6 in chromatin remodeling.[31]

9.2.4 SCHIMKE IMMUNO-OSSEOUS DYSPLASIA (SIOD)

SIOD (OMIM 242900) is an autosomal recessive disorder with the diagnostic features of spondyloepiphyseal dysplasia, renal dysfunction, skin pigmentation, episodic cerebral ischemia, and T-cell immunodeficiency commencing in childhood.[32] It is associated with mutations in a SWI/SNF2-related, matrix-associated, actin-dependent regulator of chromatin, subfamily a-like 1 (*SMARCAL1*), which is a member of the SNF2 family and hence is a putative chromatin remodeling enzyme.[33] However, SIOD patients do not exhibit hypersensitivity to ultraviolet radiation, genomic instability, increased cancer incidence, or defective DNA repair following exposure to gamma radiation. This suggests that SMARCAL1 is not a regulator of DNA replication, repair, or recombination.

9.3 ESTABLISHING DNA METHYLATION
9.3.1 IMMUNODEFICIENCY, CENTROMERIC INSTABILITY, AND FACIAL ANOMALIES (ICF) SYNDROME

ICF syndrome (OMIM 242860) is a rare autosomal recessive condition that usually presents with a tendency to recurrent infections and unusual facies. The chromosomes from an affected individual's lymphocytes are characterized by decondensation and instability in the pericentromeric heterochromatin of chromosomes 1 and 16, and to a lesser extent.[34,35]

The facial dysmorphism is subtle, and the most common features are widely spaced eyes, a broad, flat nasal bridge, and epicanthic folds. Approximately one third of affected children have mild to moderate developmental delay affecting both cognition and motor function. The immunodeficiency is common and results in severe recurrent infections affecting the respiratory and gastrointestinal systems from early childhood. Immunoglobulins are low despite the presence of B cells, and in half the cases levels of T cells are reduced (reviewed by Ehrlich et al., 2006).[36]

The chromosomal abnormalities observed in ICF syndrome are diagnostic. Whole-arm chromosomal deletions, pericentromeric breaks of chromosomes 1, 9, and 16, and multibranched formations involving these chromosomes may be seen. There also may be decondensation (stretching) of the pericentromeric regions of chromosomes 1 and 16. Curiously, these abnormalities are usually seen only in mitogen-treated lymphocytes. A similar pattern of decondensation may be seen after treating cells with demethylating agents, which was the clue that there is a defect in DNA methylation in ICF syndrome.[37] This predominantly affects the GC-rich classical satellites 2 and 3, although some affected individuals have hypomethylation of other genomic sequences (e.g., α satellites, centromeric DNA, Alu, D4Z4, and NBL2 repeats). Inactive X chromosomes are globally undermethylated in ICF syndrome, but this rarely involves CpG islands and is not accompanied by significant biallelic reactivation.[38]

In approximately 40% of ICF cases mutations are detected in the highly conserved catalytic domain of the DNA methyltransferase 3B (*DNMT3B*) gene, which is involved in de novo DNA methylation.[39] The etiology of ICF syndrome in the other 60% of cases is not known and does not appear to involve the other de novo DNA methylase DNMT3A. Hypomethylation of α satellite DNA appears to be seen only in individuals negative for *DNMT3B* mutations, strengthening the supposition that there is genetic heterogeneity in ICF syndrome.[40]

Although it seems most likely that the pathophysiology of ICF syndrome is a consequence of hypomethylation, the genes involved and the mechanism by which this occurs, given that methylation changes predominantly affect gene-poor heterochromatic sequences, are unknown. Recently a murine model for ICF has been generated in which ICF-associated mutations have been introduced into the mouse by homologous recombination.[41] Whereas mice homozygous for a null mutation of *Dnmt3b* die at late gastrula stage, mice carrying missense mutations of two human ICF syndrome alleles develop to term. These mice show phenotypes reminiscent of those seen in individuals affected by ICF, including hypomethylation of repetitive sequences, low body weight, distinctive cranial facial anomalies, and T-cell death by apoptosis. The ICF mutations in mice result in partial loss of function of Dnmt3b, which is consistent with the notion that no ICF patients are homozygous for nonsense alleles. Hopefully these mice will serve as good models for understanding the etiology of ICF syndrome and aid the identification of target genes that are regulated by DNA methylation during development.

9.4   READING DNA METHYLATION
9.4.1   RETT SYNDROME

Rett syndrome (RTT, OMIM 312750) is a severe neurodevelopmental disorder affecting females, with an incidence of approximately 1:15,000.[42] Usually there are no problems at birth and onset occurs in early childhood (6–18 mos). There is then progressive loss of intellectual and motor function, slowing of head growth, loss of purposeful hand movement and the development of stereotypic hand movements, autonomic nervous system dysfunction, and autistic features. In the later stages of the disorder the individual experiences severe growth failure, epileptic seizures, dystonia, and scoliosis (reviewed by Francke, 2006).[43]

This syndrome is associated with mutations in the X-linked gene *MECP₂* (Xq28), which encodes the methyl-CpG-binding protein 2.[44] There are eight common mutations, which are missense or truncating and lead to loss of function. In general, truncating mutations that lead to loss of the methyl-binding domain (MBD) or transcriptional repression domain (TRD) lead to a more severe failure of head growth than missense mutations.[45] C-terminal mutations that leave the MBD and TRD intact are often associated with a milder phenotype characterized by preserved speech and motor function.[46] In classic RTT the X chromosomal inactivation (XCI) pattern is random and usually balanced. However, skewing of XCI, depending on its direction, can give rise to a severe form with neonatal onset or a milder variant.[47]

Classical RTT in males is very rare and is associated with mosaicism due to either the presence of a somatic mutation of *MeCP₂*[48] or the presence of an additional X chromosome.[49] Males with a single X chromosome who have mutations in MeCP₂ present in two ways: those with inactivating mutations (associated with RTT), though apparently normal at birth, soon develop a form of encephalopathy that is lethal within a few months; individuals with mutations may also present with nonsyndromic mental retardation (reviewed by Bienvenu and Chelly, 2006).[50] In these latter cases the mutations found do not overlap with those associated with RTT. It has recently been shown that mutations in the X-linked gene cyclin-dependent kinase-like 5 (*CDKL5*) give rise to an RTT-like disease.[51,52] The possibility that MeCP₂ and CDLK5 are involved in the same pathway has been strengthened by observation that the expression pattern of these two genes in the developing mouse is similar. Furthermore, an interaction between these proteins has been observed in vitro and in vivo.[53]

Male mice with germline mutations in *Mecp2* are born normal but after a few weeks develop an RTT-like condition.[54,55] They develop tremors, have reduced mobility, and become growth restricted with small heads. They eventually die by 10 weeks. In contrast, female heterozygotes appear normal until about 6 months when they begin to develop an RTT-like condition. Male mice with a brain-specific knockout followed a course similar to those with the germline knockout, and though having a later onset, mice with a postnatal brain-specific knockout were similarly affected. The phenotype of *Mecp2* knockout mice was, however, prevented by the presence of a transgene that was expressed only in post-mitotic neurons.[56] This is consistent with the finding that Mecp2 is normally most highly expressed in brain and it is highest of all in post-mitotic, post-migratory neurons.[57] This expression pattern and the postnatal onset in RTT-affected individuals and Mecp2-deficient mice are consistent with the notion that MeCP₂ plays a role in the maturation of neurons and the formation of dendritic branches. Remarkably, in the mouse these abnormalities are reversible, and resumption of endogenous MeCp2 production in a Mecp2-deficient mouse completely corrects the neurological deficit.[58]

MeCP₂ was first identified through its ability to bind a single methyl-CpG dinucleotide in DNA via its MBD.[59] More recently it has been shown that high-affinity binding requires [A/T]₄ sequences adjacent to the methyl-CpG, which confers some sequence specificity.[60] It is thought that MeCP₂, through binding of the methyl-CpG dinucleotide, affects nearby genes by repressing transcription and modulating chromatin structure through the recruitment of corepressor complexes. Although there is some controversy as to whether MeCP₂ forms any stable complexes with other proteins, a growing list of interactions has been put forth. These include the corepressors Sin3a-HDAC,[61] c-SKI, and N-CoR[62] as well as DNMT1,[63] Suv39H1,[64] ATRX,[12] and the RNA-binding protein YB1.[65] The contribution of these factors to MeCP₂-mediated repression is still uncertain, as is the possible role of MeCP₂ in other cellular activities.

Given that the defect in RTT appears principally to affect post-mitotic neurons, it is perhaps not surprising that no widespread changes in gene expression occur in MeCP$_2$-deficient cells. In the majority of microarray-based expression studies only subtle changes were observed and these were inconsistent between studies. In one experiment, however, comparing brains from wild-type and MeCP$_2$-deficient mice, changes were observed in a group of glucocorticoid-regulated genes.[66] Those perhaps of most significance were Sgk and Fkbp5, in which changes in expression were seen prior to the onset of RTT-like signs in the animals. A candidate gene approach has led to the identification of other Mecp2-regulated genes. It was shown that in *Xenopus laevis*, MeCP$_2$ binds upstream of the promoter of *xHairy2* and represses transcription.[67] The protein product of *xHairy2* inhibits neuronal cell differentiation and in the absence of MeCP$_2$ there was enhanced expression of Hairy2a. Brain-derived neurotrophic factor (BDNF) is involved in neuronal survival and plasticity and has been shown to be a target in mice.[68,69] Chromatin immunoprecipitation (ChIP) has shown that MeCP$_2$ is bound to methyl-CpG near the gene promoter and the binding site shows the presence of the $[A/T]_4$ motif, which appears to confer specificity in vitro. In resting neurons, Ca^{2+} signaling leads to activation of BDNF, which is associated with the loss of MeCP$_2$ from the promoter. Two explanations put forward for the dissociation of MeCP$_2$ are loss of promoter methylation or the phosphorylation of MeCP$_2$ triggered by the neuron depolarization. Furthermore, in resting neurons, BDNF expression is doubled in MeCP$_2$-deficient mice. Additional evidence for a functional interaction between MeCP$_2$ and BDNF is the finding that BDNF levels can modulate the progression of disease in the MeCP$_2$-deficient mice: forebrain-specific knockout of BDNF leads to exacerbation of the RTT-like phenotype whereas overexpression of BDNF in the same cells rescues some of the phenotype.[70] Curiously, however, these effects are the opposite of what might be expected if MeCP$_2$ was repressing BDNF expression.

There has been particular interest in studying whether MeCP$_2$ plays a role in the methylation-associated gene silencing in X inactivation and imprinting. Silencing of X-inactivated alleles is maintained in clonal RTT cell lines, indicating that it is unlikely that this methyl-binding protein plays an important role. However, there is some evidence, albeit tentative, that imprinting of some genes does depend on MeCP$_2$, though there may be at least two different mechanisms at work. The *UBE3a* locus lies within a cluster of imprinted genes on human chromosome 15 and its activity depends on a complex differentially methylated imprinting control region (ICR). In brain, UBE3a is normally expressed only from the maternally derived allele. The paternal allele is silenced by a *cis*-acting UBE3a antisense RNA, which is under the control of the unmethylated ICR; on the maternal chromosome the ICR is methylated, the antisense RNA is absent, and UBE3a is expressed. In MeCP$_2$-deficient cells there is biallelic expression of the UBE3a antisense RNA, which leads to twofold downregulation of the maternal allele.[71] In this case regulation of the antisense RNA appears to depend on MeCP$_2$ reading and effecting a response to the methylation status of the ICR.

The imprinted genes *Dlx5* and *Dlx6* lie adjacent to each other. Normally, they are both silenced on the paternally derived chromosome by a looping mechanism that is mediated by MeCP$_2$ and is associated with the presence of local repressive histone marks. In MeCP$_2$-deficient mice the looping is disrupted and there is a two-fold increase in the expression of Dlx5 and Dlx6, which represents activation of the paternal allele.[72] A similar loss of imprinting was observed in human RTT cell lines. It remains unclear, however, whether aberrant expression of these genes contributes to the disease phenotype.

9.5 MODIFYING HISTONES

9.5.1 COFFIN–LOWRY SYNDROME (CLS)

CLS (OMIM 303600) is an X-linked form of severe mental retardation associated with characteristic facies (widely spaced, down-slanting eyes and full, fleshy lips), microcephaly, short stature, and skeletal abnormalities (reviewed by Young, 1988).[73] In particular, affected individuals have large, soft hands with tapering digits. Female carriers exhibit mild mental retardation, short stature, and coarse facies. CLS is associated with mutations in the ribosomal protein S6 kinase 2 (*RSK2*) gene.[74] The pattern of mutations found and the consequent phenotype are consistent with loss of function. Delaunoy et al.[75] stated that of the 128 CLS mutations reported to date, 33% are missense mutations, 15% nonsense mutations, 20% splicing errors, and 29% short deletion or insertion events, and four large deletions have been reported. The mutations were distributed throughout the *RSK2* gene and showed no obvious clustering.

In humans, the RSK family contains four closely regulated members: RSK1, RSK2, RSK3, and RSK4. They form a family of growth factor-regulated serine/threonine kinases that contain two kinase domains, one at the C terminus and one at the N terminus.[74] RSKs are implicated in the activation of the mitogen-activated kinase cascade and the stimulation of cell proliferation (at the transition between phases G_0 and G_1 of the cell cycle) and differentiation. RSK2 is prominent in the cortex and hippocampus, regions of the brain important for learning and memory.[76]

When mammalian cells are stimulated by mitogens, histone H3 is rapidly and transiently phosphorylated by one or more kinases. Sassone-Corsi et al.[77] demonstrated that RSK2 was required for epidermal growth factor (EGF)-stimulated phosphorylation of H3. Fibroblasts derived from a CLS patient failed to exhibit EGF-stimulated phosphorylation of H3 at serine 10, although H3 was phosphorylated normally during mitosis. Introduction of the wild-type *Rsk2* gene restored the activity. This function of RSK2 appears to be conserved in mice, as disruption of the *Rsk2* gene by homologous recombination in murine embryonic stem cells abolished EGF-stimulated phosphorylation of H3. It appears, therefore, that H3 may be a direct or indirect target of RSK2. Phosphorylation of histone H3S10 occurs in two circumstances. Extensive phosphorylation of H3S10 occurs as cells enter mitosis, and this is associated with chromosome condensation and the suppression of transcription. Phosphorylation of the same residue occurs when cells are stimulated by growth factors from quiescence to cell division. In the latter case the phosphorylation is limited to the immediate-early genes and is followed by a wave of H3 acetylation that is associated with the transcriptional activation of these genes.

However, RSK2 has many potential substrates and it is not clear that the abnormal phenotype is due to a defect in histone phosphorylation. RSK2 is known to phosphorylate CREB (cAMP response element-binding) proteins,[78] the Shank family of proteins that plays a role in the neuronal synapse,[79] and the transcription factor ATF4.[80] For example, phosphorylation of CREB at serine 133 leads to its activation. Indeed, in a cell line derived from an individual with CLS, growth-factor stimulated CREB phosphorylation was reduced, and when RSK2 was immunoprecipitated from patient cell lines, the capacity to phosphorylate a CREB-like peptide was reduced and there was a linear relationship between RSK2 activation of CREB and cognitive levels in patients.[81] It is clear then that although CLS is a candidate for the title of a chromatin disease, evidence is still lacking.

9.5.2   RUBENSTEIN–TAYBI SYNDROME (RSTS)

RSTS (OMIM 180849) is an autosomal dominant congenital malformation and mental retardation syndrome that occurs in approximately 1 of 125,000 births. These babies often have cardiac defects, broad thumbs, big toes, and characteristic facies, a hooked nose being a prominent feature.[82] The affected individuals also have a 350-fold increased risk of tumor formation, usually childhood cancers of neural crest origin.[83] Mutations in either CREB-binding protein (*CBP*)[84] or *EP300*[85] can give rise to the syndrome. About 20% of *CBP* mutations are deletions and protein-truncating mutations, which suggests that the haploinsufficiency of protein function is a common cause of the syndrome. CBP and EP300 are highly related proteins and act as transcriptional regulators in the control of gene expression through at least two pathways. They act as coactivators and recruit polII complex to the promoter, but also they remodel chromatin by histone acetylation.[86] Murine *Cbp* heterozygous mutants that have behavioral and cognitive deficits have been used to determine the relative contributions of these two pathways to the pathogenesis of RSTS.[87]

It was found that these aspects of the phenotype in the *Cbp* heterozygous mutants could be ameliorated using inhibitors of enzymes that compensate for a reduction in CBP's function as a CREB coactivator. Nevertheless, the majority of missense mutations in CBP lie in the HAT domain of the protein; some of these largely preserve coactivator activity although abolish HAT activity. Thus it seems likely that a defect in HAT activity contributes to the pathogenesis of this condition. This is consistent with the finding that there is reduced histone H2B acetylation in a mouse model heterozygous for a null *Cbp* allele. Furthermore, treatment of these mutants with HDAC inhibitors led to a strong increase in acetylated histone H2B, and this was associated with the reversal of a memory defect exhibited by the mutant mice. Analysis of histone modifications at specific promoters of target genes is an obvious priority. However, it must be borne in mind that CBP can acetylate other proteins such as the basic transcription apparatus, and it remains to be confirmed that perturbed histone acetylation is the cause of RSTS.

9.5.3   SOTOS SYNDROME

Sotos syndrome (OMIM 117550) is an autosomal dominant disorder that usually occurs sporadically and is characterized by excessively rapid growth in childhood, advanced bone age, characteristic facial appearance, macrocephaly, and mild learning difficulties. The facial appearance is most distinctive between 1 and 6 years and consists of a high, broad forehead, sparsity of fronto-temporal hair, malar flushing, down-slanting palpebral fissures, and a pointed chin.[88] Babies are born long and thin and neonatal hypotonia is common. Although 90% of children with Sotos have height and head circumference >2 SD above the mean,[89] the velocity of growth slows postpuberty and thus the height of many adults with Sotos may not be significantly above the norm.[88] Affected individuals are at an increased risk (170-fold) of developing a range of diverse neoplasms, especially neural crest tumors, sacrococcygeal teratomas, and some hematological malignancies, but the absolute risk of tumor development in Sotos syndrome is still <3%.[90,91]

The cloning of the breakpoints of a de novo t(5:8)(q35;q24.1) translocation in an individual with Sotos syndrome led to the identification of the disease gene, Nuclear receptor Set Domain containing protein 1 (*NSD1*) gene.[92] Various different mutational mechanisms lead to haploinsufficiency as a common pathway; these include truncating mutations, missense mutations, splice-site mutations, intragenic deletions, and 5q35 microdeletions. Truncating mutations are found throughout the gene, and missense mutations cluster in the 3' region of the gene where most of the functional domains lie. No clear genotype–phenotype correlation is evident and there is considerable variability in the phenotypic features associated with a particular mutation. *NSD1* mutations are found in >90% of Sotos cases and most are sporadic and probably represent new dominant mutations.[89]

NSD1 protein contains a SET (su(var)3-9, enhancer-of-zeste, trithorax) domain, which is a feature of histone methyltransferases. In vitro NSD1 has been shown to methylate H3K36 and H4K20 (features of silent heterochromatin), but its in vivo substrate specificity is yet to be determined.[93] NSD1 is also implicated in silencing through its interaction with NIZP1, which represses transcription in an NSD1-dependent fashion.[94]

Mice that are homozygous for a null *Nsd1* mutation can initiate mesoderm formation, but they fail to complete gastrulation.[93] The mesoderm is sparse and there is a high incidence of apoptosis in the embryonic ectoderm from which the mesoderm is generated. Heterozygous mice are normal at birth, display a normal growth rate, and are fertile. It is possible that a Sotos phenotype in mice is subtler than in man. The expression pattern of NSD1 is consistent with it playing a similar role in mouse and man, as it is expressed in brain and in the region of ossification in the developing bone. However, it is conceivable that the target genes in mouse differ from those in man.

9.6 SILENCING BY POSITION EFFECT

9.6.1 FACIOSCAPULOHUMERAL DYSTROPHY (FSHD)

FSHD (OMIM 158900) is an autosomal dominant myopathy mainly characterized by a progressive and highly variable atrophy of the facial, shoulder, and upper arm muscles. Muscle involvement is often asymmetrical and in a pair of identical twins the disease manifested itself discordantly. Weakness may appear from infancy to old age but typically is seen in the second decade. Other commonly observed features are high-tone sensorineural deafness and abnormalities of the retinal capillaries which can lead to retinopathy (reviewed by Tawil and Van der Maarel[95]).

Almost all patients carry deletions of tandem 3.3-kb heterochromatic repeats, termed *D4Z4* on chromosome 4q35.[96] The number of repeat units varies from 11 to more than 100 in the normal population, and in FSHD patients an allele of 1–10 residual units is observed because of the deletion of an integral number of these units. The number of *D4Z4* repeats is a critical determinant of the age of onset and the clinical severity of FSHD. In general, 1–3 repeats is associated with a severe form of the disease of childhood onset, 4–7 repeats with the most common form, and 8–10 repeats with a milder disease with reduced penetrance. Although the repeat contains a putative open-reading frame, there is no evidence that this is expressed and so it appears that FSHD does not arise due to the loss of protein-coding genes, but rather that the deletion of *D4Z4* affects the organization of the chromatin in this subtelomeric region and consequently perturbs the expression of nearby genes. Of significance, loss of the entire *D4Z4* repeat and adjacent 200-kb of DNA via an unbalanced translocation does not give rise to FSHD.[97] This suggests that in FSHD there is a gain of function. This hypothesis was supported by the finding that FSHD region gene 1 (*FRG1*), FSHD region gene 2 (*FRG2*), and adenine nucleotide translocator-1 gene (*ANT1*), which lie 0.1 Mb, 0.2 Mb, and 5Mb, respectively, proximal to the *D4Z4* repeat were overexpressed in FSHD muscle.[98] Of note, however, the expression of other genes within this 5-Mb interval was not perturbed in FSHD (Tupler, personal communication), which suggests some discontinuous mechanism of action.

To identify the gene responsible for FSHD pathogenesis transgenic mice were generated which selectively over-expressed FRG1, FRG2, or ANT1 in skeletal muscle.[99] FRG1 transgenic mice develop a muscular dystrophy with features characteristic of the human disease whereas FRG2 and ANT1 transgenic mice appeared normal. FRG1 is thought to play a role in RNA splicing: it exhibits a nuclear speckled pattern characteristic of mammalian splicing factors and it is a component of purified spliceosomes. The abnormal pattern of splicing seen in the affected muscles of FRG1 transgenic mice and individuals with FSHD is consistent with this role.

It appears therefore that D4Z4 normally suppresses the expression of certain nearby genes, and deletions within this repeat lead to their derepression. Electrophoretic mobility shift assay and in vitro DNaseI footprinting were used to identify a 27-bp element within the D4Z4 repeat which bound a complex of nuclear factors.[98] This complex comprised YY1, HMGB2, and nucleolin and sequence-specific binding depended on YY1. It is not yet clear how these factors might affect the expression of genes in *cis*; there is no evidence, in the wild-type allele, for spreading of heterochromatin from *D4Z4* into the adjacent chromosomal region.[100] Recent experiments have shown that YY1 recruits HDAC1 and EZH2, an H3K27 methyltransferase involved in Polycomb silencing, to the genomic region of silent muscle-specific genes in mice.[101] Given that only some genes adjacent to *D4Z4* are upregulated in FSHD,[98] it raises the possibility that the wild-type suppression of the target genes may be via a looping mechanism involving the *D4Z4* region and the target genes.

The *D4Z4* deletion in FSHD is associated with hypomethylation in the DNA of the repeat in both affected individuals and nonpenetrant deletion carriers. Of particular interest, in FSHD phenocopies (which have an identical phenotype to FSHD but no deletion within *D4Z4*) there is hypomethylation of both *D4Z4* alleles.[102] These results strongly suggest that although hypomethylation of *D4Z4* is part of a common pathway in a cascade of epigenetic events, since hypomethylation is also seen in nonpenetrant carriers, it is not sufficient to cause FSHD. Consistent with this is the observation that methylation of *D4Z4* is reduced in individuals with ICF syndrome yet they do not have FSHD.[103]

It appears that FSHD represents a fascinating example of position effect, and the epigenetic status of certain genes is dictated by the nature of the heterochromatin juxtaposing them. Given that there is considerable variability in the degree to which different muscles are affected in a given individual, this may even be an example of position effect variegation.

9.7 CONCLUDING REMARKS

The last 10 years have seen a rapid growth in the number of genetic syndromes in which the disease gene has been identified as a component of chromatin or a chromatin-modifying enzyme (Figure 9.1). As the molecular characterization of genetic disease becomes ever easier, it is likely that this list will grow.

The recognizable phenotypes presumably indicate that in each case there is a unique repertoire of genes that are perturbed. However, even with the advent of whole-genome expression profiling, identifying these genes has remained a challenging task, and it is sobering that a major disease can result from changes in gene expression that we may struggle to detect.

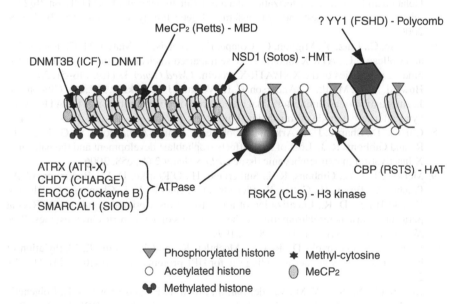

FIGURE 9.1 Components of chromatin and enzymes that modify chromatin that are affected by human disease-causing mutations. Closely packed nucleosomes on the left represent closed heterochromatin, whereas the more widely spaced nucleosomes on the right represent open euchromatin. A key to the symbols used is shown below. For each affected component or modifying enzyme, the name of the protein is followed in parenthesis by the disease, followed by the classification of the protein involved. Although it is not known how FSHD is caused, one possible model is that target genes are derepressed through a failure to recruit the Polycomb complex (see text for explanation).

It remains an intriguing observation that mutations in such general factors give rise to such specific phenotypes. This point, however, is central to understanding the role of these factors and their place in establishing and regulating the human epigenome.

REFERENCES

1. Gibbons, R. J., Picketts, D. J., and Higgs, D. R., Syndromal mental retardation due to mutations in a regulator of gene expression, *Hum Molecular Genet* 4, 1705–1709, 1995.
2. Gibbons, R. J. and Higgs, D. R., The molecular-clinical spectrum of the ATR-X syndrome, *Am J Medical Genet (Semin. Med. Genet.)* 97, 204–212, 2000.
3. Gibbons, R. J., Pellagatti, A., Garrick, D., Wood, W. G., Malik, N., Ayyub, H., Langford, C., Boultwood, J., Wainscoat, J. S., and Higgs, D. R., Identification of acquired somatic mutations in the gene encoding chromatin-remodeling factor ATRX in the alpha-thalassemia myelodysplasia syndrome (ATMDS), *Nat Genet* 34 (4), 446–9, 2003.
4. Gibbons, R. J., Suthers, G. K., Wilkie, A. O. M., Buckle, V. J., and Higgs, D. R., X-linked α thalassemia/mental retardation (ATR-X) syndrome: Localisation to Xq12-21.31 by X-inactivation and linkage analysis, *Am J Human Genet* 51, 1136–1149, 1992.
5. Aapola, U., Kawasaki, K., Scott, H. S., Ollila, J., Vihinen, M., Heino, M., Shintani, A., Minoshima, S., Krohn, K., Antonarakis, S. E., Shimizu, N., Kudoh, J., and Peterson, P., Isolation and initial characterization of a novel zinc finger gene, DNMT3L, on 21q22.3, related to the cytosine-5-methyltransferase 3 gene family, *Genomics* 65 (3), 293–8, 2000.
6. Cardoso, C., Lutz, Y., Mignon, C., Compe, E., Depetris, D., Mattei, M. G., Fontes, M., and Colleaux, L., ATR-X mutations cause impaired nuclear location and altered DNA binding properties of the XNP/ATR-X protein, *J Med Genet* 37 (10), 746–51, 2000.
7. Howard, M. T., Malik, N., Anderson, C. B., Voskuil, J. L., Atkins, J. F., and Gibbons, R. J., Attenuation of an amino-terminal premature stop codon mutation in the ATRX gene by an alternative mode of translational initiation, *J Med Genet* 41 (12), 951–6, 2004.
8. Garrick, D., Sharpe, J. A., Arkell, R., Dobbie, L., Smith, A. J., Wood, W. G., Higgs, D. R., and Gibbons, R. J., Loss of Atrx affects trophoblast development and the pattern of X-inactivation in extraembryonic tissues, *PLoS Genet* 2 (4), e58, 2006.
9. McDowell, T. L., Gibbons, R. J., Sutherland, H., O'Rourke, D. M., Bickmore, W. A., Pombo, A., Turley, H., Gatter, K., Picketts, D. J., Buckle, V. J., Chapman, L., Rhodes, D., and Higgs, D. R., Localization of a putative transcriptional regulator (ATRX) at pericentromeric heterochromatin and the short arms of acrocentric chromosomes, *Proc Natl Acad Sci U S A* 96 (24), 13983–8, 1999.
10. Lachner, M., O'Carroll, D., Rea, S., Mechtler, K., and Jenuwein, T., Methylation of histone H3 lysine 9 creates a binding site for HP1 proteins, *Nature* 410 (6824), 116–20, 2001.
11. Kourmouli, N., Sun, Y. M., van der Sar, S., Singh, P. B., and Brown, J. P., Epigenetic regulation of mammalian pericentric heterochromatin in vivo by HP1, *Biochem Biophys Res Commun* 337 (3), 901–7, 2005.
12. Nan, X., Hou, J., Maclean, A., Nasir, J., Lafuente, M. J., Shu, X., Kriaucionis, S., and Bird, A., Interaction between chromatin proteins MECP2 and ATRX is disrupted by mutations that cause inherited mental retardation, *Proc Natl Acad Sci U S A*, 2007.
13. Xue, Y., Gibbons, R., Yan, Z., Yang, D., McDowell, T. L., Sechi, S., Qin, J., Zhou, S., Higgs, D., and Wang, W., The ATRX syndrome protein forms a chromatin-remodeling complex with Daxx and localizes in promyelocytic leukemia nuclear bodies, *Proc Natl Acad Sci U S A* 100 (19), 10635–40, 2003.

14. Gibbons, R. J., McDowell, T. L., Raman, S., O'Rourke, D. M., Garrick, D., Ayyub, H., and Higgs, D. R., Mutations in ATRX, encoding a SWI/SNF-like protein, cause diverse changes in the pattern of DNA methylation, *Nat Genet* 24 (4), 368–71, 2000.

15. Blake, K. D. and Prasad, C., CHARGE syndrome, *Orphanet J Rare Dis* 1 (1), 34, 2006.

16. Vissers, L. E., van Ravenswaaij, C. M., Admiraal, R., Hurst, J. A., de Vries, B. B., Janssen, I. M., van der Vliet, W. A., Huys, E. H., de Jong, P. J., Hamel, B. C., Schoenmakers, E. F., Brunner, H. G., Veltman, J. A., and van Kessel, A. G., Mutations in a new member of the chromodomain gene family cause CHARGE syndrome, *Nat Genet* 36 (9), 955–7, 2004.

17. Sanlaville, D., Etchevers, H. C., Gonzales, M., Martinovic, J., Clement-Ziza, M., Delezoide, A. L., Aubry, M. C., Pelet, A., Chemouny, S., Cruaud, C., Audollent, S., Esculpavit, C., Goudefroye, G., Ozilou, C., Fredouille, C., Joye, N., Morichon-Delvallez, N., Dumez, Y., Weissenbach, J., Munnich, A., Amiel, J., Encha-Razavi, F., Lyonnet, S., Vekemans, M., and Attie-Bitach, T., Phenotypic spectrum of CHARGE syndrome in fetuses with CHD7 truncating mutations correlates with expression during human development, *J Med Genet* 43 (3), 211–217, 2006.

18. Jones, D. O., Cowell, I. G., and Singh, P. B., Mammalian chromodomain proteins: their role in genome organisation and expression, *Bioessays* 22 (2), 124–37, 2000.

19. Bouazoune, K., Mitterweger, A., Langst, G., Imhof, A., Akhtar, A., Becker, P. B., and Brehm, A., The dMi-2 chromodomains are DNA binding modules important for ATP-dependent nucleosome mobilization, *Embo J* 21 (10), 2430–40, 2002.

20. Akhtar, A., Zink, D., and Becker, P. B., Chromodomains are protein-RNA interaction modules, *Nature* 407 (6802), 405–9, 2000.

21. Min, J., Zhang, Y., and Xu, R. M., Structural basis for specific binding of Polycomb chromodomain to histone H3 methylated at Lys 27, *Genes Dev* 17 (15), 1823–8, 2003.

22. Feng, Q. and Zhang, Y., The NuRD complex: linking histone modification to nucleosome remodeling, *Curr Top Microbiol Immunol* 274, 269–90, 2003.

23. Nance, M. A. and Berry, S. A., Cockayne syndrome: review of 140 cases, *Am J Med Genet* 42 (1), 68–84, 1992.

24. Schmickel, R. D., Chu, E. H., Trosko, J. E., and Chang, C. C., Cockayne syndrome: a cellular sensitivity to ultraviolet light, *Pediatrics* 60 (2), 135–9, 1977.

25. Mayne, L. V. and Lehmann, A. R., Failure of RNA synthesis to recover after UV irradiation: an early defect in cells from individuals with Cockayne's syndrome and xeroderma pigmentosum, *Cancer Res* 42 (4), 1473–8, 1982.

26. Tanaka, K., Kawai, K., Kumahara, Y., Ikenaga, M., and Okada, Y., Genetic complementation groups in cockayne syndrome, *Somatic Cell Genet* 7 (4), 445–55, 1981.

27. Henning, K. A., Li, L., Iyer, N., McDaniel, L. D., Reagan, M. S., Legerski, R., Schultz, R. A., Stefanini, M., Lehmann, A. R., Mayne, L. V., and Friedberg, E. C., The Cockayne syndrome group A gene encodes a WD repeat protein that interacts with CSB protein and a subunit of RNA polymerase II TFIIH, *Cell* 82 (4), 555–64, 1995.

28. Troelstra, C., van Gool, A., de Wit, J., Vermeulen, W., Bootsma, D., and Hoeijmakers, J. H., ERCC6, a member of a subfamily of putative helicases, is involved in Cockayne's syndrome and preferential repair of active genes, *Cell* 71 (6), 939–53, 1992.

29. Beerens, N., Hoeijmakers, J. H., Kanaar, R., Vermeulen, W., and Wyman, C., The CSB protein actively wraps DNA, *J Biol Chem* 280 (6), 4722–9, 2005.

30. Citterio, E., Van Den Boom, V., Schnitzler, G., Kanaar, R., Bonte, E., Kingston, R. E., Hoeijmakers, J. H., and Vermeulen, W., ATP-dependent chromatin remodeling by the Cockayne syndrome B DNA repair-transcription-coupling factor, *Mol Cell Biol* 20 (20), 7643–53, 2000.

31. Newman, J. C., Bailey, A. D., and Weiner, A. M., Cockayne syndrome group B protein (CSB) plays a general role in chromatin maintenance and remodeling, *Proc Natl Acad Sci U S A* 103 (25), 9613–8, 2006.

32. Boerkoel, C. F., O'Neill, S., Andre, J. L., Benke, P. J., Bogdanovic, R., Bulla, M., Burguet, A., Cockfield, S., Cordeiro, I., Ehrich, J. H., Frund, S., Geary, D. F., Ieshima, A., Illies, F., Joseph, M. W., Kaitila, I., Lama, G., Leheup, B., Ludman, M. D., McLeod, D. R., Medeira, A., Milford, D. V., Ormala, T., Rener-Primec, Z., Santava, A., Santos, H. G., Schmidt, B., Smith, G. C., Spranger, J., Zupancic, N., and Weksberg, R., Manifestations and treatment of Schimke immuno-osseous dysplasia: 14 new cases and a review of the literature, *Eur J Pediatr* 159 (1–2), 1–7, 2000.

33. Boerkoel, C. F., Takashima, H., John, J., Yan, J., Stankiewicz, P., Rosenbarker, L., Andre, J. L., Bogdanovic, R., Burguet, A., Cockfield, S., Cordeiro, I., Frund, S., Illies, F., Joseph, M., Kaitila, I., Lama, G., Loirat, C., McLeod, D. R., Milford, D. V., Petty, E. M., Rodrigo, F., Saraiva, J. M., Schmidt, B., Smith, G. C., Spranger, J., Stein, A., Thiele, H., Tizard, J., Weksberg, R., Lupski, J. R., and Stockton, D. W., Mutant chromatin remodeling protein SMARCAL1 causes Schimke immuno-osseous dysplasia, *Nat Genet* 30 (2), 215–20, 2002.

34. Tiepolo, L., Maraschio, P., Gimelli, G., Cuoco, C., Gargani, G. F., and Romano, C., Multibranched chromosomes 1, 9, and 16 in a patient with combined IgA and IgE deficiency, *Hum Genet* 51 (2), 127–37, 1979.

35. Maraschio, P., Zuffardi, O., Dalla Fior, T., and Tiepolo, L., Immunodeficiency, centromeric heterochromatin instability of chromosomes 1, 9, and 16, and facial anomalies: the ICF syndrome, *J Med Genet* 25 (3), 173–80, 1988.

36. Ehrlich, M., Jackson, K., and Weemaes, C., Immunodeficiency, centromeric region instability, facial anomalies syndrome (ICF), *Orphanet J Rare Dis* 1, 2, 2006.

37. Jeanpierre, M., Turleau, C., Aurias, A., Prieur, M., Ledeist, F., Fischer, A., and Viegas-Pequignot, E., An embryonic-like methylation pattern of classical satellite DNA is observed in ICF syndrome, *Hum Mol Genet* 2 (6), 731–5, 1993.

38. Bourc'his, D., Miniou, P., Jeanpierre, M., Molina Gomes, D., Dupont, J., De Saint-Basile, G., Maraschio, P., Tiepolo, L., and Viegas-Pequignot, E., Abnormal methylation does not prevent X inactivation in ICF patients, *Cytogenet Cell Genet* 84 (3–4), 245–52, 1999.

39. Xu, G. L., Bestor, T. H., Bourc'his, D., Hsieh, C. L., Tommerup, N., Bugge, M., Hulten, M., Qu, X., Russo, J. J., and Viegas-Pequignot, E., Chromosome instability and immunodeficiency syndrome caused by mutations in a DNA methyltransferase gene, *Nature* 402 (6758), 187–91, 1999.

40. Jiang, Y. L., Rigolet, M., Bourc'his, D., Nigon, F., Bokesoy, I., Fryns, J. P., Hulten, M., Jonveaux, P., Maraschio, P., Megarbane, A., Moncla, A., and Viegas-Pequignot, E., DNMT3B mutations and DNA methylation defect define two types of ICF syndrome, *Hum Mutat* 25 (1), 56–63, 2005.

41. Ueda, Y., Okano, M., Williams, C., Chen, T., Georgopoulos, K., and Li, E., Roles for Dnmt3b in mammalian development: a mouse model for the ICF syndrome, *Development* 133 (6), 1183–92, 2006.

42. Hagberg, B., Goutieres, F., Hanefeld, F., Rett, A., and Wilson, J., Rett syndrome: criteria for inclusion and exclusion, *Brain Dev* 7 (3), 372–3, 1985.

43. Francke, U., Mechanisms of Disease: neurogenetics of MeCP2 deficiency, *Nat Clin Pract Neurol* 2 (4), 212–21, 2006.

44. Amir, R. E., Van den Veyver, I. B., Wan, M., Tran, C. Q., Francke, U., and Zoghbi, H. Y., Rett syndrome is caused by mutations in X-linked MECP2, encoding methyl-CpG-binding protein 2, *Nat Genet* 23 (2), 185–8, 1999.

45. Huppke, P., Held, M., Hanefeld, F., Engel, W., and Laccone, F., Influence of mutation type and location on phenotype in 123 patients with Rett syndrome, *Neuropediatrics* 33 (2), 63–8, 2002.

46. Smeets, E., Terhal, P., Casaer, P., Peters, A., Midro, A., Schollen, E., van Roozendaal, K., Moog, U., Matthijs, G., Herbergs, J., Smeets, H., Curfs, L., Schrander-Stumpel, C., and Fryns, J. P., Rett syndrome in females with CTS hot spot deletions: a disorder profile, *Am J Med Genet A* 132 (2), 117–20, 2005.

47. Weaving, L. S., Williamson, S. L., Bennetts, B., Davis, M., Ellaway, C. J., Leonard, H., Thong, M. K., Delatycki, M., Thompson, E. M., Laing, N., and Christodoulou, J., Effects of MECP2 mutation type, location and X-inactivation in modulating Rett syndrome phenotype, *Am J Med Genet A* 118 (2), 103–14, 2003.

48. Clayton-Smith, J., Watson, P., Ramsden, S., and Black, G. C., Somatic mutation in MECP2 as a non-fatal neurodevelopmental disorder in males, *Lancet* 356 (9232), 830–2, 2000.

49. Maiwald, R., Bonte, A., Jung, H., Bitter, P., Storm, Z., Laccone, F., and Herkenrath, P., De novo MECP2 mutation in a 46,XX male patient with Rett syndrome, *Neurogenetics* 4 (2), 107–8, 2002.

50. Bienvenu, T. and Chelly, J., Molecular genetics of Rett syndrome: when DNA methylation goes unrecognized, *Nat Rev Genet* 7 (6), 415–26, 2006.

51. Tao, J., Van Esch, H., Hagedorn-Greiwe, M., Hoffmann, K., Moser, B., Raynaud, M., Sperner, J., Fryns, J. P., Schwinger, E., Gecz, J., Ropers, H. H., and Kalscheuer, V. M., Mutations in the X-linked cyclin-dependent kinase-like 5 (CDKL5/STK9) gene are associated with severe neurodevelopmental retardation, *Am J Hum Genet* 75 (6), 1149–54, 2004.

52. Weaving, L. S., Christodoulou, J., Williamson, S. L., Friend, K. L., McKenzie, O. L., Archer, H., Evans, J., Clarke, A., Pelka, G. J., Tam, P. P., Watson, C., Lahooti, H., Ellaway, C. J., Bennetts, B., Leonard, H., and Gecz, J., Mutations of CDKL5 cause a severe neurodevelopmental disorder with infantile spasms and mental retardation, *Am J Hum Genet* 75 (6), 1079–93, 2004.

53. Mari, F., Azimonti, S., Bertani, I., Bolognese, F., Colombo, E., Caselli, R., Scala, E., Longo, I., Grosso, S., Pescucci, C., Ariani, F., Hayek, G., Balestri, P., Bergo, A., Badaracco, G., Zappella, M., Broccoli, V., Renieri, A., Kilstrup-Nielsen, C., and Landsberger, N., CDKL5 belongs to the same molecular pathway of MeCP2 and it is responsible for the early-onset seizure variant of Rett syndrome, *Hum Mol Genet* 14 (14), 1935–46, 2005.

54. Chen, R. Z., Akbarian, S., Tudor, M., and Jaenisch, R., Deficiency of methyl-CpG binding protein-2 in CNS neurons results in a Rett-like phenotype in mice, *Nat Genet* 27 (3), 327–31, 2001.

55. Guy, J., Hendrich, B., Holmes, M., Martin, J. E., and Bird, A., A mouse Mecp2-null mutation causes neurological symptoms that mimic Rett syndrome, *Nat Genet* 27 (3), 322–6, 2001.

56. Luikenhuis, S., Giacometti, E., Beard, C. F., and Jaenisch, R., Expression of MeCP2 in postmitotic neurons rescues Rett syndrome in mice, *Proc Natl Acad Sci U S A* 101 (16), 6033–8, 2004.

57. Kishi, N. and Macklis, J. D., MECP2 is progressively expressed in post-migratory neurons and is involved in neuronal maturation rather than cell fate decisions, *Mol Cell Neurosci* 27 (3), 306–21, 2004.

58. Guy, J., Gan, J., Selfridge, J., Cobb, S., and Bird, A., Reversal of neurological defects in a mouse model of Rett syndrome, *Science* 315 (5815), 1143–7, 2007.

59. Lewis, J. D., Meehan, R. R., Henzel, W. J., Maurer-Fogy, I., Jeppesen, P., Klein, F., and Bird, A., Purification, sequence, and cellular localization of a novel chromosomal protein that binds to methylated DNA, *Cell* 69 (6), 905–14, 1992.

60. Klose, R. J., Sarraf, S. A., Schmiedeberg, L., McDermott, S. M., Stancheva, I., and Bird, A. P., DNA binding selectivity of MeCP2 due to a requirement for A/T sequences adjacent to methyl-CpG, *Mol Cell* 19 (5), 667–78, 2005.

61. Nan, X., Ng, H. H., Johnson, C. A., Laherty, C. D., Turner, B. M., Eisenman, R. N., and Bird, A., Transcriptional repression by the methyl-CpG-binding protein MeCP2 involves a histone deacetylase complex, *Nature* 393 (6683), 386–9, 1998.

62. Kokura, K., Kaul, S. C., Wadhwa, R., Nomura, T., Khan, M. M., Shinagawa, T., Yasukawa, T., Colmenares, C., and Ishii, S., The Ski protein family is required for MeCP2-mediated transcriptional repression, *J Biol Chem* 276 (36), 34115–21, 2001.

63. Kimura, H. and Shiota, K., Methyl-CpG-binding protein, MeCP2, is a target molecule for maintenance DNA methyltransferase, Dnmt1, *J Biol Chem* 278 (7), 4806–12, 2003.

64. Fuks, F., Hurd, P. J., Wolf, D., Nan, X., Bird, A. P., and Kouzarides, T., The methyl-CpG-binding protein MeCP2 links DNA methylation to histone methylation, *J Biol Chem* 278 (6), 4035–40, 2003.

65. Young, J. I., Hong, E. P., Castle, J. C., Crespo-Barreto, J., Bowman, A. B., Rose, M. F., Kang, D., Richman, R., Johnson, J. M., Berget, S., and Zoghbi, H. Y., Regulation of RNA splicing by the methylation-dependent transcriptional repressor methyl-CpG binding protein 2, *Proc Natl Acad Sci U S A* 102 (49), 17551–8, 2005.

66. Nuber, U. A., Kriaucionis, S., Roloff, T. C., Guy, J., Selfridge, J., Steinhoff, C., Schulz, R., Lipkowitz, B., Ropers, H. H., Holmes, M. C., and Bird, A., Up-regulation of glucocorticoid-regulated genes in a mouse model of Rett syndrome, *Hum Mol Genet* 14 (15), 2247–56, 2005.

67. Stancheva, I., Collins, A. L., Van den Veyver, I. B., Zoghbi, H., and Meehan, R. R., A mutant form of MeCP2 protein associated with human Rett syndrome cannot be displaced from methylated DNA by notch in Xenopus embryos, *Mol Cell* 12 (2), 425–35, 2003.

68. Chen, W. G., Chang, Q., Lin, Y., Meissner, A., West, A. E., Griffith, E. C., Jaenisch, R., and Greenberg, M. E., Derepression of BDNF transcription involves calcium-dependent phosphorylation of MeCP2, *Science* 302 (5646), 885–9, 2003.

69. Martinowich, K., Hattori, D., Wu, H., Fouse, S., He, F., Hu, Y., Fan, G., and Sun, Y. E., DNA methylation-related chromatin remodeling in activity-dependent BDNF gene regulation, *Science* 302 (5646), 890–3, 2003.

70. Chang, Q., Khare, G., Dani, V., Nelson, S., and Jaenisch, R., The disease progression of Mecp2 mutant mice is affected by the level of BDNF expression, *Neuron* 49 (3), 341–8, 2006.

71. Makedonski, K., Abuhatzira, L., Kaufman, Y., Razin, A., and Shemer, R., MeCP2 deficiency in Rett syndrome causes epigenetic aberrations at the PWS/AS imprinting center that affects UBE3A expression, *Hum Mol Genet* 14 (8), 1049–58, 2005.

72. Horike, S., Cai, S., Miyano, M., Cheng, J. F., and Kohwi-Shigematsu, T., Loss of silent-chromatin looping and impaired imprinting of DLX5 in Rett syndrome, *Nat Genet* 37 (1), 31–40, 2005.

73. Young, I. D., The Coffin-Lowry syndrome, *J Med Genet* 25 (5), 344–8, 1988.

74. Trivier, E., De Cesare, D., Jacquot, S., Pannetier, S., Zackai, E., Young, I., Mandel, J. L., Sassone-Corsi, P., and Hanauer, A., Mutations in the kinase Rsk-2 associated with Coffin-Lowry syndrome, *Nature* 384 (6609), 567–70, 1996.

75. Delaunoy, J. P., Dubos, A., Marques Pereira, P., and Hanauer, A., Identification of novel mutations in the RSK2 gene (RPS6KA3) in patients with Coffin-Lowry syndrome, *Clin Genet* 70 (2), 161–6, 2006.

76. Zeniou, M., Ding, T., Trivier, E., and Hanauer, A., Expression analysis of RSK gene family members: the RSK2 gene, mutated in Coffin-Lowry syndrome, is prominently expressed in brain structures essential for cognitive function and learning, *Hum Mol Genet* 11 (23), 2929–40, 2002.

77. Sassone-Corsi, P., Mizzen, C. A., Cheung, P., Crosio, C., Monaco, L., Jacquot, S., Hanauer, A., and Allis, C. D., Requirement of Rsk-2 for epidermal growth factor-activated phosphorylation of histone H3, *Science* 285 (5429), 886–91, 1999.

78. Xing, J., Ginty, D. D., and Greenberg, M. E., Coupling of the RAS-MAPK pathway to gene activation by RSK2, a growth factor-regulated CREB kinase, *Science* 273 (5277), 959–63, 1996.

79. Thomas, G. M., Rumbaugh, G. R., Harrar, D. B., and Huganir, R. L., Ribosomal S6 kinase 2 interacts with and phosphorylates PDZ domain-containing proteins and regulates AMPA receptor transmission, *Proc Natl Acad Sci U S A* 102 (42), 15006–11, 2005.

80. Yang, X., Matsuda, K., Bialek, P., Jacquot, S., Masuoka, H. C., Schinke, T., Li, L., Brancorsini, S., Sassone-Corsi, P., Townes, T. M., Hanauer, A., and Karsenty, G., ATF4 is a substrate of RSK2 and an essential regulator of osteoblast biology; implication for Coffin-Lowry Syndrome, *Cell* 117 (3), 387–98, 2004.

81. Harum, K. H., Alemi, L., and Johnston, M. V., Cognitive impairment in Coffin-Lowry syndrome correlates with reduced RSK2 activation, *Neurology* 56 (2), 207–14, 2001.

82. Hennekam, R. C., Van Den Boogaard, M. J., Sibbles, B. J., and Van Spijker, H. G., Rubinstein-Taybi syndrome in The Netherlands, *Am J Med Genet Suppl* 6, 17–29, 1990.

83. Miller, R. W. and Rubinstein, J. H., Tumors in Rubinstein-Taybi syndrome, *Am J Med Genet* 56 (1), 112–5, 1995.

84. Petrij, F., Giles, R. H., Dauwerse, H. G., Saris, J. J., Hennekam, R. C. M., Masuno, M., Tommerup, N., van Ommen, G.-J. B., Goodman, R. H., Peters, D. J. M., and Breuning, M. H., Rubinstein-Taybi syndrome caused by mutations in the transcriptional co-activator CBP, *Nature* 376, 348–351, 1995.

85. Roelfsema, J. H., White, S. J., Ariyurek, Y., Bartholdi, D., Niedrist, D., Papadia, F., Bacino, C. A., den Dunnen, J. T., van Ommen, G. J., Breuning, M. H., Hennekam, R. C., and Peters, D. J., Genetic heterogeneity in Rubinstein-Taybi syndrome: mutations in both the CBP and EP300 genes cause disease, *Am J Hum Genet* 76 (4), 572–80, 2005.

86. Chan, H. M. and La Thangue, N. B., p300/CBP proteins: HATs for transcriptional bridges and scaffolds, *J Cell Sci* 114 (Pt 13), 2363–73, 2001.

87. Alarcon, J. M., Malleret, G., Touzani, K., Vronskaya, S., Ishii, S., Kandel, E. R., and Barco, A., Chromatin acetylation, memory, and LTP are impaired in CBP+/− mice: a model for the cognitive deficit in Rubinstein-Taybi syndrome and its amelioration, *Neuron* 42 (6), 947–59, 2004.

88. Cole, T. R. and Hughes, H. E., Sotos syndrome: a study of the diagnostic criteria and natural history, *J Med Genet* 31 (1), 20–32, 1994.

89. Tatton-Brown, K., Douglas, J., Coleman, K., Baujat, G., Cole, T. R., Das, S., Horn, D., Hughes, H. E., Temple, I. K., Faravelli, F., Waggoner, D., Turkmen, S., Cormier-Daire, V., Irrthum, A., and Rahman, N., Genotype-phenotype associations in Sotos syndrome: an analysis of 266 individuals with NSD1 aberrations, *Am J Hum Genet* 77 (2), 193–204, 2005.

90. Hersh, J. H., Cole, T. R., Bloom, A. S., Bertolone, S. J., and Hughes, H. E., Risk of malignancy in Sotos syndrome, *J Pediatr* 120 (4 Pt 1), 572–4, 1992.

91. Rahman, N., Mechanisms predisposing to childhood overgrowth and cancer, *Curr Opin Genet Dev* 15 (3), 227–33, 2005.

92. Kurotaki, N., Imaizumi, K., Harada, N., Masuno, M., Kondoh, T., Nagai, T., Ohashi, H., Naritomi, K., Tsukahara, M., Makita, Y., Sugimoto, T., Sonoda, T., Hasegawa, T., Chinen, Y., Tomita Ha, H. A., Kinoshita, A., Mizuguchi, T., Yoshiura Ki, K., Ohta, T., Kishino, T., Fukushima, Y., Niikawa, N., and Matsumoto, N., Haploinsufficiency of NSD1 causes Sotos syndrome, *Nat Genet* 30 (4), 365–6, 2002.

93. Rayasam, G. V., Wendling, O., Angrand, P. O., Mark, M., Niederreither, K., Song, L., Lerouge, T., Hager, G. L., Chambon, P., and Losson, R., NSD1 is essential for early post-implantation development and has a catalytically active SET domain, *Embo J* 22 (12), 3153–63, 2003.

94. Nielsen, A. L., Jorgensen, P., Lerouge, T., Cervino, M., Chambon, P., and Losson, R., Nizp1, a novel multitype zinc finger protein that interacts with the NSD1 histone lysine methyltransferase through a unique C2HR motif, *Mol Cell Biol* 24 (12), 5184–96, 2004.

95. Tawil, R. and Van der Maarel, S. M., Facioscapulohumeral muscular dystrophy, *Muscle Nerve* 34 (1), 1–15, 2006.

96. van Deutekom, J. C., Wijmenga, C., van Tienhoven, E. A., Gruter, A. M., Hewitt, J. E., Padberg, G. W., van Ommen, G. J., Hofker, M. H., and Frants, R. R., FSHD associated DNA rearrangements are due to deletions of integral copies of a 3.2 kb tandemly repeated unit, *Hum Mol Genet* 2 (12), 2037–42, 1993.

97. Tupler, R., Berardinelli, A., Barbierato, L., Frants, R., Hewitt, J. E., Lanzi, G., Maraschio, P., and Tiepolo, L., Monosomy of distal 4q does not cause facioscapulohumeral muscular dystrophy, *J Med Genet* 33 (5), 366–70, 1996.

98. Gabellini, D., Green, M. R., and Tupler, R., Inappropriate gene activation in FSHD: a repressor complex binds a chromosomal repeat deleted in dystrophic muscle, *Cell* 110 (3), 339–48, 2002.

99. Gabellini, D., D'Antona, G., Moggio, M., Prelle, A., Zecca, C., Adami, R., Angeletti, B., Ciscato, P., Pellegrino, M. A., Bottinelli, R., Green, M. R., and Tupler, R., Facioscapulohumeral muscular dystrophy in mice overexpressing FRG1, *Nature* 439 (7079), 973–7, 2006.

100. Jiang, G., Yang, F., van Overveld, P. G., Vedanarayanan, V., van der Maarel, S., and Ehrlich, M., Testing the position-effect variegation hypothesis for facioscapulohumeral muscular dystrophy by analysis of histone modification and gene expression in subtelomeric 4q, *Hum Mol Genet* 12 (22), 2909–21, 2003.

101. Caretti, G., Di Padova, M., Micales, B., Lyons, G. E., and Sartorelli, V., The Polycomb Ezh2 methyltransferase regulates muscle gene expression and skeletal muscle differentiation, *Genes Dev* 18 (21), 2627–38, 2004.

102. van Overveld, P. G., Lemmers, R. J., Sandkuijl, L. A., Enthoven, L., Winokur, S. T., Bakels, F., Padberg, G. W., van Ommen, G. J., Frants, R. R., and van der Maarel, S. M., Hypomethylation of D4Z4 in 4q-linked and non-4q-linked facioscapulohumeral muscular dystrophy, *Nat Genet* 35 (4), 315–7, 2003.

103. Ehrlich, M., The ICF syndrome, a DNA methyltransferase 3B deficiency and immunodeficiency disease, *Clin Immunol* 109 (1), 17–28, 2003.

10 Epigenetics and Immunity

Esteban Ballestar and Bruce Richardson

CONTENTS

10.1 Introduction 175
10.2 Epigenetics in Immune Differentiation and the Immune Response 176
    10.2.1 Transcription Factors and Epigenetic Modifications in
        Lymphocyte Differentiation 177
    10.2.2 DNA Methylation Changes Associated with Lymphocyte
        Differentiation and Activation 178
10.3 Epigenetics in Autoimmunity 179
    10.3.1 Epigenetics and Systemic Lupus Erythematosus 179
        10.3.1.1 DNA Methylation and T-Cell Autoreactivity 180
        10.3.1.2 DNA Demethylation and Autoimmunity 180
        10.3.1.3 DNA Methylation and Drug-Induced Lupus 181
        10.3.1.4 DNA Methylation and Idiopathic Lupus 182
        10.3.1.5 Histone Modifications in Lupus 183
        10.3.1.6 Summary 183
    10.3.2 Epigenetic Changes in Other Autoimmune Disorders 183
        10.3.2.1 Rheumatoid Arthritis 183
        10.3.2.2 Progressive Systemic Sclerosis 184
        10.3.2.3 X Chromosome Skewing in Autoimmunity 184
        10.3.2.4 Aging 184
10.4 Conclusions 185
References 185

10.1 INTRODUCTION

The immune system is unique in that cells such as T and B lymphocytes differentiate throughout life and can be readily obtained, making them useful models for differentiation and gene expression studies. As in other tissues, epigenetic mechanisms are essential to both differentiation and gene expression in these cells. Further, a failure to maintain epigenetic homeostasis in the immune response leads to aberrant gene expression, contributing to autoimmunity. This chapter reviews how epigenetic mechanisms contribute to immune differentiation and function and how epigenetic abnormalities contribute to human autoimmune disease.

10.2   EPIGENETICS IN IMMUNE DIFFERENTIATION
AND THE IMMUNE RESPONSE

The differentiation of hematopoietic precursor cells into lymphocytes and their subsequent activation by foreign antigens are key processes determining immune function. Lymphocytes develop through a stepwise process in which a progressive narrowing of developmental potential results in commitment to lineages such as B or T cells (Figure 10.1). During this process, lineage-specific genes are sequentially activated, while lineage-inappropriate genes are turned off. Each lymphocyte also rearranges its antigen receptor genes during development to generate a molecule specifically binding antigens on foreign pathogens such as microbes.

The resultant B and T lymphocytes are immature when leaving the bone marrow and thymus, respectively, as they are capable of responding to foreign antigen but have not yet been activated. Prior to antigenic exposure, the lymphocytes are referred to as *naive*. When a naive cell encounters its antigen, it responds and undergoes division with further differentiation. The progeny are clonally distinguished by their unique antigen receptor, but differ from the naive cell by expressing a different repertoire of effector molecules. For example, developing T cells commit first to the CD4+ *helper* or CD8+ *cytotoxic* subsets, and CD4+ cells differentiate further into T helper 1 (Th1) or T helper 2 (Th2) cells. These latter subsets express and secrete distinct patterns of cytokines, which play an important role in activating subpopulations of macrophages, B cells, and other T cells. The Th1 subset expresses pro-inflammatory cytokines such as interferon-gamma (IFN-γ), while the Th2 subset expresses cytokines such as IL-4, which promotes B-cell antibody production (see Figure 10.1).

Each differentiation step is mediated by transcription factors, cell-specific or ubiquitous. However, activation and suppression of genes during differentiation are determined not only by activated transcription factors binding regulatory elements, but also by the chromatin configuration around the genes, which is determined by epigenetic mechanisms such as histone modifications, DNA methylation, and nucleosome positioning. Further, some transcription factors cause regional modifications of the chromatin structure that affect gene expression. For example, differentiation results in modifications to the chromatin structure at the regulatory regions of the effector-cytokine genes [1], including changes in DNA methylation [2] and histone acetylation that modulate accessibility to NFAT1 (nuclear factor of activated T cells 1) [3]. These modifications contribute to polarization of the developing Th1 and Th2 cells, such that IL-4 and IFN-γ expression is mutually exclusive. The silencing of IL-4 or IFN-γ expression may be stabilized further by the recruitment of these loci to specific heterochromatin domains in the nucleus [4]. Interestingly, with successive divisions, the progeny lose the ability to revert to the alternative lineage, an observation consistent with lineage restriction being sustained by progressive gene silencing [5]. The end result is a variety of lymphocyte subsets with distinct repertoires of gene expression, reflecting distinct patterns of transcription factor expression and distinct epigenetic chromatin modifications.

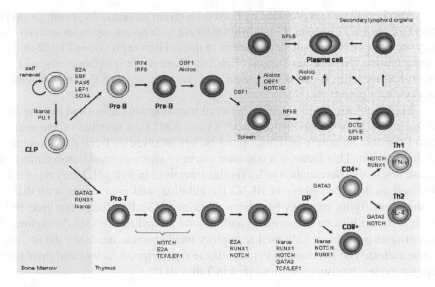

FIGURE 10.1 (See color figure following page 52.) *Transcriptional regulation of T- and B-lymphocyte differentiation.* Bone marrow stem cells differentiate first into common lymphoid precursor (CLP) cells, which then either migrate to the thymus where they differentiate first into pro-T cells (Pro-T), then serially into CD4+ CD8+ (DP) cells, CD4+, or CD8+ cells, then Th1 or Th2 cells. Early B-cell development starts in the bone marrow with pro-B cells maturing to pre-B cells, which migrate to the spleen then secondary lymphoid tissues, where they differentiate into further B cells and plasma cells. The transcription factors involved are named at each step.

10.2.1 Transcription Factors and Epigenetic Modifications in Lymphocyte Differentiation

Gene targeting studies have identified several transcription factors essential for lymphocyte development. For example, B-lymphoid differentiation is regulated in part by PU.1, Ikaros, E2A, EBF, and Pax5 during early stages of lineage commitment, and Aiolos, Spi-B, Oct2, OBF1, NF-κB, or Notch at later stages. In contrast, transcription factors involved in early T-cell differentiation include GATA3, Notch, TCF-1/LEF1, or Runx (see Figure 10.1). In many cases, these transcription factors operate through mechanisms involving the recruitment of enzymes that modify histones or nucleosome positioning. This is exemplified by the Ikaros proteins, a family of zinc-finger transcriptional regulators that includes Ikaros, Aiolos, and Helios. These play an important role in the control of B-lymphocyte differentiation and proliferation and are essential for the proper development of the lymphoid arm of the hematopoietic system [6]. Several reports provide clues about how Ikaros proteins regulate gene transcription. Aiolos and Ikaros recruit the Mi-2/NuRD and SIN3 histone deacetylase (HDAC)-containing repressor complexes [7,8], and also the nucleosome-remodeling complex SWI/SNF to transcriptionally silenced regions of the genome [8,9]. Additionally, Aiolos and Ikaros associate with CtBP-interacting protein (CtIP), which has been linked with HDAC-independent mechanisms of repression [10].

Another example is provided by Polycomb group proteins. Polycomb group protein Ezh2 is a K27 histone H3 methyltransferase with potent repressor activity and a key regulator of embryonic development in mice. High expression of Ezh2 in developing murine lymphocytes suggests Ezh2 involvement in lymphopoiesis, and Ezh2 is critical for early B-cell development and rearrangement of the immunoglobulin heavy-chain gene (Igh) [11].

The complexity of interactions between transcription factors and histone-modifying enzymes is exemplified by RUNX1 (or AML1), a member of the mammalian Runt domain family of transcription factors involved in T-cell development and differentiation. This factor is a common target of chromosomal translocations that cause arrest of differentiation in leukemias [reviewed in Ref. 12]. Many reports have focused on the mechanisms of RUNX1 regulation, and interactions with different histone-modifying enzymes have been described [13]. It is likely that interactions with different histone deacetylases and histone methyltransferases throughout differentiation provide an additional regulatory mechanism to modulate the activity of these factors. The relationship between these transcription factors and their related epigenetic mechanisms is summarized in Table 10.1.

10.2.2 DNA METHYLATION CHANGES ASSOCIATED WITH LYMPHOCYTE DIFFERENTIATION AND ACTIVATION

Genes not required for the function of specific lymphoid cell types may also be suppressed by DNA methylation. Several mammalian DNA methyltransferases account for this activity. While the extent to which methylation physiologically regulates the expression of tissue-specific genes is unclear, recent examples suggest an important role in silencing unnecessary genes in specific subsets. When naive CD4+ T cells

TABLE 10.1
Relationship between Transcriptional Regulators and Epigenetic Machinery in Lymphocyte Differentiation

Factor	Expression	Interaction with Epigenetic Machinery	Ref.
Aiolos	B cells	HDACs: HDAC1	7
		Corepresssor complex: Sin3A, Sin3B (contains HDAC1)	
GATA3	T cells	Nucleosome remodeling complex: Mi-2/NuRD (contains HDAC1)	66, 67
		HDACs: HDAC3, HDAC4, HDAC5	
Ikaros	T cells	HDACs: HDAC1	
		Corepressor complex: Sin3A, Sin3B (contains HDAC1)	
LEF1	B, T cells	HDAC1	68
NFκB	B cells	HATs: CBP/P300 (HAT	69, 70
		HACs: HDAC1	
NOTCH	T cells	HAT: P300, PCAF	71
		HDACs: HDAC1, HDAC2	
RUNXI	T cells	HMTs: SUV39H1 (K9 specific)	72, 73, 13

HAT: histone acetyltransferase, HDAC: hisone deacetylase; HMT: histone methyltransferase.

are activated they produce IL-2 and proliferate. As noted above, they also undergo a differentiation event that makes them competent to produce other cytokines such as IFN-γ and IL-4. This differentiation correlates with demethylation of CpGs in the promoters of IFN-γ and IL-4 [14,15]. Further differentiation into Th1 or Th2 cells is associated with methylation of IL-4 in Th1 cells, and IFN-γ in Th2, and treatment of Th1 or Th2 cells with DNA methylation inhibitors can reactivate expression of the suppressed gene [16].

Interestingly, a small region in the promoter-enhancer of the IL-2 gene demethylates in T lymphocytes following activation, but before cell division, and remains demethylated thereafter. These epigenetic changes are necessary and sufficient to enhance transcription [17]. Further, IL-2 demethylation allows binding of the transcription factor Oct-1 [18]. These processes imply that demethylation can occur by an active enzymatic mechanism. Although DNA methyltransferases are well characterized, the molecular mechanisms causing DNA demethylation, aside from passive demethylation caused by Dnmt1 inhibition during mitosis, are poorly understood. Recent evidence suggests that Gadd45a (growth arrest and DNA-damage-inducible protein 45 alpha), a nuclear protein involved in maintenance of genomic stability, DNA repair, and suppression of cell growth, may participate in active DNA demethylation [19]. Whether this mechanism is involved in the IL-2 demethylation remains to be determined.

10.3 EPIGENETICS IN AUTOIMMUNITY

The importance of epigenetic mechanisms in immune differentiation and function is evidenced by immune diseases caused at least in part by epigenetic abnormalities. Perhaps the clearest example is ICF syndrome (immunodeficiency, centromeric instability, and facial anomalies), caused by mutations of the *DNMT3B* gene. This disorder, discussed in Chapter 15 of this book, is characterized by a B-cell immunodeficiency and indicates that genetic abnormalities in DNA methylation may contribute to immune disorders. However, epigenetic marks in mature cells must be replicated during mitosis. The maintenance of epigenetic marks on DNA and histones is susceptible to modification by environmental factors, and these changes can lead to aberrant gene expression and abnormal immune function. At present the role of acquired epigenetic abnormalities in the development of autoimmunity is best studied in human systemic lupus erythematosus (SLE), although there are indications that other diseases may also have an epigenetic basis. The role of epigenetics, and in particular DNA methylation, in human autoimmunity is summarized below.

10.3.1 EPIGENETICS AND SYSTEMIC LUPUS ERYTHEMATOSUS

Human lupus is a systemic autoimmune disease primarily affecting women. Lupus is characterized by autoantibody formation to nuclear antigens and immune complex deposition in tissues such as the kidney. The cause of human lupus is unknown, but current models propose that both genetic and environmental factors are required. Familial studies provide evidence for multiple genetic loci predisposing to lupus [20]. However, incomplete concordance in identical twins [21], and observations that some drugs cause a lupus-like disease and that sunlight triggers lupus flares [22], indicates a role for environmental factors. Current models thus postulate that lupus occurs in genetically predisposed individuals when exposed to one or more environmental triggers. While persuasive evidence for environmental agents other than sunlight and drugs has yet to be shown, environmentally induced epigenetic changes and, in particular, altered DNA methylation, are likely to be involved. Much of this work derives from studies on the role of DNA methylation in regulating gene expression in mature T cells.

10.3.1.1 DNA Methylation and T-Cell Autoreactivity

Early studies used the irreversible DNA methyltransferase (Dnmt) inhibitor 5-azacytidine (5-azaC) to probe for functional changes caused by DNA methylation inhibition in mature T cells. One observation was that CD4+ T cells become autoreactive following 5-azaC treatment. CD4+ T cells normally respond to peptides presented in the antigen-binding cleft of "self" class II MHC molecules on antigen-presenting cells. Following 5-azaC treatment, antigen-specific CD4+ T cells lose the requirement for specific antigen and respond to antigen-presenting cells alone. The response is specific for self class II MHC molecules and is reversible, in that antigen responsiveness is slowly recovered once the Dnmt inhibitor is removed. The autoreactivity has been demonstrated with cloned and polyclonal human and murine CD4+ T cells [23]. CD8+ T cells do not become autoreactive; the reason is unknown [24,25].

Mechanistic studies revealed that the autoreactivity correlates with increased expression of LFA-1 (CD11a/CD18), caused by increased levels of CD11a (*ITGAL*) transcripts [26], and LFA-1 overexpression caused by transfection results in identical autoreactivity in human [27] and murine [28] T cells. Bisulfite sequencing revealed demethylation of *alu* repeats upstream of the *ITGAL* promoter, and cassette methylation of the region suppressed promoter function in transfection studies [29], indicating transcriptional relevance. LFA-1 is an adhesion molecule and surrounds the T-cell antigen receptor (TCR) to form the "immunologic synapse" during activation by antigen-presenting cells, providing both increased stability to the TCR–MHC interaction as well as co-stimulatory signals [30]. Increased LFA-1 expression may cause a T-cell response to MHC molecules presenting inappropriate antigens by overstabilizing the lower affinity interaction between the TCR and class II MHC molecules bearing inappropriate peptide fragments, by providing increased co-stimulatory signals, or by a combination of both.

10.3.1.2 DNA Demethylation and Autoimmunity

The response of demethylated CD4+ cells to self class II MHC molecules demonstrates that normal, antigen-reactive T cells may be modified by exogenous agents to become autoreactive, potentially contributing to an autoimmune disease. CD4+ T cells similarly responding to host class II MHC molecules cause chronic graft-vs-host disease (GVHD), with many features of human lupus, including antinuclear antibodies and an immune complex kidney disease [31], suggesting that the 5-azaC modified cells may cause a disease resembling SLE. Pathogenicity of the autoreactive cells was demonstrated by injecting cloned or polyclonal 5-azaC-treated CD4+

T cells into syngeneic mice. The recipients developed anti-DNA antibodies and an immune complex glomerulonephritis as well as other histologic features of autoimmunity, depending on the strain [32,33]. LFA-1-transfected CD4+ T cells caused a similar lupus-like disease in the same system, indicating that LFA-1 overexpression contributes to the autoimmunity induced by demethylated T cells [28].

T-cell effector functions are also modified by DNA demethylation. Coculture of 5-azaC-treated T cells with autologous B cells results in IgG hypersecretion [25], due to both increased expression of Th1 and Th2 cytokines including IFN-γ, IL-4, and IL-6 [33] as well as overexpression of B-cell co-stimulatory molecules including CD70 and CD40L [34,35]. The IgG overstimulation may contribute to the increased autoantibody titers found in mice receiving demethylated T cells.

In contrast to the B-cell stimulation, coculture of demethylated T cells with autologous monocytes/macrophages (M$\emptyset$) results in M$\emptyset$ death by apoptosis [27]. Antigen-presenting M$\emptyset$ normally undergo apoptosis after stimulating T cells [36]. However, hypomethylated autoreactive CD4+ T cells respond to all M$\emptyset$ bearing autologous class II MHC molecules, resulting in promiscuous M$\emptyset$ killing. Increased M$\emptyset$ apoptosis may initiate an anti-DNA response by providing a source of antigenic nucleosomes, since injecting apoptotic cells into normal mice also results in anti-DNA antibodies [37]. Further, transgenic mice lacking one or more of the molecules involved in clearance of apoptotic debris develop a lupus-like disease with similar anti-DNA antibodies [38]. The consequence of increased M$\emptyset$ apoptosis has been tested by injecting control and lupus-prone mice with clodronate vesicles. These vesicles are phagocytosed by M$\emptyset$, causing release of the clodronate and apoptotic death of the phagocyte. Clodronate vesicles cause antinucleosome antibodies when injected into normal mice and accelerate autoimmunity in lupus-prone mice [39].

10.3.1.3 DNA Methylation and Drug-Induced Lupus

The experiments summarized above imply that inhibiting CD4+ T-cell DNA methylation may cause a lupus-like disease. More than 100 drugs have been reported to cause a lupus-like disease with antinuclear antibodies in the patients receiving them, typically in a small subset of the people [40]. However, a majority of the people receiving the antiarrhythmic agent procainamide, or the antihypertensive drug hydralazine, eventually develop antinuclear antibodies, and a subset develop an illness resembling lupus [41,42]. These considerations have prompted studies testing if procainamide and hydralazine were DNA methylation inhibitors. In vitro studies showed that procainamide and hydralazine decreased total T-cell deoxymethylcytosine (dmC) content and induced LFA-1 overexpression and autoreactivity similar to 5-azaC, although 5-azaC was considerably more potent [43]. Procainamide- and hydralazine-treated CD4+ T cells also increased B-cell antibody production similar to 5-azaC-treated T cells [25].

Procainamide was found to be a competitive inhibitor of DNA methyltransferase enzymatic activity and had no effect on intracellular S-adenosylmethionine or S-adenosylhomocysteine pools [44]. Confirming studies demonstrated that procainamide is a selective DNA methyltransferase 1 inhibitor, reducing the affinity of the enzyme for its substrates, hemimethylated DNA, and S-adenosylmethionine [45]. In contrast, hydralazine selectively inhibits T-cell ERK pathway signaling, preventing upregulation of DNA methyltransferase 1 and 3a during mitosis, resulting in hypomethylation of the daughter cells [46]. Pathogenicity of decreased ERK pathway signaling was confirmed by treating CD4+ T cells with U0126, a selective MEK inhibitor that decreases ERK pathway signaling, and injecting the cells into syngeneic mice. The T cells overexpressed LFA-1 and became autoreactive, and mice receiving the treated cells developed anti-DNA antibodies, similar to procainamide-treated cells [46].

10.3.1.4 DNA Methylation and Idiopathic Lupus

Similar mechanisms may contribute to idiopathic human lupus. Early studies showed that T cells from patients with active lupus have decreased total deoxymethylcytosine (dmC) content relative to patients with inactive lupus and normal controls. Decreased dmC content was associated with decreased DNA methyltransferase enzyme activity [47]. Later studies demonstrated decreased Dnmt1 transcripts in lupus T cells [48]. Since lupus T cells have multiple signaling abnormalities [49], and Dnmt1 expression is regulated by the ERK and JNK pathways [46], T-cell signaling was examined in human lupus. Patients with active but not inactive lupus had decreased ERK phosphorylation in response to stimulation that was identical to hydralazine-treated cells [48], while signaling through the JNK and p38 pathways was intact [50].

Other studies revealed functional and epigenetic similarities between lupus and experimentally demethylated T cells. Functional studies revealed that lupus T cells overstimulate autologous B-cell antibody production, similar to 5-azaC-treated T cells [51]. A subset of lupus T cells also overexpresses LFA-1, and this subset spontaneously kills autologous Mø in an MHC-restricted, autoreactive fashion identical to experimentally demethylated cells [26]. Further, patients with active lupus have circulating apoptotic monocytes in their peripheral blood, suggesting that a similar killing occurs in vivo [52].

Epigenetic similarities between 5-azaC-treated and lupus T cells were sought using bisulfite sequencing. As noted above, 5-azaC causes LFA-1 overexpression by demethylating a series of *alu* repeats 5′ to the *ITGAL* (CD11a) gene. Bisulfite sequencing of DNA from CD4+ lupus T cells demonstrated demethylation of the same sequences, and the degree of demethylation was proportional to disease activity [29].

The similarities between *ITGAL* demethylation in 5-azaC-treated and lupus T cells has raised the possibility that 5-azaC treatment may be used to identify methylation-sensitive T-cell genes overexpressed in lupus and to predict the sequences demethylated. Normal T cells were treated with 5-azaC, affected transcripts identified using oligonucleotide arrays, and genes relevant to B-cell overstimulation and macrophage killing sought. Of the >100 genes affected, perforin (*PRF1*), a cytotoxic molecule normally expressed by natural killer (NK) cells and cytotoxic CD8+ cells but not CD4+ cells, and CD70 (*TNFSF7*), a B-cell costimulatory molecule, were selected for further study. Confirming studies determined that 5-azaC induced perforin mRNA and protein expression in the CD4+ T-cell subset, and that expression was due to demethylation of a conserved region located between the *PRF1* promoter and upstream enhancer [53]. Studies in lupus patients demonstrated identical demethylation of the same sequence and aberrant perforin expression in CD4+ T cells [54]. Concanamycin, a perforin antagonist, prevented the autoreactive Mø killing by lupus T cells, implicating perforin in this phenomenon [39].

Similar studies compared *TNFSF7* methylation, expression, and function in 5-azaC-treated and lupus T cells. Bisulfite sequencing revealed that the core *TNFSF7* promoter is normally demethylated in CD4+ T cells and that 5-azaC extends the demethylated region upstream by ~300 bp. Cassette methylation confirmed transcriptional suppression when the region is methylated. Interestingly, other DNA methylation inhibitors including procainamide, hydralazine, and the MEK inhibitor U0126 all demethylated the same sequence and increased CD70 expression as 5-azaC. Studies in CD4+ T cells from lupus patients demonstrated CD70 overexpression and demethylation of the same sequence [34,55]. Thus, for at least three genes (*ITGAL, TNFSF7,* and *PRF1*), identical changes in methylation and expression are found in experimentally demethylated and lupus T cells.

10.3.1.5   Histone Modifications in Lupus

The role of histone modifications in lupus is less well understood. Treating lupus T cells with histone deacetylase inhibitors including trichostatin A and SAHA restores aberrant expression of some genes [56]. However, these drugs also modify acetylation of other proteins including transcription factors [57], and confirmatory studies at the chromatin level still need to be performed.

10.3.1.6   Summary

T-cell DNA methylation abnormalities in human lupus appear to be predicted by experimental demethylation of T cells in vitro. Identical changes occur at the DNA, mRNA, protein, and functional levels for at least three genes that contribute to T-cell autoreactivity, B-cell overstimulation, and macrophage killing (*ITGAL, TNFSF7,* and *PRF1*), respectively. Experimentally demethylated T cells cause a lupus-like disease in vivo. Finally, at least two lupus-inducing drugs are DNA methylation inhibitors with identical effects on T cells as 5-azaC. In lupus, DNA demethylation appears to be caused by a failure to upregulate Dnmt1 during mitosis, due to a defect in ERK pathway signaling. It seems reasonable to propose that defective T-cell DNA methylation may contribute to the pathogenesis of lupus in genetically predisposed individuals. This model is illustrated in Figure 10.2.

10.3.2   Epigenetic Changes in Other Autoimmune Disorders
10.3.2.1   Rheumatoid Arthritis

Rheumatoid arthritis, a systemic autoimmune disease that causes a deforming and disabling arthritis, is also associated with genome-wide T-cell DNA hypomethylation [47]. However, the genes affected and mechanism causing the demethylation are unknown. Methotrexate, used to treat rheumatoid arthritis, has been reported to increase dmC levels in rheumatoid arthritis T cells [58]. Since methotrexate is a folate antagonist, this effect is paradoxical and may be secondary to other effects

DECREASED T CELL SIGNALING
↓
DECREASED DNA METHYLTRANSFERASES
↓
DNA HYPOMETHYLATION
↓
ITGAL, PRF1, TNFSF7 OVEREXPRESSION
↓
T CELL AUTOREACTIVITY,
INCREASED Mø KILLING AND B CELL HELP
↓
ANTI-DNA ANTIBODIES

FIGURE 10.2 *Relationship between T-cell signaling abnormalities, DNA hypomethylation, and lupus.* The sequence of events leading to generation of autoreactive T cells and the development of lupus is shown. (Mø: macrophage)

on the disease process. Nonetheless, the increased methylation is interesting and deserves further study.

10.3.2.2 Progressive Systemic Sclerosis

Progressive systemic sclerosis (PSS), also known as scleroderma, is a rare and poorly understood condition characterized in part by collagen deposition in skin as well as other organs. PSS is considered an autoimmune disease because of the frequent presence of autoantibodies to nuclear as well as other autoantigens. The fibrosis is due to aberrant activation and collagen secretion by fibroblasts. The *Fli1* gene, encoding a regulatory protein suppressing collagen synthesis, has been reported to be aberrantly methylated in PSS fibroblasts, raising the possibility that fibroblast DNA methylation abnormalities may contribute to this disorder [59].

10.3.2.3 X Chromosome Skewing in Autoimmunity

Women tend to develop autoimmune disease more frequently than men. The mechanisms causing this dimorphism are unknown, but some have proposed that skewed X chromosome inactivation may contribute. As discussed in other chapters in this book, DNA methylation contributes to the silencing of one X chromosome in women. During early development both X chromosomes are active, then one or the other is silenced on a random basis, resulting in mosaicism with a 50:50 distribution of the inactivated chromosomes. In some conditions, however, one or the other X chromosome is preferentially inactivated, resulting in an aberrant relative overexpression of the genes on the active X. This aberrant inactivation is referred to as skewing. Conditions associated with skewing include scleroderma [60] and autoimmune thyroiditis [61].

10.3.2.4 Aging

Most people develop antinuclear antibodies (ANAs) with age. Total genomic dmC content decreases with age in most vertebrate tissues, and T cells similarly demethylate in the elderly [62]. The demethylation is associated with decreased Dnmt1 expression and with LFA-1 overexpression due to demethylation of the same sequences affected by 5-azaC [63]. These observations suggest that age-dependent DNA demethylation may contribute to the development of ANAs and lupus-like diseases in aging. While lupus is thought to be a genetically predisposed condition [20], the average age of onset is ~40 [64]. It is possible that age-dependent decreases in T-cell Dnmt1 and dmC content contribute to lupus in older individuals. T-cell function also declines with age, and epigenetic changes may contribute to this process as well [65].

10.4   CONCLUSIONS

T lymphocytes differentiate throughout life, and epigenetic mechanisms play an essential role in the regulation of subset-specific genes. The expression of some of these genes, such as IL-4 in Th1 cells, IFN-γ in Th2 cells, and perforin in CD4+ cells, involves changes in DNA methylation, and pharmacologic demethylation of these genes can lead to their reexpression. The level of expression of some genes, like *ITGAL* and *TNFSF7*, is also modified by the methylation status of regions flanking their promoters. Current models indicate that failure to maintain these methylation patterns can modify T-cell gene expression and hence immune function, contributing to the development of lupus-like diseases and perhaps other forms of autoimmunity. Further studies are needed to extend these observations and identify ways to correct these abnormalities.

REFERENCES

1. Agarwal, S., and A. Rao. Modulation of chromatin structure regulates cytokine gene expression during T cell differentiation. *Immunity* 9:765.1998.
2. Lee, D. U., S. Agarwal, and A. Rao. Th2 lineage commitment and efficient IL-4 production involves extended demethylation of the IL-4 gene. *Immunity* 16:649.2002.
3. Avni, O., et al. T(H) cell differentiation is accompanied by dynamic changes in histone acetylation of cytokine genes. *Nat Immunol* 3:643.2002.
4. Grogan, J. L., and R. M. Locksley. T helper cell differentiation: on again, off again. *Curr Opin Immunol* 14:366.2002.
5. Grogan, J. L., et al. Early transcription and silencing of cytokine genes underlie polarization of T helper cell subsets. *Immunity* 14:205.2001.
6. Cortes, M., et al. Control of lymphocyte development by the Ikaros gene family. *Curr Opin Immunol* 11:167.1999.
7. Koipally, J., et al. Repression by Ikaros and Aiolos is mediated through histone deacetylase complexes. *Embo J* 18:3090.1999.
8. Kim, J., et al. Ikaros DNA-binding proteins direct formation of chromatin remodeling complexes in lymphocytes. *Immunity* 10:345.1999.
9. Georgopoulos, K. Haematopoietic cell-fate decisions, chromatin regulation and ikaros. *Nat Rev Immunol* 2:162.2002.
10. Koipally, J., and K. Georgopoulos. Ikaros-CtIP interactions do not require C-terminal binding protein and participate in a deacetylase-independent mode of repression. *J Biol Chem* 277:23143.2002.
11. Su, I. H., et al. Ezh2 controls B cell development through histone H3 methylation and Igh rearrangement. *Nat Immunol* 4:124.2003.
12. Ichikawa, M., et al. Runx1/AML-1 ranks as a master regulator of adult hematopoiesis. *Cell Cycle* 3:722.2004.

13. Reed-Inderbitzin, E., et al. RUNX1 associates with histone deacetylases and SUV39H1 to repress transcription. *Oncogene* 25:*5777*.2006.
14. Fitzpatrick, D. R., et al. Distinct methylation of the interferon gamma (IFN-gamma) and interleukin 3 (IL-3) genes in newly activated primary CD8+ T lymphocytes: regional IFN-gamma promoter demethylation and mRNA expression are heritable in CD44(high)CD8+ T cells. *J Exp Med* 188:*103*.1998.
15. Fitzpatrick, D. R., K. M. Shirley, and A. Kelso. Cutting edge: stable epigenetic inheritance of regional IFN-gamma promoter demethylation in CD44highCD8+ T lymphocytes. *J Immunol* 162:*5053*.1999.
16. Sanders, V. M. Epigenetic regulation of Th1 and Th2 cell development. *Brain Behav Immun* 20:*317*.2006.
17. Bruniquel, D., and R. H. Schwartz. Selective, stable demethylation of the interleukin-2 gene enhances transcription by an active process. *Nat Immunol* 4:*235*.2003.
18. Murayama, A., et al. A specific CpG site demethylation in the human interleukin 2 gene promoter is an epigenetic memory. *Embo J* 25:*1081*.2006.
19. Barreto, G., et al. Gadd45a promotes epigenetic gene activation by repair-mediated DNA demethylation. *Nature* 445:*671*.2007.
20. Namjou, B., J. Kelly, and J. Harley. 2007. The Genetics of Lupus. In *Systemic Lupus Erythematosus*. G. Tsokos, C. Gordon, and J. Smolen, eds. Mosby Elsevier, Phiadelphia, p. 74.
21. Jarvinen, P., et al. Systemic lupus erythematosus and related systemic diseases in a nationwide twin cohort: an increased prevalence of disease in MZ twins and concordance of disease features. *J Intern Med* 231:*67*.1992.
22. Sawalha, A., and B. Richardson. 2007. The Environment in the Pathogenesis of Systemic Lupus Erythematosus. In *Systemic Lupus Erythematosus*. G. Tsokos, C. Gordon, and J. Smolen, eds. Mosby Elsevier, Philadephia, p. 64.
23. Richardson, B. DNA methylation and autoimmune disease. *Clin Immunol* 109:*72*.2003.
24. Richardson, B. Effect of an inhibitor of DNA methylation on T cells. II. 5-Azacytidine induces self-reactivity in antigen-specific T4+ cells. *Hum Immunol* 17:*456*.1986.
25. Richardson, B. C., M. R. Liebling, and J. L. Hudson. CD4+ cells treated with DNA methylation inhibitors induce autologous B cell differentiation. *Clin Immunol Immunopathol* 55:*368*.1990.
26. Richardson, B. C., et al. Phenotypic and functional similarities between 5-azacytidine-treated T cells and a T cell subset in patients with active systemic lupus erythematosus. *Arthritis Rheum* 35:*647*.1992.
27. Richardson, B., et al. Lymphocyte function-associated antigen 1 overexpression and T cell autoreactivity. *Arthritis Rheum* 37:*1363*.1994.
28. Yung, R., et al. Mechanisms of drug-induced lupus. II. T cells overexpressing lymphocyte function-associated antigen 1 become autoreactive and cause a lupuslike disease in syngeneic mice. *J Clin Invest* 97:*2866*.1996.
29. Lu, Q., et al. Demethylation of ITGAL (CD11a) regulatory sequences in systemic lupus erythematosus. *Arthritis Rheum* 46:*1282*.2002.
30. Wulfing, C., et al. Costimulation and endogenous MHC ligands contribute to T cell recognition. *Nat Immunol* 3:*42*.2002.
31. Rolink, A. G., S. T. Pals, and E. Gleichmann. Allosuppressor and allohelper T cells in acute and chronic graft-vs.-host disease. II. F1 recipients carrying mutations at H-2K and/or I-A. *J Exp Med* 157:*755*.1983.
32. Yung, R. L., et al. Mechanism of drug-induced lupus. I. Cloned Th2 cells modified with DNA methylation inhibitors in vitro cause autoimmunity in vivo. *J Immunol* 154:*3025*.1995.
33. Quddus, J., et al. Treating activated CD4+ T cells with either of two distinct DNA methyltransferase inhibitors, 5-azacytidine or procainamide, is sufficient to cause a lupus-like disease in syngeneic mice. *J Clin Invest* 92:*38*.1993.

34. Oelke, K., et al. Overexpression of CD70 and overstimulation of IgG synthesis by lupus T cells and T cells treated with DNA methylation inhibitors. *Arthritis Rheum* 50:*1850*.2004.

35. Lu, Q., et al. Women and lupus: the inactive X awakens. *Arthritis Rheum* 54:*S775*.2006.

36. Richardson, B. C., et al. Evidence that macrophages are programmed to die after activating autologous, cloned, antigen-specific, CD4+ T cells. *Eur J Immunol* 23:*1450*.1993.

37. Mevorach, D., et al. Systemic exposure to irradiated apoptotic cells induces autoantibody production. *J Exp Med* 188:*387*.1998.

38. Walport, M. J. Lupus, DNase and defective disposal of cellular debris. *Nat Genet* 25:*135*.2000.

39. Denny, M. F., et al. Accelerated macrophage apoptosis induces autoantibody formation and organ damage in systemic lupus erythematosus. *J Immunol* 176:*2095*.2006.

40. Yung, R., and B. Richardson. 2003. Drug-induced Lupus. In *Rheumatology*. M. Hochberg, A. Silman, J. Smolen, M. Weinblatt, and M. Weisman, eds. Mosby, New York, p. 1385.

41. Henningsen, N. C., et al. Effects of long-term treatment with procaine amide. A prospective study with special regard to ANF and SLE in fast and slow acetylators. *Acta Med Scand* 198:*475*.1975.

42. Blomgren, S. E., J. J. Condemi, and J. H. Vaughan. Procainamide-induced lupus erythematosus. Clinical and laboratory observations. *Am J Med* 52:*338*.1972.

43. Cornacchia, E., et al. Hydralazine and procainamide inhibit T cell DNA methylation and induce autoreactivity. *J Immunol* 140:*2197*.1988.

44. Scheinbart, L. S., et al. Procainamide inhibits DNA methyltransferase in a human T cell line. *J Rheumatol* 18:*530*.1991.

45. Lee, B. H., et al. Procainamide is a specific inhibitor of DNA methyltransferase 1. *J Biol Chem* 280:*40749*.2005.

46. Deng, C., et al. Hydralazine may induce autoimmunity by inhibiting extracellular signal-regulated kinase pathway signaling. *Arthritis Rheum* 48:*746*.2003.

47. Richardson, B., et al. Evidence for impaired T cell DNA methylation in systemic lupus erythematosus and rheumatoid arthritis. *Arthritis Rheum* 33:*1665*.1990.

48. Deng, C., et al. Decreased Ras-mitogen-activated protein kinase signaling may cause DNA hypomethylation in T lymphocytes from lupus patients. *Arthritis Rheum* 44:*397*.2001.

49. Kammer, G. M., et al. Abnormal T cell signal transduction in systemic lupus erythematosus. *Arthritis Rheum* 46:*1139*.2002.

50. Oelke, K., and B. Richardson. Decreased T cell ERK pathway signaling may contribute to the development of lupus through effects on DNA methylation and gene expression. *Int Rev Immunol* 23:*315*.2004.

51. Crow, M. 1999. Mechanisms of T-Helper Cell Activation and Function in Systemic Lupus Erythematosus. In *Lupus: Molecular and Cellular Pathogenesis*. G. M. Kammer and G. Tsokos, eds. Humana Press, Totowa NJ, p. 231.

52. Richardson, B. C., et al. Monocyte apoptosis in patients with active lupus. *Arthritis Rheum* 39:*1432*.1996.

53. Lu, Q., et al. DNA methylation and chromatin structure regulate T cell perforin gene expression. *J Immunol* 170:*5124*.2003.

54. Kaplan, M. J., et al. Demethylation of promoter regulatory elements contributes to perforin overexpression in CD4+ lupus T cells. *J Immunol* 172:*3652*.2004.

55. Lu, Q., A. Wu, and B. C. Richardson. Demethylation of the same promoter sequence increases CD70 expression in lupus T cells and T cells treated with lupus-inducing drugs. *J Immunol* 174:*6212*.2005.

56. Mishra, N., et al. Trichostatin A reverses skewed expression of CD154, interleukin-10, and interferon-gamma gene and protein expression in lupus T cells. *Proc Natl Acad Sci U S A* 98:*2628*.2001.

57. Marks, P. A., and M. Dokmanovic. Histone deacetylase inhibitors: discovery and development as anticancer agents. *Expert Opin Investig Drugs* 14:*1497.*2005.
58. Kim, Y. I., et al. DNA hypomethylation in inflammatory arthritis: reversal with methotrexate. *J Lab Clin Med* 128:*165.*1996.
59. Wang, Y., P. S. Fan, and B. Kahaleh. Association between enhanced type I collagen expression and epigenetic repression of the FLI1 gene in scleroderma fibroblasts. *Arthritis Rheum* 54:*2271.*2006.
60. Ozbalkan, Z., et al. Skewed X chromosome inactivation in blood cells of women with scleroderma. *Arthritis Rheum* 52:*1564.*2005.
61. Ozcelik, T., et al. Evidence from autoimmune thyroiditis of skewed X-chromosome inactivation in female predisposition to autoimmunity. *Eur J Hum Genet* 14:*791.*2006.
62. Golbus, J., T. D. Palella, and B. C. Richardson. Quantitative changes in T cell DNA methylation occur during differentiation and ageing. *Eur J Immunol* 20:*1869.*1990.
63. Zhang, Z., et al. Age-dependent DNA methylation changes in the ITGAL (CD11a) promoter. *Mech Ageing Dev* 123:*1257.*2002.
64. Cooper, G. S., and B. C. Stroehla. The epidemiology of autoimmune diseases. *Autoimmun Rev* 2:*119.*2003.
65. Richardson, B. Impact of aging on DNA methylation. *Ageing Res Rev* 2:*245.*2003.
66. Chen, G. Y., et al. Interaction of GATA-3/T-bet transcription factors regulates expression of sialyl Lewis X homing receptors on Th1/Th2 lymphocytes. *Proc Natl Acad Sci U S A* 103:*16894.*2006.
67. Han, S., et al. Recruitment of histone deacetylase 4 by transcription factors represses interleukin-5 transcription. *Biochem J* 400:*439.*2006.
68. Henderson, B. R., et al. Lymphoid enhancer factor-1 blocks adenomatous polyposis coli-mediated nuclear export and degradation of beta-catenin. Regulation by histone deacetylase 1. *J Biol Chem* 277:*24258.*2002.
69. Zhong, H., R. E. Voll, and S. Ghosh. Phosphorylation of NF-κB p65 by PKA stimulates transcriptional activity by promoting a novel bivalent interaction with the coactivator CBP/p300. *Mol Cell* 1:*661.*1998.
70. Zhong, H., et al. The phosphorylation status of nuclear NF-κB determines its association with CBP/p300 or HDAC-1. *Mol Cell* 9:*625.*2002.
71. Wallberg, A. E., et al. p300 and PCAF act cooperatively to mediate transcriptional activation from chromatin templates by notch intracellular domains in vitro. *Mol Cell Biol* 22:*7812.*2002.
72. Vaute, O., et al. Functional and physical interaction between the histone methyl transferase Suv39H1 and histone deacetylases. *Nucleic Acids Res* 30:*475.*2002.
73. Chakraborty, S., et al. SUV39H1 interacts with AML1 and abrogates AML1 transactivity. AML1 is methylated in vivo. *Oncogene* 22:*5229.*2003.

11 Etiology of Major Psychosis
Why Do We Need Epigenetics?

Gabriel Oh and Arturas Petronis

CONTENTS

11.1 Introduction 189
11.2 Three Fundamental Concepts behind the Epigenetic Model of Major
Psychosis 190
11.3 Traditional Theories of Mental Health Disorder and Epigenetics 192
    11.3.1 Genetic Theory 192
    11.3.2 Neurodevelopmental Theory 193
    11.3.3 Neurochemical Theory 193
    11.3.4 Traditional Theories of Major Psychosis from the Epigenetic
    Perspective 194
11.4 The Non-Mendelian Features of Psychiatric Disease and Their
Epigenetic Interpretation... 195
11.5 The Epigenetic Model of Major Psychosis... 197
11.6 Experimental Considerations: New Complexities 197
11.7 Final Notes 200
References 200

11.1 INTRODUCTION

In the past three decades, significant effort has been dedicated to the development of new experimental methods and computational strategies for the identification of primary molecular causes of human diseases. Many series of genes that cause or contribute to various human diseases have already been identified. The vast majority of such disease genes, however, are those that cause simple Mendelian diseases, such as cystic fibrosis and Huntington's disease. In contrast, the genes responsible for complex diseases—those that exhibit a heritable component but do not follow Mendel's laws—have proven to be more elusive than previously thought. This could be due to the limitations of current research strategies, which are typically based on searching for DNA sequence variations (i.e., mutations and polymorphisms). Although there were some limited successes in identifying DNA-based origins of complex diseases (e.g., colon cancer, breast cancer, and Alzheimer's diseases), the molecular origins for the overwhelming majority of complex diseases remain unexplained. This is the case with major psychosis (i.e., schizophrenia and bipolar disorder). Schizophrenia is characterized by psychotic features, including delusions and hallucinations, as well as disturbances in normal affective responses and altered cognitive functions. Bipolar disorder presents with severe disturbances in mood, ranging from extreme elation to severe depression, and is usually accompanied by psychotic features. Major psychoses, unlike simple-Mendelian diseases, exhibit relatively late onset, discordance between monozygotic (MZ) twins, sexual dimorphism, parent-of-origin effect, fluctuating disease course, and sometimes, partial recovery, none of which are consistent with DNA sequence-based disease mechanisms. New opportunities may arise from epigenetic interpretations of major psychosis by providing alternative research strategies, which work in unison with traditional research methods.

Epigenetics by definition refers to the regulation of gene expression that is not based on DNA sequence but, rather, is controlled by heritable and potentially reversible changes to the DNA (methylation of cytosines) and/or the chromatin structure (acetylation, methylation, phosphorylation, and other modifications of histones).[1] Increasing evidence tells us that in order for cells to operate "correctly," both the DNA sequence and the epigenetic factors must be normal. Thus, individuals with epigenetic misregulation may suffer from a disease, even if they are carrying functionally impeccable DNA sequences (no disease mutations or predisposing DNA sequence variants). Consequently, the current research strategies, which focus on DNA sequence-based analysis, might not be sufficient and efficient for studying complex diseases such as schizophrenia and bipolar disorder. Epigenetic research may open new opportunities for understanding the various non-Mendelian features of complex psychiatric diseases. This chapter is dedicated to the brief introduction of the epigenetic concepts and mechanisms, as well as a discussion on the putative role of epigenetic factors in etiopathogenesis of major psychiatric diseases.

11.2   THREE FUNDAMENTAL CONCEPTS BEHIND THE EPIGENETIC MODEL OF MAJOR PSYCHOSIS

The epigenetic model of major psychosis rests on three fundamental concepts. Each concept is based on observable facts and enables explanation of the peculiar nature of non-Mendelian characteristics of complex psychiatric diseases. The three fundamental concepts are as follows:

1. *Compared to the DNA sequence, the epigenetic status is considerably more dynamic.* Epigenetic patterns, like DNA sequences, are transmitted from maternal chromatids to daughter chromatids during mitosis. The mode of epigenetic status transmission is called the epigenetic inheritance system.[2] Unlike DNA sequences, which are almost flawlessly reproduced in each cell division, epigenetic patterns usually exhibit partial stability (sometimes

called epigenetic metastability). After mitosis, the daughter chromosomes do not necessarily carry epigenetic patterns identical to their parental chromosomes. For example, in mammalian cell cultures, the maintenance of methylation status was detected to be between 97 and 99.9% and de novo methylation activity was as high as 3–5% per mitosis.[3] Although promoter regions of functionally important genes showed a substantially higher fidelity of methylation patterns, with a rate of 99.85–99.92% site/generation,[4] substantial epigenetic differences may be accumulated over time across the cells of the same cell type or tissue, despite the fact that they all have the same DNA sequence. Several groups of factors contribute to epigenetic metastability. Changes in epigenetic patterns may depend on the internal and external environments of the individual, as well as various stages of development during the individual's life.[5-8] In tissue cultures, however, neither genetic nor environmental factors can be blamed for epigenetic changes. Hence, it can be explained only as stochastic events in the maintenance of epigenetic profiles.

2. *Some epigenetic signals can be transmitted from one generation to another, along with the DNA sequence.* Until recently, it has been generally accepted that during the maturation of the germline cells, the epigenetic patterns are erased and new epigenetic patterns are established.[9] However, it is now becoming apparent that some epigenetic patterns "survive" gametogenesis and can be transmitted from one generation to another, sequentially passing on the epigenetically determined traits from the parents to the offspring.[10-13] Therefore, DNA sequence is not the only molecular substrate of heritable traits (Figure 11.1).

3. *Epigenetic signals are critical for the regulation of various genomic functions, and epigenetic misregulation may be detrimental to an individual in the same way mutant genes are.* Both epigenetic studies of individual genes and manipulations of the key players in epigenetic machinery demonstrate the role of epigenetic regulation in a functionally normal cell. One of the most thoroughly investigated genes, β globin, from both the DNA sequence and epigenetic perspectives, revealed that complex interactions of DNA methylation and histone modifications regulate the activity of this gene across various developmental stages of the organism.[14-18] Numerous other examples demonstrate the critical role of epigenetic factors in the regulation of gene activity.[19-24]

Knocking out the gene encoding for DNA methyltransferase 1 (Dnmt1) led to developmental arrest at an early stage of embryogenesis.[9] The conditional disruption of Dnmt3a in germ cells, with their preservation in somatic cells, resulted in the termination of female offspring in utero, while Dnmt3a conditional mutant males showed impaired spermatogenesis and lack of methylation at two of the three paternally imprinted loci examined in spermatogonia.[25] In addition to the regulation of gene activity, epigenetic factors also play an important role in numerous other genomic functions, including genetic recombination and DNA mutability.[3,26-35]

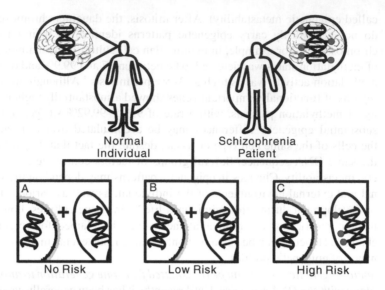

FIGURE 11.1 (See color figure following page 52.) Partial meiotic epigenetic stability. In this hypothetical family, the father is affected with schizophrenia and has an epimutation on the gene predisposing to schizophrenia. (*A*) The epimutation is completely erased in the father's germline and the offspring has no disease. (*B*) There is partial erasure of epimutation, which results in a higher risk of developing schizophrenia. (*C*) The epimutation is meiotically stable; in which case, the offspring has a high chance of developing schizophrenia.

11.3 TRADITIONAL THEORIES OF MENTAL HEALTH DISORDER AND EPIGENETICS

A number of traditional theories have tried to explain the various aspects of schizophrenia and bipolar disease. However, such theories focus on isolated features of a disease, rather than attempting to identify common molecular mechanisms that would be consistent with the wide variety of epidemiological, clinical, and molecular findings. This section will provide a brief overview of the main traditional theories of psychiatric disorders.

11.3.1 GENETIC THEORY

Psychiatric disorders are complex non-Mendelian diseases that show relatively high degrees of heritability. For example, heritability in schizophrenia is estimated to be ~70%, although neither the mode of inheritance nor the number of involved genes and their interactions are clear. Traditional genetic theory assumes that DNA sequence variation accounts for the differential susceptibility to schizophrenia or bipolar disease. A number of genes, which may increase the risk of getting schizophrenia and bipolar disorder, have been identified, including: the genes encoding for dystro-brevin binding protein 1 (*DTNBP1*),[36] neuregulin 1 (*NRG1*),[37] d-amino-acid oxidase (*DAO*) and DAO activator (*DAOA*),[38] regulator of G-protein signaling 4 (*RGS4*),[39] catechol-O-methyltransferase (*COMT*),[40] and the gene "disrupted in schizophrenia 1" (*DISC 1*),[41] to list a few.[42] Very large samples and linkage disequilibrium-based mapping are required to identify single-nucleotide polymorphisms (SNPs), which exhibit sufficient statistical evidence for association with the disease phenotype. Identification of SNP variants associated with the disease, however, raise numerous questions and difficulties that have been observed in molecular genetic studies of other complex diseases.[43] In an overwhelming majority of cases, evidence for association to a disease derives from haplotypes that are built on SNPs of unknown functional significance. Identification of a causative mutation (as in simple genetic diseases) or a functional polymorphism is often difficult. Furthermore, replication studies often reveal a different risk allele or haplotype, which complicates the interpretation of such findings in terms of disease etiopathogenesis.

11.3.2 Neurodevelopmental Theory

The neurodevelopmental theory of schizophrenia states that predisposition to schizophrenia is caused by early brain insults, which prevent normal brain development and eventually result in the dysfunction of the mature brain. The foundation for this theory was formed in the early twentieth century by clinical psychiatrists who speculated that schizophrenia is caused by cerebral maldevelopment.[44] In the present day, researchers have found a wide range of neurodevelopmental defects attributed to individuals affected with schizophrenia, including physical abnormalities, neuropsychological deficits, and anatomical abnormalities in the brain.[45-50] Although the results of the various studies were not always consistent, there is converging evidence that at least some of the individuals who suffer from schizophrenia have subtle developmental abnormalities that most likely occurred during the early stages of development. Differences in neuronal cell migration and clustering, synaptic formation, cytoarchitectural organization, neuronal density in various regions of the brain, and brain asymmetry have been detected.[51-53] Morphological differences associated with schizophrenia have also been documented at the macroscopic level, including changes in the size of lateral brain ventricles, subcortical structures, and cortical volume. Prospective and retrospective studies have shown that individuals affected with schizophrenia exhibit higher incidences of delayed attainment of developmental milestones and various other subtle behavioral and intellectual abnormalities, long before they are affected by the disease.[51-53] Individuals with schizophrenia are also reported to have a higher incidence of minor physical anomalies, including low-set ears, furrowed tongue, high-arched palate, curved fingers, and greater distance between the eyes. Since both the skin and the brain originate from the ectoderm, it is possible that physical anomalies observed in schizophrenics are caused by damage to the ectoderm during early stages of development, which is indirect support for abnormal neurodevelopment.[54]

11.3.3 Neurochemical Theory

In the past 30 years, researchers have accumulated a substantial amount of information regarding the dysregulation of various neurotransmitter systems in the brains of individuals affected with schizophrenia and bipolar disease. This has provided the basis for the neurochemical theory of major psychosis. In support of the theory, it was identified that various psychiatric medications target the dopaminergic, serotonergic, or norepinephric systems.[55-59] For example, antipsychotic drugs that are used to treat schizophrenia exhibit high affinity for dopamine receptors, and the antipsychotic potency of these drugs correlate with their affinity for the dopamine D2 receptors.[60-62] Furthermore, the density of dopamine D2 receptors is elevated in the postmortem brains of schizophrenia patients.[63-65] These pharmacological findings stimulated molecular genetic studies of the genes for dopamine receptors in schizophrenic patients, particularly the dopamine D2 receptor gene, *DRD2*.[66] While there is no doubt that psychosis is accompanied by neurochemical changes in the brain, it remains unclear why and how the changes come about.

11.3.4 TRADITIONAL THEORIES OF MAJOR PSYCHOSIS FROM THE EPIGENETIC PERSPECTIVE

Since some epigenetic signals can be transmitted from one generation to another along with DNA sequence, DNA sequence is not the only molecular substrate of heritable traits. The partial meiotic epigenetic metastability may shed a new light not only on heritability, but also on the issue of familiality and sporadicity in major psychosis. The traditional DNA sequence-based theory assumes that the difference between familial and sporadic cases of psychiatric disorder is caused by the difference in genes that are involved; however, the clinical phenotypes in sporadic and familial cases of schizophrenia are indistinguishable. From the epigenetic point of view, it is possible that meiotically stable epimutations of schizophrenia or bipolar disease might present themselves as quasi-Mendelian in familial cases of the disease, while epimutations that are unstable and are partially or completely fixed during gametogenesis may be the cause for sporadic cases. In other words, it can be hypothesized that disease epimutations may develop in two possible ways. The overwhelming majority of them regress toward the norm in the germline of an affected individual, and his/her offspring will not be affected, while the minority of epimutations may persist across generations. The reason behind why some epimutations are more stable during the meiosis than others remains unclear.

Since epigenetic factors are directly involved in tissue differentiation, development, and overall regulation of gene activities, inherited and acquired epigenetic problems may also be the primary causes of the myriad of neurodevelopmental changes detected in schizophrenia and bipolar disease. As mentioned before, epigenetic modifications are closely linked with the neurodevelopmental processes. DNA methylation patterns in mammals are subjected to major changes during gametogenesis, tissue differentiation, and in early and later stages of development.[67-70] The crucial role of DNA methylation in development is demonstrated by the presence of developmental defects, resulting from the disruption of normal genomic imprinting and changes in the pattern of gene expression induced by the demethylating agent, 5-azacytidine.[71-74] Therefore, it is entirely conceivable that epigenetic factors may lead to neurodevelopmental defects.

Since epigenetic modifications of DNA and chromatin contribute to the regulation of gene expression in differentiated tissues and organs, putative epigenetic misregulation may also be a candidate mechanism for the explanation of the neurochemical changes in the brains of psychiatric patients. It is equally possible that the changes in densities of dopamine D2 and numerous other receptors in schizophrenic patients are due to epigenetic misregulation of their respective genes or the structural changes in their DNA sequences.

In summary, epigenetics may shed new light on the experimental findings and complement the traditional theories of major psychiatric disease. The value of the epigenetic model of psychiatric disease lies in the possibility of integrating a variety of epidemiological, clinical, and molecular data, which seem to be unrelated, into a new theoretical framework. In addition, epigenetics may provide a new interpretation of various non-Mendelian features of major psychosis, most of which have not been explored from either the theoretical or the experimental points of view.

11.4 THE NON-MENDELIAN FEATURES OF PSYCHIATRIC DISEASE AND THEIR EPIGENETIC INTERPRETATION

Identical, or monozygotic (MZ), twins share identical DNA sequences. Phenotypic differences (i.e., MZ twin discordance) have been one of the key features of complex diseases. For example, MZ concordance (phenotypic similarity) for bipolar disorder[75] and schizophrenia[76] were 62–79% and 41–65%, respectively. The differential susceptibility to disease in MZ co-twins was identified decades ago, but the causes of such differences remain unknown. Several attempts have been made to try and identify DNA sequence differences in MZ twins who are discordant for psychiatric disorders. However, differences in systemic DNA sequences in MZ twins are yet to be identified.[77-80] The traditional explanation for MZ twin discordance in disease state is based on so-called non-shared environmental effects.[81] Identification of specific environmental factors that would predispose an individual to a disease, however, is very difficult. Thus far, not a single environmental risk factor for major psychosis has been identified. The epigenetic model of complex disease provides a novel molecular interpretation of the environment-induced phenotypic differences in MZ twins to their epigenetic differences. Differential exposure to a wide variety of environmental factors may reflect on the differences in epigenetic patterns of MZ twins.[82,83] The list of environmental factors affecting the epigenetic status of the genome and individual genes is growing larger every year.[6,84,85] Studies have shown that the intake of folic acid affects both the global methylation level in the genome and the regulation of imprinted genes.[86,87] Animal studies have shown that an increased intake of maternal dietary methyl supplements during pregnancy, which increase DNA methylation, can lead to changes in methylation-dependent epigenetic phenotypes in their offspring.[5,7] Recreational drugs may also alter epigenetic regulation: methamphetamine, which can cause psychosis in humans, alters the DNA methylation profile and the expression of genes that are thought to be involved in schizophrenia.[88] Of greater interest, it was recently discovered that exposure to certain behaviors (or the lack thereof) alone could alter the epigenetic pattern. Specifically, the lack of pup licking, grooming, and arched-back nursing (i.e., nurturing behaviors) by mother rats, which are thought to increase levels of stress in the offspring, induced histone modifications and changes in DNA methylation at a glucocorticoid receptor gene promoter in the hippocampus of the offspring.[8] However, changes in epigenetic patterns are not entirely due to differential environmental exposures. A significant portion of epigenetic changes may occur even in the absence of evident environmental differences. As mentioned before, a substantial degree of epigenetic variation can be accumulated over the lifespan of an individual, even in genetically identical individuals, simply due to the epigenetic metastability in somatic cells. This is well illustrated in the epigenetic studies of inbred animals.[12]

Parent-of-origin effect, which is differential risk to the offspring depending on which of the two parents was affected with the disease, is commonly observed in major psychiatric disorders. For example, the risk of developing bipolar disorder is higher in individuals with mothers, rather than fathers, who are affected by the disease.[89] This effect is also observed in other studies of major psychosis.[90-94] An epigenetic phenomenon, called genomic imprinting, is one of the most common mechanisms behind the parent-of-origin effect.[95] The principle of genomic imprinting is based on differential epigenetic modification of genes, which depend on their parental origin.[96,97] As a rule, disruption of normal imprinting patterns affects growth, development, and behavior, which results in severe health problems.[98,99] Animal studies using chimeric mice have shown the detrimental impact of disrupted imprinting patterns on the development of the brain. Mice expressing normal and uniparental cells showed that cells that display complete paternal disomy (i.e., androgenetic [Ag] cells) or complete maternal disomy (i.e., parthenogenetic [Pg] cells) have differential effects on specific regions of the brain.[100] They found that Ag cells proliferated extensively in the medio-basal forebrain and were essential for the proper development of the preoptic area, the hypothalamus, and the septum, which are regions in the brain that are important for primary motivated behavior. Pg cells, however, proliferated in areas where Ag cells did not. These areas included the neocortex, striatum, and telencephalic structures.[101] These results offer some insight into the importance of epigenetic processes and genomic imprinting in brain development. Altered epigenetic processes are known to cause aberrations in brain development. For example, individuals with Angelman syndrome have abnormal cortical growth, resulting in cortical atrophy, microencephaly, and ventricular dilation. The inadequate brain development results in behavioral dysfunction involving seizures, attention deficit, hyperactivity, and aggressive behavior.[102,103]

Another feature of major psychosis, which cannot be explained using a simple Mendelian model, is the sex effect, or sexual dimorphism. The sex effect refers to differential susceptibility to a disease that is dependent on the gender of the individual, and nearly all complex diseases exhibit some degree of sexual dimorphism.[56,104] It is interesting to note that such effects cannot be explained by genetic risk factors on the sex chromosomes. Some genetic association studies have found that numerous autosomal genes exhibit sex effects. For example, in schizophrenia studies of *DISC1*, a common haplotype (HEP3) was found to be significantly undertransmitted to affected individuals.[105] This haplotype also displayed sex differences in transmission distortion; the undertransmission being significant only in affected females (p = .00024) (undertransmission in affected males p = .38), suggesting that this haplotype may confer a protective effect against the disease in women.[105] A Scottish study of patients with schizophrenia and bipolar disease found that an extended HEP3 haplotype was overrepresented in males affected with bipolar disease (in the male group p = .008, while in the females p = .25).[106] Another frequently analyzed gene in psychiatric research, *COMT*, revealed an association to schizophrenia only in women (p = 9.1×10^{-6}) while men showed no evidence for such an association (p = .10). While sex hormones have been the usual "culprit" in the explanation of gender effects in complex diseases (based on the myriad of data associating hormonal differences with disease states and their critical involvement in human biology),[107] no underlying mechanisms have been proposed to explain how such hormones predispose or protect individuals from a disease relating to the specific molecular mechanisms of hormone action. The gender-specific effects in genetic linkage and association studies suggest that chromosomes and individual genes can be targets of sex hormones. While such hormones cannot change DNA sequence, they can be potent modifiers of epigenetic status, which controls gene expression and various other genomic activities. One of the mechanisms by which sex hormones can lead to epigenetic misregulation is by altering the epigenetic signatures of chromosomal regions that modulate the access of transcription factors to the target sequence.[108] The epigenetic modifications mediated through the sex hormones and their corresponding receptors will likely vary by gene and tissue between the sexes.[108] This mechanism may contribute to the clinical differences of schizophrenia in males and females. In conclusion, it can be hypothesized that differential susceptibility to complex psychiatric diseases in males and females is mediated by sex hormone-induced changes in the epigenetic regulation of genes.

11.5 THE EPIGENETIC MODEL OF MAJOR PSYCHOSIS

The epigenetic model of major psychosis could be thought as the result of a chain of deviant epigenetic events, beginning with a pre-epimutation, an epigenetic change that takes place during gametogenesis or embryogenesis. Pre-epimutation increases the risk for major psychosis, but it is not sufficient to cause the disease. The phenotypic outcome is dependent on the overall effect of a series of pre- and postnatal factors on the pre-epimutation. The multidirectional effects of tissue differentiation, stochastic factors, hormones, and external environmental factors (nutrition, infections, medications, addictions, etc.) further amplify the epigenetic misregulation, but this could be tolerated to some extent (Figure 11.2). It may take decades for the epigenetic misregulation to reach a critical mass, beyond which the cell (or the tissue) is no longer functionally normal. Only a fraction of the predisposed individuals may reach the threshold of epigenetic misregulation that causes the phenotypic changes, which meet the criteria for major psychosis. The severity of epigenetic misregulation may fluctuate over time, which could lead to remission and relapse (Figure 11.3). In some cases, "aging" epimutations may start slowly regressing back to the normal state. This is seen as partial recovery, which is consistent with age-dependent epigenetic changes in the genome.[109]

11.6 EXPERIMENTAL CONSIDERATIONS: NEW COMPLEXITIES

Despite the significant theoretical value of the epigenetic model of major psychosis, experimental epigenetic and epigenomic studies are not straightforward. Here are

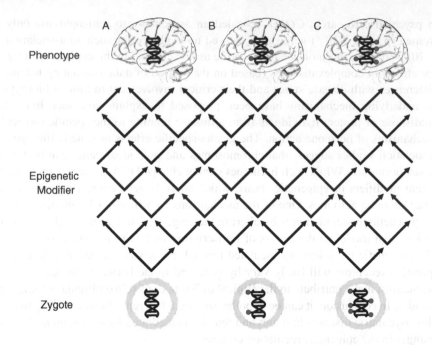

FIGURE 11.2 Epigenetic model of major psychosis. An inherited epigenetic status in disease genes may vary from severe epimutation to a perfectly normal epigenetic profile. The multidirectional effects of tissue differentiation, stochastic factors, hormones, and external environmental factors, such as nutrition, infections, medications, addictions, etc. (indicated by arrows), can further either increase or decrease the degree of epigenetic misregulation. The more severe the inherited epigenetic defect, the higher the chance that such an individual will develop psychosis. Vice versa, the chance of contracting the disease is slim for individuals with normal inherited epigenotype. Exceptions in both cases are also possible.

some of the confounding factors that should be taken into account when designing experiments and interpreting data.

1. Since epigenetic patterns, unlike DNA sequences, are cell type specific, tissues from the primary site of disease manifestation are required for an epigenetic analysis. Therefore, when studying psychiatric diseases, the ideal sample for epigenetic analysis is brain tissue. It is also important to note that there is increasing evidence that many epimutations are not limited to the affected tissue or cell type, but can also be detected in other tissues.[110,111]

2. It should also be noted that a complex organ, such as the brain, is a concoction of many different types of cells (astrocytes, glial cells, pyramidal cells, etc.). To make matters more complicated, specific regions of the brain have cells that are specific to that region only (i.e., not all neurons express the same receptors). Once again, different types of cells have unique epigenetic patterns that are specific to each cell type. For example, in an ideal situation, one could isolate dopaminergic neurons from the ventral tegmental

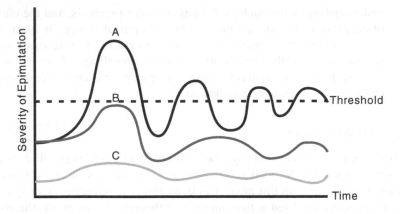

FIGURE 11.3   Severity of epimutation and "threshold" for major psychosis. Lines *A* and *B* represent identical twins who share the same epigenetic predisposition for major psychosis, while line *C* represents a normal individual. The level of epimutation is identical at birth for the twins but the epigenetic differences grow larger as time progresses. Differential exposure to factors such as hormone-induced epigenetic changes, environmental effects, and stochastic events may cause one of the twins to go over the threshold while the other twin stays below the threshold. The first peak over the threshold for twin A represents the primary psychotic episode, which is most likely to occur shortly after puberty. The subsequent oscillation above and below the threshold represents disease relapse and remission.

    area or striatum using laser capture microdissection to study the epigenetic regulation of dopamine D2 receptor gene in schizophrenia.

3. Since histone modifications are not stable in postmortem samples, studies in human postmortem tissues have to be carefully designed and numerous control experiments performed. Since DNA methylation is more stable in the postmortem tissues, DNA methylation studies are more feasible.

4. The cause-and-effect relationship between the epigenetic changes and disease may not be evident because disease phenotypes might be caused by epigenetic changes or the phenotype itself might cause epigenetic changes. Associations between epigenetic changes in the affected tissue and disease phenotype do not reveal anything about which component of this dyad is primary and which one is secondary. Prospective studies, analyses of unaffected tissues of the affected individuals including the germline, and animal models may help to establish the nature of the relationship between epimutations and disease.

5. Although twin and inbred animal studies have clearly shown that epigenetic patterns are at least partially autonomous from DNA sequence, interactions of DNA sequence and epigenetic profiles are possible in some cases. There is an increasing list of observations showing that, in some cases, DNA sequences may partially determine epigenetic peculiarities (e.g., Beckwith–Wiedemann syndrome[112] and the gene encoding *CDH13*[113]).

6. Identification of the targets—genes and genomic loci—as candidates for epigenetic and epigenomic studies in psychiatric disease is not trivial. The traditionally investigated genes may be a reflection of our (limited)

understanding of the etiological basis of major psychosis, and the choice of candidates from this list may not be the optimal strategy in identification of etiological epimutations. Unless the evidence to investigate some specific genes and their regulatory regions is compelling,[114-116] an unbiased, large-scale, microarray-based search for epigenetic changes across entire chromosomes or genome is indicated.

11.7   FINAL NOTES

The epigenetic model of major psychiatric disorders can explain peculiar characteristics of complex non-Mendelian diseases, which traditional theories of psychosis cannot. The true value of this model lies in its ability to integrate a variety of theories, which seem unrelated at first, into a new theoretical framework, thereby providing the basis for new experimental approaches. However, the theoretical notions about a role for epigenetic processes in the development of psychiatric disorders do not prove such a link. There is a need to move from theory into molecular research in order to test the role of epigenetics in complex psychiatric diseases. Although it is premature to conclude that epigenetics will lead to revolutionary discoveries in non-Mendelian biology, it has the potential to transform understanding about the molecular etiology of complex diseases. The goal of this chapter is not to discredit DNA sequence variation-based analysis, but rather, to suggest that in complex diseases such as psychiatric disorders, the contribution of epigenetic factors may be substantial, and that DNA sequence variations of genes should be investigated in parallel with the epigenetic regulation of genes.

REFERENCES

1. Henikoff, S. and Matzke, M. A., Exploring and explaining epigenetic effects, *Trends Genet* 13 (8), 293-5, 1997.
2. Maynard Smith, J., Models of a dual inheritance system, *J Theor Biol* 143 (1), 41-53, 1990.
3. Riggs, A. D., Xiong, Z., Wang, L., and LeBon, J. M., Methylation dynamics, epigenetic fidelity and X chromosome structure, *Novartis Found Symp* 214, 214-25; discussion 225-32, 1998.
4. Ushijima, T., Watanabe, N., Okochi, E., Kaneda, A., Sugimura, T., and Miyamoto, K., Fidelity of the methylation pattern and its variation in the genome, *Genome Res* 13 (5), 868-74, 2003.
5. Cooney, C. A., Dave, A. A., and Wolff, G. L., Maternal methyl supplements in mice affect epigenetic variation and DNA methylation of offspring, *J Nutr* 132 (8 Suppl), 2393S-2400S, 2002.
6. Sutherland, J. E. and Costa, M., Epigenetics and the environment, *Ann N Y Acad Sci* 983, 151-60, 2003.
7. Waterland, R. A. and Jirtle, R. L., Transposable elements: targets for early nutritional effects on epigenetic gene regulation, *Mol Cell Biol* 23 (15), 5293-300, 2003.
8. Weaver, I. C., Cervoni, N., Champagne, F. A., D'Alessio, A. C., Sharma, S., Seckl, J. R., Dymov, S., Szyf, M., and Meaney, M. J., Epigenetic programming by maternal behavior, *Nat Neurosci* 7 (8), 847-54, 2004.
9. Li, E., Chromatin modification and epigenetic reprogramming in mammalian development, *Nat Rev Genet* 3 (9), 662-73, 2002.

10. Rakyan, V. and Whitelaw, E., Transgenerational epigenetic inheritance, *Curr Biol* 13 (1), R6, 2003.

11. Rakyan, V. K., Preis, J., Morgan, H. D., and Whitelaw, E., The marks, mechanisms and memory of epigenetic states in mammals, *Biochem J* 356 (Pt 1), 1-10, 2001.

12. Rakyan, V. K., Blewitt, M. E., Druker, R., Preis, J. I., and Whitelaw, E., Metastable epialleles in mammals, *Trends Genet* 18 (7), 348-51, 2002.

13. Richards, E. J., Inherited epigenetic variation—revisiting soft inheritance, *Nat Rev Genet* 7 (5), 395-401, 2006.

14. Litt, M. D., Simpson, M., Gaszner, M., Allis, C. D., and Felsenfeld, G., Correlation between histone lysine methylation and developmental changes at the chicken beta-globin locus, *Science* 293 (5539), 2453-5, 2001.

15. Singal, R. and Ginder, G. D., DNA methylation, *Blood* 93 (12), 4059-70, 1999.

16. Kim, A., Kiefer, C. M., and Dean, A., Distinctive signatures of histone methylation in transcribed coding and noncoding human beta-globin sequences, *Mol Cell Biol* 27 (4), 1271-9, 2007.

17. Chakalova, L., Carter, D., Debrand, E., Goyenechea, B., Horton, A., Miles, J., Osborne, C., and Fraser, P., Developmental regulation of the beta-globin gene locus, *Prog Mol Subcell Biol* 38, 183-206, 2005.

18. Robertson, G., Garrick, D., Wilson, M., Martin, D. I., and Whitelaw, E., Age-dependent silencing of globin transgenes in the mouse, *Nucleic Acids Res* 24 (8), 1465-71, 1996.

19. Robertson, K. D. and Wolffe, A. P., DNA methylation in health and disease, *Nat Rev Genet* 1 (1), 11-9, 2000.

20. Robertson, K. D., DNA methylation and human disease, *Nat Rev Genet* 6 (8), 597-610, 2005.

21. Dolinoy, D. C., Weidman, J. R., and Jirtle, R. L., Epigenetic gene regulation: Linking early developmental environment to adult disease, *Reprod Toxicol*, 2006.

22. D'Alessio, A. C. and Szyf, M., Epigenetic tete-a-tete: the bilateral relationship between chromatin modifications and DNA methylation, *Biochem Cell Biol* 84 (4), 463-76, 2006.

23. Wright, K. L. and Ting, J. P., Epigenetic regulation of MHC-II and CIITA genes, *Trends Immunol* 27 (9), 405-12, 2006.

24. Razin, A. and Kantor, B., DNA methylation in epigenetic control of gene expression, *Prog Mol Subcell Biol* 38, 151-67, 2005.

25. Kaneda, M., Okano, M., Hata, K., Sado, T., Tsujimoto, N., Li, E., and Sasaki, H., Essential role for de novo DNA methyltransferase Dnmt3a in paternal and maternal imprinting, *Nature* 429 (6994), 900-3, 2004.

26. Yang, A. S., Jones, P. A., and Shibata, A., The mutational burden of 5-methylcytosine, in *Epigenetic mechanisms of gene regulation*, Martienssen, R. A., Riggs, A. D., and Russo, V. E. A. Cold Spring Harbor Laboratory Press, Plainview, N.Y., 1996, pp. 77-94.

27. Constancia, M., Pickard, B., Kelsey, G., and Reik, W., Imprinting mechanisms, *Genome Res* 8 (9), 881-900, 1998.

28. Ehrlich, M. and Ehrlich, K. C., Effect of DNA methylation on the binding of vertebrate and plant proteins to DNA, *Exs* 64, 145-68, 1993.

29. Jones, P. L., Veenstra, G. J., Wade, P. A., Vermaak, D., Kass, S. U., Landsberger, N., Strouboulis, J., and Wolffe, A. P., Methylated DNA and MeCP2 recruit histone deacetylase to repress transcription, *Nat Genet* 19 (2), 187-91, 1998.

30. Nan, X., Ng, H. H., Johnson, C. A., Laherty, C. D., Turner, B. M., Eisenman, R. N., and Bird, A., Transcriptional repression by the methyl-CpG-binding protein MeCP2 involves a histone deacetylase complex, *Nature* 393 (6683), 386-9, 1998.

31. Razin, A. and Shemer, R., Epigenetic control of gene expression, *Results Probl Cell Differ* 25, 189-204, 1999.

32. Siegfried, Z., Eden, S., Mendelsohn, M., Feng, X., Tsuberi, B. Z., and Cedar, H., DNA methylation represses transcription in vivo, *Nat Genet* 22 (2), 203-6, 1999.
33. Bestor, T. H., Chandler, V. L., and Feinberg, A. P., Epigenetic effects in eukaryotic gene expression, *Dev Genet* 15 (6), 458-62, 1994.
34. Riggs, A. D. and Porter, T., Overview of epigenetic mechanisms, in *Epigenetic mechanisms of gene regulation*, Martienssen, R. A., Riggs, A. D., andRusso, V. E. A. Cold Spring Harbor Laboratory Press, Plainview, N.Y., 1996, pp. 29-45.
35. Petronis, A., Genomic imprinting in unstable DNA diseases, *Bioessays* 18 (7), 587-90, 1996.
36. Straub, R. E., Jiang, Y., MacLean, C. J., Ma, Y., Webb, B. T., Myakishev, M. V., Harris-Kerr, C., Wormley, B., Sadek, H., Kadambi, B., Cesare, A. J., Gibberman, A., Wang, X., O'Neill, F. A., Walsh, D., and Kendler, K. S., Genetic variation in the 6p22.3 gene DTNBP1, the human ortholog of the mouse dysbindin gene, is associated with schizophrenia, *Am J Hum Genet* 71 (2), 337-48, 2002.
37. Stefansson, H., Sarginson, J., Kong, A., Yates, P., Steinthorsdottir, V., Gudfinnsson, E., Gunnarsdottir, S., Walker, N., Petursson, H., Crombie, C., Ingason, A., Gulcher, J. R., Stefansson, K., and St Clair, D., Association of neuregulin 1 with schizophrenia confirmed in a Scottish population, *Am J Hum Genet* 72 (1), 83-7, 2003.
38. Chumakov, I., Blumenfeld, M., Guerassimenko, O., Cavarec, L., Palicio, M., Abderrahim, H., Bougueleret, L., Barry, C., Tanaka, H., La Rosa, P., Puech, A., Tahri, N., Cohen-Akenine, A., Delabrosse, S., Lissarrague, S., Picard, F. P., Maurice, K., Essioux, L., Millasseau, P., Grel, P., Debailleul, V., Simon, A. M., Caterina, D., Dufaure, I., Malekzadeh, K., Belova, M., Luan, J. J., Bouillot, M., Sambucy, J. L., Primas, G., Saumier, M., Boubkiri, N., Martin-Saumier, S., Nasroune, M., Peixoto, H., Delaye, A., Pinchot, V., Bastucci, M., Guillou, S., Chevillon, M., Sainz-Fuertes, R., Meguenni, S., Aurich-Costa, J., Cherif, D., Gimalac, A., Van Duijn, C., Gauvreau, D., Ouellette, G., Fortier, I., Raelson, J., Sherbatich, T., Riazanskaia, N., Rogaev, E., Raeymaekers, P., Aerssens, J., Konings, F., Luyten, W., Macciardi, F., Sham, P. C., Straub, R. E., Weinberger, D. R., Cohen, N., and Cohen, D., Genetic and physiological data implicating the new human gene G72 and the gene for D-amino acid oxidase in schizophrenia, *Proc Natl Acad Sci U S A* 99 (21), 13675-80, 2002.
39. Chowdari, K. V., Mirnics, K., Semwal, P., Wood, J., Lawrence, E., Bhatia, T., Deshpande, S. N., B, K. T., Ferrell, R. E., Middleton, F. A., Devlin, B., Levitt, P., Lewis, D. A., and Nimgaonkar, V. L., Association and linkage analyses of RGS4 polymorphisms in schizophrenia, *Hum Mol Genet* 11 (12), 1373-80, 2002.
40. Badner, J. A. and Gershon, E. S., Meta-analysis of whole-genome linkage scans of bipolar disorder and schizophrenia, *Mol Psychiatry* 7 (4), 405-11, 2002.
41. Millar, J. K., Wilson-Annan, J. C., Anderson, S., Christie, S., Taylor, M. S., Semple, C. A., Devon, R. S., Clair, D. M., Muir, W. J., Blackwood, D. H., and Porteous, D. J., Disruption of two novel genes by a translocation co-segregating with schizophrenia, *Hum Mol Genet* 9 (9), 1415-23, 2000.
42. Craddock, N., O'Donovan, M. C., and Owen, M. J., The genetics of schizophrenia and bipolar disorder: dissecting psychosis, *J Med Genet* 42 (3), 193-204, 2005.
43. Harrison, P. J. and Owen, M. J., Genes for schizophrenia? Recent findings and their pathophysiological implications, *Lancet* 361 (9355), 417-9, 2003.
44. Kraepelin, E. and Robertson, G. M., *Dementia praecox and paraphrenia*, Livingstone, Edinburgh, 1919.
45. Green, M. F., Satz, P., Gaier, D. J., Ganzell, S., and Kharabi, F., Minor physical anomalies in schizophrenia, *Schizophr Bull* 15 (1), 91-9, 1989.
46. Guy, J. D., Majorski, L. V., Wallace, C. J., and Guy, M. P., The incidence of minor physical anomalies in adult male schizophrenics, *Schizophr Bull* 9 (4), 571-82, 1983.

47. Bracha, H. S., Etiology of structural asymmetry in schizophrenia: an alternative hypothesis, *Schizophr Bull* 17 (4), 551-3, 1991.
48. Harrison, P. J., The neuropathology of schizophrenia. A critical review of the data and their interpretation, *Brain* 122 (Pt 4), 593-624, 1999.
49. Lieberman, J. A., Is schizophrenia a neurodegenerative disorder? A clinical and neurobiological perspective, *Biol Psychiatry* 46 (6), 729-39, 1999.
50. Barkataki, I., Kumari, V., Das, M., Taylor, P., and Sharma, T., Volumetric structural brain abnormalities in men with schizophrenia or antisocial personality disorder, *Behav Brain Res* 169 (2), 239-47, 2006.
51. Lobato, M. I., Belmonte-de-Abreu, P., Knijnik, D., Teruchkin, B., Ghisolfi, E., and Henriques, A., Neurodevelopmental risk factors in schizophrenia, *Braz J Med Biol Res* 34 (2), 155-63, 2001.
52. Lewis, D. A. and Levitt, P., Schizophrenia as a disorder of neurodevelopment, *Annu Rev Neurosci* 25, 409-32, 2002.
53. Woods, B. T., Is schizophrenia a progressive neurodevelopmental disorder? Toward a unitary pathogenetic mechanism, *Am J Psychiatry* 155 (12), 1661-70, 1998.
54. Quan, X. J. and Hassan, B. A., From skin to nerve: flies, vertebrates and the first helix, *Cell Mol Life Sci* 62 (18), 2036-49, 2005.
55. Seeman, P., Lee, T., Chau-Wong, M., and Wong, K., Antipsychotic drug doses and neuroleptic/dopamine receptors, *Nature* 261 (5562), 717-9, 1976.
56. Seeman, M. V., Psychopathology in women and men: focus on female hormones, *Am J Psychiatry* 154 (12), 1641-7, 1997.
57. Van Tol, H. H., Bunzow, J. R., Guan, H. C., Sunahara, R. K., Seeman, P., Niznik, H. B., and Civelli, O., Cloning of the gene for a human dopamine D4 receptor with high affinity for the antipsychotic clozapine, *Nature* 350 (6319), 610-4, 1991.
58. Emrich, H. M., Berger, M., Riemann, D., and von Zerssen, D., Serotonin reuptake inhibition vs. norepinephrine reuptake inhibition: a double-blind differential-therapeutic study with fluvoxamine and oxaprotiline in endogenous and neurotic depressives, *Pharmacopsychiatry* 20 (2), 60-3, 1987.
59. Pedersen, O. L., Kragh-Sorensen, P., Bjerre, M., Overo, K. F., and Gram, L. F., Citalopram, a selective serotonin reuptake inhibitor: clinical antidepressive and long-term effect—a phase II study, *Psychopharmacology (Berl)* 77 (3), 199-204, 1982.
60. Creese, I., Burt, D. R., and Snyder, S. H., Dopamine receptor binding predicts clinical and pharmacological potencies of antischizophrenic drugs, *Science* 192 (4238), 481-3, 1976.
61. Seeman, P., Dopamine receptor sequences. Therapeutic levels of neuroleptics occupy D2 receptors, clozapine occupies D4, *Neuropsychopharmacology* 7 (4), 261-84, 1992.
62. Kapur, S. and Mamo, D., Half a century of antipsychotics and still a central role for dopamine D2 receptors, *Prog Neuropsychopharmacol Biol Psychiatry* 27 (7), 1081-90, 2003.
63. Seeman, P., Bzowej, N. H., Guan, H. C., Bergeron, C., Reynolds, G. P., Bird, E. D., Riederer, P., Jellinger, K., and Tourtellotte, W. W., Human brain D1 and D2 dopamine receptors in schizophrenia, Alzheimer's, Parkinson's, and Huntington's diseases, *Neuropsychopharmacology* 1 (1), 5-15, 1987.
64. Seeman, P., Guan, H. C., Nobrega, J., Jiwa, D., Markstein, R., Balk, J. H., Picetti, R., Borrelli, E., and Van Tol, H. H., Dopamine D2-like sites in schizophrenia, but not in Alzheimer's, Huntington's, or control brains, for [3H]benzquinoline, *Synapse* 25 (2), 137-46, 1997.
65. Seeman, P. and Niznik, H. B., Dopamine receptors and transporters in Parkinson's disease and schizophrenia, *Faseb J* 4 (10), 2737-44, 1990.
66. Noble, E. P., D2 dopamine receptor gene in psychiatric and neurologic disorders and its phenotypes, *Am J Med Genet B Neuropsychiatr Genet* 116 (1), 103-25, 2003.
67. Razin, A. and Cedar, H., DNA methylation and embryogenesis, *Exs* 64, 343-57, 1993.

68. Razin, A. and Shemer, R., DNA methylation in early development, *Hum Mol Genet* 4 Spec No, 1751-5, 1995.

69. Surani, M. A., Hayashi, K., and Hajkova, P., Genetic and epigenetic regulators of pluripotency, *Cell* 128 (4), 747-62, 2007.

70. Reik, W., Dean, W., and Walter, J., Epigenetic reprogramming in mammalian development, *Science* 293 (5532), 1089-93, 2001.

71. Franklin, G. C., Adam, G. I., and Ohlsson, R., Genomic imprinting and mammalian development, *Placenta* 17 (1), 3-14, 1996.

72. Jones, P. A., Altering gene expression with 5-azacytidine, *Cell* 40 (3), 485-6, 1985.

73. Zagris, N. and Podimatas, T., 5-Azacytidine changes gene expression and causes developmental arrest of early chick embryo, *Int J Dev Biol* 38 (4), 741-4, 1994.

74. Doerksen, T. and Trasler, J. M., Developmental exposure of male germ cells to 5-azacytidine results in abnormal preimplantation development in rats, *Biol Reprod* 55 (5), 1155-62, 1996.

75. Bertelsen, A., Harvald, B., and Hauge, M., A Danish twin study of manic-depressive disorders, *Br J Psychiatry* 130, 330-51, 1977.

76. Cardno, A. G. and Gottesman, II, Twin studies of schizophrenia: from bow-and-arrow concordances to star wars Mx and functional genomics, *Am J Med Genet* 97 (1), 12-7, 2000.

77. Lavrentieva, I., Broude, N. E., Lebedev, Y., Gottesman, II, Lukyanov, S. A., Smith, C. L., and Sverdlov, E. D., High polymorphism level of genomic sequences flanking insertion sites of human endogenous retroviral long terminal repeats, *FEBS Lett* 443 (3), 341-7, 1999.

78. Polymeropoulos, M. H., Xiao, H., Torrey, E. F., DeLisi, L. E., Crow, T., and Merril, C. R., Search for a genetic event in monozygotic twins discordant for schizophrenia, *Psychiatry Res* 48 (1), 27-36, 1993.

79. Tsujita, T., Niikawa, N., Yamashita, H., Imamura, A., Hamada, A., Nakane, Y., and Okazaki, Y., Genomic discordance between monozygotic twins discordant for schizophrenia, *Am J Psychiatry* 155 (3), 422-4, 1998.

80. Vincent, J. B., Kalsi, G., Klempan, T., Tatuch, Y., Sherrington, R. P., Breschel, T., McInnis, M. G., Brynjolfsson, J., Petursson, H., Gurling, H. M., Gottesman, II, Torrey, E. F., Petronis, A., and Kennedy, J. L., No evidence of expansion of CAG or GAA repeats in schizophrenia families and monozygotic twins, *Hum Genet* 103 (1), 41-7, 1998.

81. Reiss, D., Plomin, R., and Hetherington, E. M., Genetics and psychiatry: an unheralded window on the environment, *Am J Psychiatry* 148 (3), 283-91, 1991.

82. Petronis, A., Gottesman, II, Kan, P., Kennedy, J. L., Basile, V. S., Paterson, A. D., and Popendikyte, V., Monozygotic twins exhibit numerous epigenetic differences: clues to twin discordance?, *Schizophr Bull* 29 (1), 169-78, 2003.

83. Fraga, M. F., Ballestar, E., Paz, M. F., Ropero, S., Setien, F., Ballestar, M. L., Heine-Suner, D., Cigudosa, J. C., Urioste, M., Benitez, J., Boix-Chornet, M., Sanchez-Aguilera, A., Ling, C., Carlsson, E., Poulsen, P., Vaag, A., Stephan, Z., Spector, T. D., Wu, Y. Z., Plass, C., and Esteller, M., Epigenetic differences arise during the lifetime of monozygotic twins, *Proc Natl Acad Sci U S A* 102 (30), 10604-9, 2005.

84. Jablonka, E. and Lamb, M. J., *Epigenetic inheritance and evolution : the Lamarckian dimension*, Oxford University Press, Oxford; New York, 1995.

85. Ross, S. A., Diet and DNA methylation interactions in cancer prevention, *Ann N Y Acad Sci* 983, 197-207, 2003.

86. Ingrosso, D., Cimmino, A., Perna, A. F., Masella, L., De Santo, N. G., De Bonis, M. L., Vacca, M., D'Esposito, M., D'Urso, M., Galletti, P., and Zappia, V., Folate treatment and unbalanced methylation and changes of allelic expression induced by hyperhomocysteinaemia in patients with uraemia, *Lancet* 361 (9370), 1693-9, 2003.

87. Wolff, G. L., Kodell, R. L., Moore, S. R., and Cooney, C. A., Maternal epigenetics and methyl supplements affect agouti gene expression in Avy/a mice, *Faseb J* 12 (11), 949-57, 1998.

88. Numachi, Y., Yoshida, S., Yamashita, M., Fujiyama, K., Naka, M., Matsuoka, H., Sato, M., and Sora, I., Psychostimulant alters expression of DNA methyltransferase mRNA in the rat brain, *Ann N Y Acad Sci* 1025, 102-9, 2004.

89. McMahon, F. J., Stine, O. C., Meyers, D. A., Simpson, S. G., and DePaulo, J. R., Patterns of maternal transmission in bipolar affective disorder, *Am J Hum Genet* 56 (6), 1277-86, 1995.

90. Crow, T. J., DeLisi, L. E., and Johnstone, E. C., Concordance by sex in sibling pairs with schizophrenia is paternally inherited. Evidence for a pseudoautosomal locus, *Br J Psychiatry* 155, 92-7, 1989.

91. Ohara, K., Xu, H. D., Mori, N., Suzuki, Y., Xu, D. S., Ohara, K., and Wang, Z. C., Anticipation and imprinting in schizophrenia, *Biol Psychiatry* 42 (9), 760-6, 1997.

92. McMahon, F. J., Hopkins, P. J., Xu, J., McInnis, M. G., Shaw, S., Cardon, L., Simpson, S. G., MacKinnon, D. F., Stine, O. C., Sherrington, R., Meyers, D. A., and DePaulo, J. R., Linkage of bipolar affective disorder to chromosome 18 markers in a new pedigree series, *Am J Hum Genet* 61 (6), 1397-404, 1997.

93. Petronis, A., Popendikyte, V., Kan, P., and Sasaki, T., Major psychosis and chromosome 22: genetics meets epigenetics, *CNS Spectr* 7 (3), 209-14, 2002.

94. Schulze, T. G., Chen, Y. S., Badner, J. A., McInnis, M. G., DePaulo, J. R., Jr., and McMahon, F. J., Additional, physically ordered markers increase linkage signal for bipolar disorder on chromosome 18q22, *Biol Psychiatry* 53 (3), 239-43, 2003.

95. Hall, J. G., Genomic imprinting: review and relevance to human diseases, *Am J Hum Genet* 46 (5), 857-73, 1990.

96. Barlow, D. P., Gametic imprinting in mammals, *Science* 270 (5242), 1610-3, 1995.

97. Reik, W. and Walter, J., Genomic imprinting: parental influence on the genome, *Nat Rev Genet* 2 (1), 21-32, 2001.

98. Pfeifer, K., Mechanisms of genomic imprinting, *Am J Hum Genet* 67 (4), 777-87, 2000.

99. Horsthemke, B., Epimutations in human disease, *Curr Top Microbiol Immunol* 310, 45-59, 2006.

100. Allen, N. D., Logan, K., Lally, G., Drage, D. J., Norris, M. L., and Keverne, E. B., Distribution of parthenogenetic cells in the mouse brain and their influence on brain development and behavior, *Proc Natl Acad Sci U S A* 92 (23), 10782-6, 1995.

101. Keverne, E. B., Genomic imprinting in the brain, *Curr Opin Neurobiol* 7 (4), 463-8, 1997.

102. Leonard, C. M., Williams, C. A., Nicholls, R. D., Agee, O. F., Voeller, K. K., Honeyman, J. C., and Staab, E. V., Angelman and Prader-Willi syndrome: a magnetic resonance imaging study of differences in cerebral structure, *Am J Med Genet* 46 (1), 26-33, 1993.

103. Williams, C. A., Hendrickson, J. E., Cantu, E. S., and Donlon, T. A., Angelman syndrome in a daughter with del(15) (q11q13) associated with brachycephaly, hearing loss, enlarged foramen magnum, and ataxia in the mother, *Am J Med Genet* 32 (3), 333-8, 1989.

104. Ostrer, H., Sex-based differences in gene transmission and gene expression, *Lupus* 8 (5), 365-9, 1999.

105. Hennah, W., Varilo, T., Kestila, M., Paunio, T., Arajarvi, R., Haukka, J., Parker, A., Martin, R., Levitzky, S., Partonen, T., Meyer, J., Lonnqvist, J., Peltonen, L., and Ekelund, J., Haplotype transmission analysis provides evidence of association for DISC1 to schizophrenia and suggests sex-dependent effects, *Hum Mol Genet* 12 (23), 3151-9, 2003.

106. Thomson, P. A., Wray, N. R., Millar, J. K., Evans, K. L., Hellard, S. L., Condie, A., Muir, W. J., Blackwood, D. H., and Porteous, D. J., Association between the TRAX/DISC locus and both bipolar disorder and schizophrenia in the Scottish population, *Mol Psychiatry* 10 (7), 657-68, 616, 2005.

107. Exploring the biological contributions to human health: does sex matter?, *J Womens Health Gend Based Med* 10 (5), 433-9, 2001.
108. Kaminsky, Z. A., Wang, S. C., and Petronis, A., Complex disease, gender and epigenetics, *Ann Med* 38 (8), 530-544, 2006.
109. Fuke, C., Shimabukuro, M., Petronis, A., Sugimoto, J., Oda, T., Miura, K., Miyazaki, T., Ogura, C., Okazaki, Y., and Jinno, Y., Age related changes in 5-methylcytosine content in human peripheral leukocytes and placentas: an HPLC-based study, *Ann Hum Genet* 68 (Pt 3), 196-204, 2004.
110. Cui, H., Cruz-Correa, M., Giardiello, F. M., Hutcheon, D. F., Kafonek, D. R., Brandenburg, S., Wu, Y., He, X., Powe, N. R., and Feinberg, A. P., Loss of IGF2 imprinting: a potential marker of colorectal cancer risk, *Science* 299 (5613), 1753-5, 2003.
111. Martin, D. I., Ward, R., and Suter, C. M., Germline epimutation: A basis for epigenetic disease in humans, *Ann N Y Acad Sci* 1054, 68-77, 2005.
112. Murrell, A., Heeson, S., Cooper, W. N., Douglas, E., Apostolidou, S., Moore, G. E., Maher, E. R., and Reik, W., An association between variants in the IGF2 gene and Beckwith-Wiedemann syndrome: interaction between genotype and epigenotype, *Hum Mol Genet* 13 (2), 247-55, 2004.
113. Flanagan, J. M., Popendikyte, V., Pozdniakovaite, N., Sobolev, M., Assadzadeh, A., Schumacher, A., Zangeneh, M., Lau, L., Virtanen, C., Wang, S. C., and Petronis, A., Intra- and interindividual epigenetic variation in human germ cells, *Am J Hum Genet* 79 (1), 67-84, 2006.
114. Costa, E., Grayson, D. R., and Guidotti, A., Epigenetic downregulation of GABAergic function in schizophrenia: potential for pharmacological intervention?, *Mol Interv* 3 (4), 220-9, 2003.
115. Noh, J. S., Sharma, R. P., Veldic, M., Salvacion, A. A., Jia, X., Chen, Y., Costa, E., Guidotti, A., and Grayson, D. R., DNA methyltransferase 1 regulates reelin mRNA expression in mouse primary cortical cultures, *Proc Natl Acad Sci U S A* 102 (5), 1749-54, 2005.
116. Costa, E., Dong, E., Grayson, D. R., Ruzicka, W. B., Simonini, M. V., Veldic, M., and Guidotti, A., Epigenetic targets in GABAergic neurons to treat schizophrenia, *Adv Pharmacol* 54, 95-117, 2006.

12 Epigenetics and Cardiovascular Disease

Gertrud Lund and Silvio Zaina

CONTENTS

12.1   Introduction 207
12.2   Atherosclerosis: An Overview 208
12.3   DNA Methylation and the Natural History of Atherosclerosis 209
      12.3.1   DNA Hypomethylation Genome Wide and at Specific Genes 209
      12.3.2   DNA Hypermethylation Genome Wide and at Specific Genes 211
      12.3.3   Mechanisms of Aberrant DNA Methylation Patterns 211
            12.3.3.1   The Folate and Homocysteine Connection 212
            12.3.3.2   Control of Chromatin Structure by Lipids 213
            12.3.3.3   DNA Demethylation in Nonreplicative Cells? 214
12.4   DNA Methylation and Stroke 214
12.5   Epigenetics and Risk Factors for CVD 215
      12.5.1   Nutritional Factors and CVD Risk 215
            12.5.1.1   Nutrition in Utero 215
            12.5.1.2   Nutrition in Infancy 216
            12.5.1.3   Transgenerational Effects of Nutrients 216
      12.5.2   Early Non-Nutritional Factors and CVD Risk 217
            12.5.2.1   Aortic Coarctation 217
            12.5.2.2   Tobacco Smoke 218
      12.5.3   Alternatives to Epigenetic Mechanisms: Are There Any? 218
12.6   Future Directions 219
Acknowledgments 220
References 220

12.1   INTRODUCTION

Sensible progress has been made in recent years to understand the epigenetic bases of cardiovascular disease (CVD) and atherosclerosis, the underlying cause of CVD. Yet, the field is still in its infancy in comparison with cancer studies. The description of the epigenome of cell types involved in CVD is a potentially crucial advance because of at least two fundamental characteristics of the disease. First, environmental and nutritional factors constitute an important portion of the complex etiology of CVD. Second, the incidence of risk factors for atherosclerosis such as obesity and diabetes is increasing worldwide. The underlying cause of the latter phenomenon is unlikely to be genetic; rather, it must reflect nongenetic mechanisms of gene expression regulation by environmental and nutritional factors. In principle, epigenetics provides unique conceptual and experimental instruments to understand how these CVD risk factors act at the molecular level to change gene expression patterns. Moreover, epigenetic mechanisms of gene regulation imply a degree of flexibility and reversibility, thus justifying hopes for "epigenomic therapies" to be developed in the future. This chapter reviews evidence supporting the case for a strong epigenetic component in the etiology of atherosclerosis and CVD.

12.2   ATHEROSCLEROSIS: AN OVERVIEW

Atherosclerosis and its complications are a major cause of death and disability in westernized countries. The disease is initiated by infiltration of plasma lipoproteins, some of which are oxidized, into the vascular wall of large arterial vessels, followed by macrophage recruitment and chronic inflammation [1]. Lipoproteins are complex particles containing lipids—triglycerides, cholesterol, phospholipids—and polypeptides (apolipoproteins) (Figure 12.1). Apolipoprotein and lipid content varies qualitatively and quantitatively in different lipoprotein types, with very-low-density lipoproteins (VLDL) being relatively lipid rich, particularly in triglycerides. According to a simplified but useful view of the biological activity of lipoproteins, VLDL and low-density lipoproteins (LDL, the product of VLDL delipidation by lipases) tend to promote atherogenesis, while high-density lipoproteins (HDL) are protective, mainly due to their role as sinks for peripheral tissue cholesterol (reverse cholesterol transport activity) (Figure 12.1). The initial lipoprotein infiltration and macrophage recruitment lead to the formation of a "fatty streak," an early vascular lesion consisting of lipoproteins and lipid-loaded macrophages known as foam cells. Subsequently, a fibrocellular, elevated plaque gradually develops through further recruitment of macrophages and lymphocytes, migration and proliferation of smooth muscle cells (SMC), and synthesis of the abundant extracellular matrix by SMC. The fibrocellular plaque is generally asymptomatic and stable over many years until its structure weakens, causing rupture, thrombosis, and the clinical complications that result in CVD—stroke, coronary heart disease, and peripheral vascular disease. It is believed that the macrophage is a pivotal cell type in events leading to plaque weakening, instability, and rupture [2].

Atherosclerosis is the consequence of diverse and partly overlapping genetic, environmental, nutritional, and lifestyle-related risk factors, some of which are diseases per se—familial hyperlipidemias, age, male gender, smoke, obesity, hypertension, diabetes, and others. Crucially, at least some risk factors for CVD are present only transiently during an individual's lifetime and act as if imposing a "hit" during a specific time window. These hits increase the risk of developing CVD many years later or even transgenerationally. One possible if not the most adequate molecular explanation for these phenomena is that CVD has a strong epigenetic basis (i.e., lifelong stable and inheritable perturbations of the epigenome are among the pivotal underlying causes of CVD). In this chapter we will review experimental evidence suggesting that aberrations in DNA methylation patterns accompany the natural

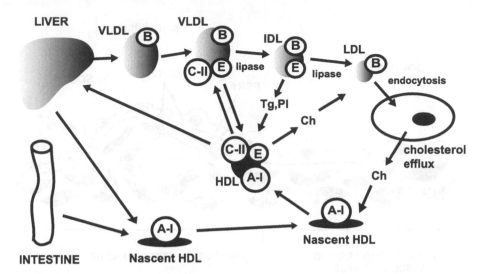

FIGURE 12.1 Schematic view of dynamic processes in lipoprotein metabolism. VLDL particles are produced by the liver as relatively large apolipoprotein B (apoB)-containing and trygliceride-rich particles. Mature VLDL are formed by exchange of apoC-II and apoE with HDL. Lipase activity in the capillary circulation removes triglycerides, which are incorporated in storage or used as cellular fuel. As a consequence of this removal, VLDL is converted into particles of increasing density and decreasing size (intermediate-density lipoprotein, or IDL, and LDL). Excess LDL and other apoB-rich particles are normally removed by endocytosis in peripheral tissue or the liver. During atherogenesis, VLDL and LDL infiltrate the vascular wall and unchain inflammatory processes. Nascent HDL produced by the liver or intestine is converted to mature HDL by incorporating cholesterol (Ch) removed from peripheral cells such as macrophages. Cholesterol is transferred from HDL to apoB-containing particles in exchange for triglycerides and phospholipids (Tg and Pl). A portion of HDL and apoB-containing particles is absorbed by the liver. Circled numbers indicate apolipoproteins (e.g., *B* indicates apoB).

history of CVD and may be critical in initiating the cascade of pathophysiological processes leading to CVD.

12.3 DNA METHYLATION AND THE NATURAL HISTORY OF ATHEROSCLEROSIS

12.3.1 DNA Hypomethylation Genome Wide and at Specific Genes

To our knowledge, the idea of an epigenetic basis for atherosclerosis was first proposed by Newman, who argued that folate, vitamin B6, and B12 deficiency causes atherosclerosis by inducing DNA hypomethylation [3]. The hypothesis is based on the sound argument that folate and vitamin B are crucial regulators of the cellular levels of S-adenosylmethionine (SAM), a universal donor of methyl groups in DNA methylation reactions (see Section 12.3.3.1). The idea is apparently backed by spectacular data obtained in animal models, showing that folate supplementation can control DNA methylation and gene expression even transgenerationally [4].

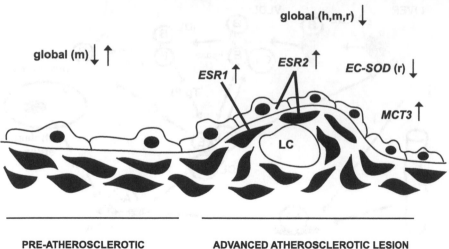

FIGURE 12.2   Synopsis of known changes in DNA methylation globally and at individual genes, in pre-atherosclerotic vessels (*left*) and advanced lesions (*right*), compared to corresponding healthy controls. Lines connect gene names to cell types where epigenetic changes have been demonstrated. Vascular tissue is represented as a monolayer of endothelial cells (white, black nuclei) and underlying SMC (black). The central mass in the lesion is a macrophage-rich lipid core (LC). Upward and downward arrows indicate DNA hyper- and hypomethylation, respectively. h, human; m, mouse; r, rabbit. No species symbol indicates data obtained only in humans.

Nonetheless, subsequent clinical studies addressing the potential anti-inflammatory and anti-atherogenic effects of folate and vitamin B supplementation yielded controversial results, indicating that the causal relationship between vitamin deficiency, atherogenesis, and aberrant DNA methylation is not as straightforward as previously thought [reviewed in Refs. 5-7]. At any rate, the basic idea of Newman's seminal paper that DNA hypomethylation is associated with the natural history of atherosclerosis has been confirmed experimentally by a number of studies. Global DNA hypomethylation has been observed in advanced atherosclerotic lesions in humans and animal models including rabbits and atherosclerosis-prone, hyperlipidemic apolipoprotein E (APOE)-deficient mice [8,9]. A relatively moderate DNA hypomethylation coexisting with DNA hypermethylation was subsequently demonstrated in aortas of young APOE-deficient mice before the appearance of histologically detectable lesions [10] (Figure 12.2).

At gene level, a CpG island in the extracellular superoxide dismutase (*ec-sod*) sequence was the first reported example of hypomethylation in a rabbit model of atherosclerosis [8]. EC-SOD protects vascular tissues against the harmful effects of superoxide (O_2-) and is abundantly produced by lesion SMC and macrophages in accordance with the hypomethylated status of the corresponding gene (Figure 12.2).

12.3.2 DNA HYPERMETHYLATION GENOME WIDE AND AT SPECIFIC GENES

We reported that at the whole-genome level similar amounts of de novo methylation and demethylation coexist in pre-atherosclerotic mouse aortas [10]. In the same study, we showed that a lipoprotein mix enriched in atherogenic lipoproteins (i.e., with a relatively high LDL- and VLDL-to-HDL ratio) caused de novo DNA methylation in human cultured macrophages, compared with a lipoprotein mix containing the same absolute amount of total lipoprotein but at relative proportions observed in normolipidemic conditions (i.e., a relatively low LDL- and VLDL-to-HDL ratio). In a subsequent preliminary study, we showed that the DNA hypermethylation response is induced mainly by VLDL, whereas the atheroprotective HDL is inactive or induces a marginal DNA hypomethylation. These observations indicate that qualitatively different lipoprotein profiles have specific epigenetic effects and that triglyceride-rich particles play a major role in these responses.

Studies at gene level focused on oestrogen receptors (ESR) α and β (*ESR1* and *ESR2* genes, respectively). The interest in ESR stems from the atheroprotective properties of estrogens and from the observation that methylation at the *ESR* genes increases with aging [11]. Two studies by the same group demonstrated hypermethylation of both *ESR1* and *ESR2* genes in human atherosclerotic lesions [12,13]. Hypermethylation of *ESR2* was detected in both cultured vascular SMC and endothelial cells, whereas *ESR1* hypermethylation was apparently restricted to the former cell type. The authors put forward the hypothesis that focal hypermethylation at *ESR* genes contributes to determining the site of atherosclerotic lesion susceptibility. Perhaps contrary to known differential susceptibility to atherosclerosis among venous and arterial venous tissue, *ESR2* was more methylated in the former than the latter tissue [13]. Importantly though, the authors point out that differences in *ESR2* methylation between internal mammary artery and saphenous vein reflect well the differential effectiveness of the two tissues in coronary artery bypass grafting (Figure 12.2).

Another study addressed the methylation pattern of the monocarboxylate transporter 3 (*MCT3*) gene in human atherosclerotic lesions and cultured vascular SMC [14]. MCT3 prevents intracellular acidification of SMC during anaerobic metabolism, by mediating the efflux of lactate. The authors observed a decrease in *MCT3* expression and concomitant hypermethylation of a CpG island located in exon 2 of the gene in atherosclerotic lesions relative to control aortas and in cultured SMC. Noticeably, the degree of DNA hypermethylation was positively correlated with the severity of atherosclerosis. Backed by siRNA-induced knockdown data, the authors conclude that downregulation of MCT3 induces SMC hyperproliferation in vascular lesions. The functional significance of the observed hypermethylation of *MCT3* exon 2 is not clear and awaits further supporting experimental data. Nonetheless, the data are in accordance with studies suggesting a significant role for differential methylation at intragenic sequences in gene expression [15,16] (Figure 12.2).

12.3.3 MECHANISMS OF ABERRANT DNA METHYLATION PATTERNS

Currently, efforts are devoted to understanding how pathological levels of homocysteine (Hcy) may impose aberrant DNA methylation patterns in atherosclerosis. In addition to this more traditional view, recent evidence suggests that lipid and lipo-proteins may have previously unappreciated chromatin-modifying functions.

12.3.3.1 The Folate and Homocysteine Connection

The biosynthesis and bioavailability of the universal methyl group donor SAM are controlled by cellular levels of Hcy and the vitamins folate, B6, and B12, acting as cofactors for crucial enzymes participating in SAM production. Accordingly, deficiency of any of the three mentioned vitamins results in reduced SAM levels [17]. Moreover, excess Hcy causes the accumulation of its own precursor S-adeno-sylhomocysteine (SAH), an inhibitor of SAM binding to methyltransferases [17]. Therefore, both vitamin deficiency and hyperhomocysteinemia have the potential for causing DNA hypomethylation. The involvement of this biochemical pathway in atherosclerosis has been studied intensely. Hyperhomocysteinemia has been long recognized as an independent risk factor for atherosclerosis. It has been hypothe-sized that the main consequence of hyperhomocysteinemia is DNA hypomethylation via blockage of SAM bioavailability by the SAH-mediated mechanisms described above. The observation that hyperhomocysteinemia is generally associated with reduced SAM levels and DNA hypomethylation is consistent with this view [18,19]. Furthermore, mice deficient in methylenetetrahydrofolate reductase, a crucial folate-dependent enzyme in the pathway generating SAM, show DNA hypomethylation and vascular lipid deposits arguably resembling fatty streaks [20]. Nonetheless, vari-ous documented effects of Hcy on cellular homeostasis and proliferation raise the question whether hyperhomocysteinemia causes cellular responses that are indepen-dent of SAH accumulation and DNA methylation [discussed in Refs. 21 and 22]. Hcy may act indirectly on DNA methylation simply by promoting cell hyperprolif-eration and defective maintenance of DNA methylation patterns in advanced ath-erosclerotic lesions [23]. Since proliferating SMC represent a significant portion of elevated fibrocellular lesions, it is not surprising that the measurement of total lesion DNA methylation status yields hypomethylation. Furthermore, elevated Hcy is not an absolute prerequisite for DNA hypomethylation, as the latter is a consistent fea-ture of advanced atherosclerotic lesions in hypohomocysteinemic APOE-null mice. Reported Hcy levels in this mouse model are in the region of 4.6 µM, compared with the 5.8–8.7 µM normal human value range [18,24]. Nonetheless, it cannot be excluded that significant differences in sensitivity to Hcy exist between humans and mice. Whether DNA hypomethylation is a feature restricted to the reported 28% of total atherosclerotic patients that are hyperhomocysteinemic is a relevant question that to our knowledge has not been addressed yet [25]. DNA hypomethylation is not a feature of all cancer types, and atherosclerosis may show a similar degree of vari-ability [26].

Taken together, the available data fail to convincingly support the attractive model that an impairment of the SAM-producing biochemical cycle—via either hyperhomocysteinemia or vitamin deficiency—is a straightforward cause of DNA hypomethylation in atherosclerosis.

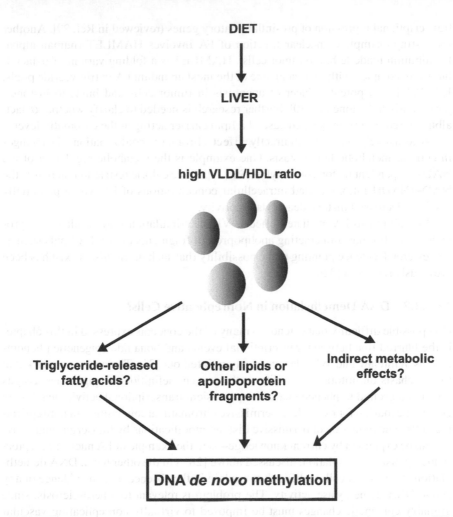

FIGURE 12.3 Possible alternative or concomitant mechanisms mediating aberrant DNA methylation by triglyceride-rich VLDL particles.

12.3.3.2 Control of Chromatin Structure by Lipids

The idea that atherogenic lipids may have direct effects on the epigenome is supported by our observation that hyperlipidemic lipoprotein profiles (i.e., with a relatively high VLDL + LDL-to-HDL ratio) produce a rapid, significant increase in DNA methylation (25%) and histone deacetylation in cultured human macrophages THP-1 cell line compared to normolipidemic lipid profiles, with VLDL being the most active component [10]. The fact that VLDL is a triglyceride-rich lipoprotein and therefore a potential source of fatty acids (FA) offers some clues on mechanisms mediating its epigenetic effects (Figure 12.3). Several FA are ligands of peroxisome proliferator-activated receptors (PPARs), a family of nuclear receptors that regulate many genes involved in lipid metabolism and inflammation [reviewed in Refs. 27 and 28]. PPARs act as transcription factors and have also been shown to mediate transcriptional repression of pro-inflammatory genes (reviewed in Ref. 29]. Another interesting example of nuclear function of FA involves HAMLET (human alpha-lactalbumin made lethal to tumor cells). HAMLET is a folding variant of α-lactal-bumin in complex with oleic acid, one of the most abundant FA in triglyceride pools. HAMLET is a potent inducer of apoptosis in tumor cells and binds to histones, particularly to histone H3 [30]. Further research is needed to clarify whether α-lact-albumin represents an isolated case of a lipid carrier acting at the chromatin level.

Alternatively, FA may indirectly affect chromatin conformation via changes in cellular metabolic homeostasis. One example is the metabolic regulation of the NAD+-dependent histone deacetylase Sir [31]. Since caloric restriction increases the NAD+/NADH ratio, elevated intracellular concentrations of FA would potentially result in decreased histone deacetylase activity.

In addition to FA-mediated effects, VLDL stimulation may result in the pro-duction of chromatin-interacting apolipoprotein fragments or non-FA lipids such as cholesterol. Evidence pointing to the possibility that such mechanisms exist has been previously reviewed [32].

12.3.3.3   DNA Demethylation in Nonreplicative Cells?

One possible difficulty that extends to many of the concepts expressed in this chapter is the blurred line between transcriptional events and bona fide epigenetic phenom-ena, as Whitelaw and Whitelaw correctly pointed out [33]. The authors remark that the emphasis on mitotic heritability in the current definition of epigenetics assigns undefined ground to phenomena such as transient transcriptional activity leaving an epigenetic mark—for example, a permissive chromatin architecture—in nonreplicat-ing cells. The induction of permissive histone modifications by transcriptional activ-ity can be explained by current knowledge—see the example of FA nuclear receptors inducing histone acetylation, discussed above [28]. On the other hand, DNA demeth-ylation in quiescent cells is difficult to explain in the absence of firm evidence of any active DNA demethylase activity. The problem is relevant to atherosclerosis, since arguably epigenetic changes must be imposed to virtually nonreplicating vascular cells by early atherogenic stimuli [34]. Indeed, we detected DNA hypomethylation during early hyperlipidemia in mice in apparently normal aortas, as judged by cell replication markers [10]. The molecular phenomena underlying these responses are currently hard to pinpoint. We anticipate that advances in this particular issue will greatly clarify mechanisms of epigenetic changes in atherosclerosis and in other relevant phenomena such as epigenetic programming in the brain.

12.4   DNA METHYLATION AND STROKE

The involvement of DNA methylation in stroke, a major complication of atheroscle-rosis, has been investigated in a mouse model of cerebral ischemia. In an initial study, DNA hypermethylation was observed in mild focal ischemia, and mice with reduced DNA methyltransferase (DNMT) activity were resistant to mild ischemic damage [35]. None of these features was observed in a severe ischemia model. The same group later refined their findings by examining the effects of conditional genetic inactivation of DNMT type 1 (DNMT1) in postnatal brain. As suggested by their previous study, reduced DNMT1 levels were protective from ischemic injury, yet complete absence of DNMT1 was not [36]. These studies are important because they identify de novo DNA methylation as a critical initial step in mild ischemic injury of the brain and demonstrate that a basal pre-injury DNA methylation at unknown loci is essential for protection. Further investigations, particularly epigenome description, will further understanding of the complex contribution of epigenetic changes to stroke.

12.5 EPIGENETICS AND RISK FACTORS FOR CVD

The hypothesis that the genome interacts with environmental factors in early life to program adult development was proposed in pioneer work by Dörner in the 1970s [37]. The fundamental framework of this idea has been applied to aspects of the epidemiology of CVD and other metabolic diseases—obesity, diabetes, metabolic syndrome [38,39]—and lends support to the idea of a strong epigenetic component of these diseases. For example, transiently acting risk factors for CVD show delayed effects on the manifestation of clinical symptoms of the disease, often decades after the apparent initial hit. This "remote control" of CVD risk may operate within one individual's lifetime or even transgenerationally. Many of the relevant studies describe complex phenomena in which maternal effects, adult nutrition, and lifestyle are not clearly distinguishable. That is to say that high adult CVD risk may be the result of continued, lifelong exposure to hyperlipidemia or other metabolic disorders that would cumulate to similar factors present in early life. In these cases, disease predisposition is probably the sum of different concomitant, lifelong acting biochemical, genetic, and epigenetic components. For example, it is clear that both childhood and adult obesity, but not the former alone, are determinant for high CVD risk [40]. On the other hand, there are examples of genuinely transient exposures to risk factors in early life that result in adult CVD, likely by imposing stable epigenetic hits with longlasting effects throughout an individual's lifetime or transgenerationally [reviewed in Ref. 41]. We propose that these are useful simplified models in which the epigenetic contribution to CVD may be addressed by molecular studies. In the following paragraphs we review selected examples and discuss them in the context of epigenetics.

12.5.1 NUTRITIONAL FACTORS AND CVD RISK

An increasing amount of evidence implicates early nutrition as an important determinant of CVD risk in adulthood [reviewed in Refs. 42 and 43]. Given the possible links between lipids and epigenetic alterations discussed above, the role of dietary lipids is particularly important in this context.

12.5.1.1 Nutrition In Utero

In human females, low energy intake during pregnancy is associated with more aggressive early atherosclerosis in the offspring [44]. This important study proves the idea that maternal dietary factors in utero control predisposition to atherosclerosis transgenerationally. Furthermore, children whose mothers were hypercholesterolemic during pregnancy have a higher atherosclerosis burden than children of normocholesterolemic women [45]. In this case, fetal exposure to high cholesterol is excluded due to the placental barrier. Nonetheless, lipid peroxidation products that can diffuse through the placenta increase in the fetus proportionally to maternal hypercholesterolemia. Interestingly, animal models show that lipid peroxidation products impose aberrant gene expression patterns in the vascular tissue that persist long after fetal exposure [41]. Gale and coauthors propose that exposure to maternal low energy intake in utero produces long-term dyslipidemia in the offspring [44]. The idea is backed by animal studies demonstrating hypercholesterolemia and endothelial dysfunction if maternal low energy intake is experimentally reproduced in rats [46-48].

Low birth weight (LBW) is another key player in CVD risk. LBW is inversely correlated with risk of adult CVD when concomitant with low weight in early childhood and subsequent fast weight gain. LBW is a surrogate measure of complex metabolic defects, yet small babies have vascular anomalies already at birth, suggesting that molecular defects predisposing to adult CVD are present from very early phases of this condition [49].

12.5.1.2   Nutrition in Infancy

A number of large population-based studies demonstrated a protective effect of breastfeeding for classical CVD risk factors such as blood pressure, insulin resistance, and obesity [reviewed in Ref. 50]. Given the reported effects of atherogenic lipoprotein profiles on DNA methylation patterns [10], it is interesting to speculate that differences in atherogenic lipoprotein levels between breastfed and formula-fed infants produces distinct epigenetic patterns during an individual's lifetime. For example, VLDL levels were shown to transiently peak and decrease to newborn values between 7 and 30 days in breastfed children, whereas infants receiving formula milk had prolonged, significantly higher VLDL and lower LDL levels [51]. Furthermore, a follow-up study of 216 children 13 to 16 years old who had been breastfed as infants showed a 14% reduction in the LDL:HDL ratio and significantly lower CRP (a marker of inflammation) compared to the formula-fed group [52]. These stable changes are probably independent of later variables, since among 405 participants in a 65-year follow-up study, breastfeeding was inversely associated with intima-media thickness and carotid and femoral plaque prevalence, even after controlling for socioeconomic and behavioral factors [53].

12.5.1.3   Transgenerational Effects of Nutrients

If the epigenetic basis of delayed effects of CVD risk factors during one individual's lifetime can reasonably be questioned, that is certainly more difficult for transgenerational transmission of CVD risk, particularly when such a transmission occurs along the male line; that is, independently of potentially confusing in utero effects (e.g., transplacental, physical constraint-related, etc.). The phenomenon has been clearly described in a Swedish study addressing the transgenerational effects of food availability during prepubertal slow growth period (SGP) in an isolated rural community in that country. The results showed a number of interesting effects along both the female and male lines. From an epigenetic point of view, the striking results were the ones relative to male line transmission. Food shortage in paternal SGP resulted in reduced CVD mortality. Moreover, abundance of food in paternal grandfather's SGP was associated with a high risk of diabetes mortality [54]. SGP is associated with a dynamic epigenetic state in the germline due to chromatin reprogramming events in emerging pools of spermatocytes [reviewed in Ref. 55]. Therefore, it has been proposed that nutritional factors could alter these reprogramming events, thus imposing aberrant epigenetic patterns [56].

A recent unpublished study provides another example of paternal transgenerational transmission of high risk of cardiovascular death and suggests potential clues on mechanisms underlying these curious phenomena. The study analyzes the metabolic profile of a cohort of Mexican normal- and low-birth weight children and their families. The data indicate that elevated paternal LDL is strongly associated with low birth weight (G. Barbosa Sabanero, personal communication, 2007), thus linking a paternal metabolic abnormality to a childhood risk factor for adult CVD. This result is important because it reinforces the idea that transgenerational transmission of CVD risk is indeed real and relevant. Furthermore, it offers the simple hypothesis that an atherogenic lipoprotein causes epigenetic, inheritable "hits" in the male germline. Given that LDL and VLDL cause aberrant DNA methylation patterns in cultured human macrophages, it is conceivable that plasma LDL would produce similar effects in the gonadal tissue [10]. Experimental work is ongoing to verify this fascinating idea.

In the Swedish studies described above, the documented effects along the female line are likely due to a number of factors, ranging from in utero nutrient availability to "pure" epigenetic components. The latter included effects of food availability during the grandmother's oocyte pool maturation phase, thus resembling the discussed possible effects in the male germline [54]. An interesting parallel to these observations is the positive association between mother's leg length, but not trunk length, and offspring's birth weight, suggesting a transgenerational effect of nutritional factors during the mother's childhood [57]. Another case of transgenerational transmission of CVD risk along the female line is a Dutch study showing that adults exposed to famine during early gestation had a more atherogenic lipid profile than those whose mothers had not faced calorie restrictions [58].

At least one documented example of transgenerational transmission of epigenetic patterns in humans exists. It provides experimental bases to the largely descriptive phenomena described above. The study describes transgenerational paternal transmission of a functional neocentromere on chromosome 4. This has to be considered a case of meiotic transmission of a chromatin state, since no underlying genetic change could be demonstrated [59].

12.5.2 EARLY NON-NUTRITIONAL FACTORS AND CVD RISK
12.5.2.1 Aortic Coarctation

Aortic coarctation is a narrowing of the aorta that can be successfully repaired by surgery. Occurring in children, this vascular structural anomaly is associated with

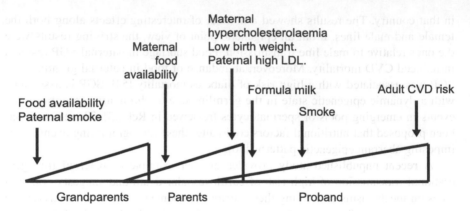

FIGURE 12.4 Schematic view of selected candidate factors imposing epigenetic "hits" in utero, during childhood or transgenerationally, that result in increased CVD risk in adulthood. Overlapping triangles represent successive generations; aging proceeds from left to right.

an increased risk of CVD. Importantly, successful surgical repair does not diminish CVD risk. Moreover, surgery survivors have local persistent functional defects in the endothelium and SMC proximal to the coarctation site and a thickening of the aortic intimal layer, a predictor of CVD [60,61]. Whether such functional defects are cause or consequence of coarctation, the data strongly suggest the occurrence of stable aberrant gene expression patterns that are associated with coarctation.

12.5.2.2 Tobacco Smoke

Smoke exposure is another example of a transient factor that imposes stable long-term alterations. Active and environmental (passive) smoking in the young are risk factors for atherosclerosis [62]. Carotid artery intimal thickness was shown to progress at the same rate in past and current smokers [63]. The same study showed that intimal thickening progression caused by exposure to environmental smoking was only partially reversible. Strikingly, transgenerational effects of early paternal smoking on obesity in sons, but not in daughters, were observed [64]. As pointed out by Pembrey [56], it is difficult to explain these phenomena with mechanisms other than epigenetic.

A synopsis of selected risk factors performing a "remote control" of adult CVD risk is presented in Figure 12.4.

12.5.3 ALTERNATIVES TO EPIGENETIC MECHANISMS: ARE THERE ANY?

Many of the phenomena described in this chapter are examples of programming, a biological phenomenon that has been well documented in experimental models. The question is whether the underlying mechanisms are epigenetic; in other words, whether stable epigenetic hits are produced in early life by nutritional factors or other stimuli, which increase CVD risk in adult life due to concomitant additional hits or accumulation of epigenetic aberrations above a critical threshold. In the absence of hard data (i.e., description of the epigenome of involved tissues in proper clinical studies), one can only reason on how defendable are alternative mechanisms to epigenetics. Hypotheses not involving the epigenome have to assume that somatic mutations play a major role, an idea that is difficult to reconcile with currently accepted somatic mutation rates [65]. Alternatively, it is possible that atherogenic factors impose purely "biochemical," gene expression-independent modifications; that is, either stable modifications of long-lived cellular structures or stable biochemical loops. One if not the only example of the former is provided by studies addressing the involvement of collagen in diabetes. One prominent feature of diabetes is nonenzymatic reactions between glucose and proteins that result in the production of advanced glycation endproducts (AGE). Collagen glycation has been implicated in major complication of diabetes such as chronic inflammation and CVD. Since collagen has an extrapolated half-life of up to 117 years in humans, it has been proposed that glycation of extracellular matrix proteins may represent a "metabolic memory" that mediates long-term, irreversible effects of diabetes, extending beyond any eventual regression of conditions favoring the formation of AGE [66]. This is one remarkable example of an entirely biochemical, genome- or epigenome-independent mechanism to explain delayed effects for a CVD risk "hit." It is not currently clear whether such mechanisms are of general significance in the pathophysiology of CVD or other diseases.

Other possible scenarios assume that atherogenic stimuli trigger abnormal, self-sustaining, stable endocrine or paracrine loops. For example, it has been suggested that a positive feedback loop involving angiotensin II and locally produced inflammatory mediators explains self-sustaining vascular inflammation during atherosclerosis [67]. Even if self-sustaining signaling loops imposed by a transient stimulus were proven to underlie a large number of pathophysiological phenomena, they would inevitably rely on changes in gene expression patterns induced one way or another by a new epigenetic state (i.e., an aberrant distribution of permissive and nonpermissive chromatin architecture).

In summary, in the absence of available alternative mechanisms, it is likely that epigenetics will prove to be a crucial conceptual and technical tool to understand the natural history of and predisposing factors to CVD.

12.6   FUTURE DIRECTIONS

The necessity to interpret the human epigenome has been clearly advocated, and efforts to complete the task are starting to produce results that are promising for understanding CVD, among other diseases [15,16,68]. A particularly challenging task unique to atherosclerosis will be to define the epigenome of every cell type participating in the vascular lesion—at least endothelial cells, SMC, macrophages, and immune cells—and at different stages of the disease. A further task will be to incorporate these technologies in large clinical studies addressing the level of adult CVD risk associated with a specific epigenome.

In parallel, technologies able to modify the epigenome in a controlled way, if implemented, are likely to revolutionize prevention and therapy of CVD. A related issue that must be resolved concerns epigenetic variability and the stability of epigenetic modifications. It is conceivable that epigenetic changes imposed by environmental and nutritional factors have an intrinsic degree of stability and reversibility that may change depending on the duration and intensity of the provoking stimulus. Long-term, stable effects of the discussed CVD risk factors may take place only if a certain threshold of epigenetic change is exceeded—for example, if cytosines in a regulatory sequence are de novo methylated above a certain density. Also, epigenetic variability is likely to explain interindividual differences in CVD outcome. We are testing these ideas in a mouse model recently created in our group, with chronic, progressive de novo methylation of CpG islands.

Epigenetics changed forever the way scientists look at tumorigenesis and cancer therapy. A similar cultural change may well take place in cardiovascular sciences in the near future, thus opening a new era in prevention and therapy of CVD.

ACKNOWLEDGMENTS

This work was supported by the Science and Technology Council of the State of Guanajuato (CONCYTEG) and the University of Guanajuato.

REFERENCES

1. Libby, P., Fat fuels the flame: triglyceride-rich lipoproteins and arterial inflammation, *Circ. Res.*, 100, 299, 2007.
2. Shah, P.K., et al., Human monocyte-derived macrophages induce collagen breakdown in fibrous caps of atherosclerotic plaques. Potential role of matrix-degrading metalloproteinases and implications for plaque rupture, *Circulation*, 92, 1565, 1995.
3. Newman, P.E., Can reduced folic acid and vitamin B12 cause deficient DNA methylation producing mutations which initiate atherosclerosis?, *Med. Hypotheses*, 53, 421, 1999.
4. Waterland, R.A., and Jirtle, R.L., Transposable elements: targets for early nutritional effects on epigenetic gene regulation, *Mol. Cell. Biol.*, 23, 5293, 2003.
5. Carlsson, C.M., Homocysteine lowering with folic acid and vitamin B supplements: effects on cardiovascular disease in older adults, *Drugs Aging*, 23, 491, 2006.
6. Mangoni, A.A., Folic acid, inflammation, and atherosclerosis: false hopes or the need for better trials?, *Clin. Chim. Acta*, 367, 11, 2006.
7. Davis, N., Katz, S., and Wylie-Rosett, J., The effect of diet on endothelial function, *Cardiol. Rev.*, 15, 62, 2007.
8. Laukkanen, M.O., et al., Local hypomethylation in atherosclerosis found in rabbit ecsod gene, *Arterioscler. Thromb. Vasc. Biol.*, 19, 2171, 1999.
9. Hiltunen, M.O., et al., DNA hypomethylation and methyltransferase expression in atherosclerotic lesions, *Vasc. Med.*, 7, 5, 2002.
10. Lund, G., et al., DNA methylation polymorphisms precede any histological sign of atherosclerosis in mice lacking apolipoprotein E., *J. Biol. Chem.*, 279, 29147, 2004.
11. Issa, J.P., Aging, DNA methylation and cancer, *Crit. Rev. Oncol. Hematol.*, 32, 31, 1999.
12. Post, W.S., et al., Methylation of the estrogen receptor gene is associated with aging and atherosclerosis in the cardiovascular system, *Cardiovasc. Res.*, 43, 985, 1999.
13. Kim, J., et al., Epigenetic changes in estrogen receptor beta gene in atherosclerotic cardiovascular tissues and in-vitro vascular senescence, *Biochim. Biophys. Acta*, 1772, 72, 2007.
14. Zhu, S., Goldschmidt-Clermont, P.J., and Dong, C., Inactivation of monocarboxylate transporter MCT3 by DNA methylation in atherosclerosis, *Circulation*, 112, 1353, 2005.

15. Weber, M., et al., Chromosome-wide and promoter-specific analyses identify sites of differential DNA methylation in normal and transformed human cells, *Nat. Genet.*, 37, 853, 2005.

16. Eckhardt, F., et al., DNA methylation profiling of human chromosomes 6, 20 and 22, *Nat. Genet.*, 38, 1378, 2006.

17. Castro. R., et al., Homocysteine metabolism, hyperhomocysteinaemia and vascular disease: an overview, *J. Inherit. Metab. Dis.*, 29, 3, 2006.

18. Yi, P., et al., Increase in plasma homocysteine associated with parallel increases in plasma S-adenosylhomocysteine and lymphocyte DNA hypomethylation, *J. Biol. Chem.*, 275, 29318, 2000.

19. Ingrosso, D., et al., Folate treatment and unbalanced methylation and changes of allelic expression induced by hyperhomocysteinaemia in patients with uraemia, *Lancet*, 361, 1693, 2003.

20. Chen, Z., et al., Mice deficient in methylenetetrahydrofolate reductase exhibit hyper-homocysteinemia and decreased methylation capacity, with neuropathology and aortic lipid deposition, *Hum. Mol. Genet.*, 10, 433, 2001.

21. Zaina, S., Lindholm, M.W., and Lund, G., Nutrition and aberrant DNA methylation patterns in atherosclerosis: more than just hyperhomocysteinemia?, *J. Nutr.*, 135, 5, 2005.

22. Ulrey, C.L., et al., The impact of metabolism on DNA methylation, *Hum. Mol. Genet.*, 14, R139, 2005.

23. Kimura, F., et al., Decrease of DNA methyltransferase 1 expression relative to cell proliferation in transitional cell carcinoma, *Int. J. Cancer*, 104, 568, 2003.

24. Troen, A.M., et al., The atherogenic effect of excess methionine intake, *Proc. Natl. Acad. Sci. USA*, 100, 15089, 2003.

25. Wilcken, D.E., Wilcken, B., The pathogenesis of coronary artery disease. A possible role for methionine metabolism, *J.Clin. Invest.*, 57, 1079, 1976.

26. Wilson, A.S., Power, B.E., and Molloy, P.L., DNA hypomethylation and human diseases, *Biochim. Biophys. Acta,* 1775, 138, 2007.

27. Hwang, D., and Rhee, S.H., Receptor-mediated signaling pathways: potential targets of modulation by dietary fatty acids, *Am. J. Clin. Nutr.*, 70, 545, 1999.

28. Pegorier, J.P., Le May, C., and Girard, J., Control of gene expression by fatty acids, *J Nutr.*, 134, 2444S, 2004.

29. Chinetti, G., Fruchart, J.-C., and Staels, B., Peroxisome proliferator-activated receptors (PPARs): Nuclear receptors at the crossroads between lipid metabolism and inflammation, *Inflamm. Res.*, 49, 497, 2000.

30. Duringer, C., et al., HAMLET interacts with histones and chromatin in tumor cell nuclei, *J. Biol. Chem.*, 278, 42131, 2003.

31. Gasser, S.M., and Cockell, M.M., The molecular biology of the SIR proteins, *Gene*, 279, 1, 2001.

32. Zaina, S., et al., Chromatin modification by lipids and lipoprotein components: an initiating event in atherogenesis?, *Curr. Opin. Lipidol.*, 16, 549, 2005.

33. Whitelaw, N.C., and Whitelaw, E., How lifetimes shape epigenotype within and across generations, *Hum. Mol. Genet.*, 15 Spec No 2, R131, 2006.

34. Schwartz, S.M., and Benditt, E.P., Cell replication in the aortic endothelium: a new method for study of the problem, *Lab. Invest.*, 28, 699, 1973.

35. Endres, M., et al., DNA methyltransferase contributes to delayed ischemic brain injury, *J. Neurosci.*, 20, 3175, 2000.

36. Endres, M., et al., Effects of cerebral ischemia in mice lacking DNA methyltransferase 1 in post-mitotic neurons, *Neuroreport*, 12, 3763, 2001.

37. Dörner, G., Perinatal hormone levels and brain organization, in *Anatomical neuroendocrinology*, Stumpf, W.E., and Grant, L.D., Eds., Karger, Basel, 245, 1975.

38. Barker, D.J.P., et al., Fetal nutrition and cardiovascular disease in adult life, *Lancet*, 341, 938, 1993.

39. Lucas, A., Programming by early nutrition in man, in *The childhood environment and adult disease (CIBA Foundation symposium 156)*, Bock, G.R., and Whelan, J., Eds., Whiley, Chichester, UK, 38, 1991.

40. Berenson, G.S., Obesity--a critical issue in preventive cardiology: the Bogalusa Heart Study, *Prev. Cardiol*, 8, 234, 2005.

41. Celermajer, D.S., and Ayer, J.G., Childhood risk factors for adult cardiovascular disease and primary prevention in childhood, *Heart*, 92, 1701, 2006.

42. Gluckman, P.D., and Hanson, M.A., Adult disease: echoes of the past. *Eur. J. Endocrinol.*, 155 Suppl 1, S47, 2006.

43. Gluckman, P.D., Hanson, M.A., and Beedle, A.S., Early life events and their consequences for later disease: a life history and evolutionary perspective, *Am. J. Hum. Biol.*,19, 1, 2007.

44. Gale, C.R., et al., Maternal Diet During Pregnancy and Carotid Intima-Media Thickness in Children, *Arterioscler. Thromb. Vasc. Biol.*, 26, 1877, 2006.

45. Napoli, C., et al., Influence of maternal hypercholesterolaemia during pregnancy on progression of early atherosclerotic lesions in childhood: Fate of Early Lesions in Children (FELIC) study, *Lancet*, 354, 1234, 1999.

46. Szitanyi, P., Hanzlova, J., and Poledne, R., Influence of intrauterine undernutrition on the development of hypercholsterolemia in an animal model, *Physiol. Res.*, 49, 721, 2000.

47. Khan, I.Y., et al., A high-fat diet during rat pregnancy or suckling induces cardiovascular dysfunction in adult offspring, *Am. J. Physiol. Regul. Integr. Comp. Physiol.*, 288, R127, 2005.

48. Koukkou, E., et al., Offspring of normal and diabetic rats fed saturated fat in pregnancy demonstrate vascular dysfunction, *Circulation*, 98, 2899, 1998.

49. Skilton. M.R., et al., Aortic wall thickness in newborns with intrauterine growth restriction, *Lancet*, 365, 1484, 2005.

50. Singhal, A., Early nutrition and long-term cardiovascular health, *Nutr. Rev.*, 64, S44, 2006.

51. Van Biervliet, J.P., Rosseneu, M., and Caster, H., Influence of dietary factors on the plasma lipoprotein composition and content in neonates, *Eur. J. Pediatr.*, 144, 489, 1986.

52. Singhal, A., et al., Breastmilk feeding and lipoprotein profile in adolescents born preterm: follow-up of a prospective randomised study, *Lancet*, 363, 1571, 2004.

53. Martin, R.M., et al., Breastfeeding and atherosclerosis: intima-media thickness and plaque at 65-year follow-up of the Boyd Orr Cohort, *Arterioscl. Thromb. Vasc. Biol.*, 25, 1482, 2005.

54. Kaati, G., Bygren, L.O., and Edvinsson, S., Cardiovascular and diabetes mortality determined by nutrition during parents' and grandparents' slow growth period, *Eur. J. Hum. Genet.*, 10, 682, 2002.

55. Gallou-Kabani, C., and Junien, C., Nutritional epigenomics of metabolic syndrome: new perspective against the epidemic, *Diabetes*, 54, 1899, 2005.

56. Pembrey, M.E., Time to take epigenetic inheritance seriously, *Eur. J. Hum. Genet.*, 10, 669, 2002.

57. Martin, R.M., et al., Parents' growth in childhood and the birth weight of their offspring, *Epidemiology*, 15, 308, 2004.

58. Roseboom, T.J., et al., Plasma lipid profiles in adults after prenatal exposure to the Dutch famine, *Am. J. Clin. Nutr.*, 72, 1101, 2000.

59. Amor, D.J., et al., Human centromere repositioning "in progress," *Proc. Natl. Acad. Sci. USA*, 101, 6542, 2004.

60. Cohen, M., et al., Coarctation of the aorta. Long-term follow-up and prediction of outcome after surgical correction, *Circulation*, 80, 840, 1989.

61. O'Leary, D.H., et al., Carotid-artery intima and media thickness as a risk factor for myocardial infarction and stroke in older adults. Cardiovascular Health Study Collaborative Research Group, *N. Engl. J. Med.*, 340, 14, 1999.
62. Berenson, G.S., et al., Association between multiple cardiovascular risk factors and atherosclerosis in children and young adults. The Bogalusa Heart Study, *N. Engl. J. Med.*, 338, 1650, 1998.
63. Howard, G., et al., Cigarette smoking and progression of atherosclerosis: The Atherosclerosis Risk in Communities (ARIC) Study, *JAMA*, 279, 119, 1998.
64. Pembrey, M.E., et al., Sex-specific, male-line transgenerational responses in humans, *Eur. J. Hum. Genet.*, 14, 159, 2006.
65. Hinds, D.A., et al., Whole-genome patterns of common DNA variation in three human populations, *Science*, 307, 1072, 2005.
66. Genuth, S., et al., DCCT Skin Collagen Ancillary Study Group. Glycation and carboxymethyllysine levels in skin collagen predict the risk of future 10-year progression of diabetic retinopathy and nephropathy in the diabetes control and complications trial and epidemiology of diabetes interventions and complications participants with type 1 diabetes, *Diabetes*, 54, 3103, 2005.
67. Brasier, A.R., Recinos, A. 3rd, and Eledrisi, M.S., Vascular inflammation and the renin-angiotensin system, *Arterioscler. Thromb. Vasc. Biol.*, 22, 1257, 2002.
68. Esteller, M., The necessity of a human epigenome project, *Carcinogenesis*, 27, 1121, 2006.

13 Plant Epigenetics

*Mónica Meijón, Luis Valledor, Jose Luis Rodríguez,
Rodrigo Hasbún, Estrella Santamaría,
Isabel Feito, Maria Jesús Cañal, Maria Berdasco,
Mario F. Fraga, and Roberto Rodríguez*

CONTENTS

13.1 Introduction: DNA Methylation in Plants 225
13.2 Implication of DNA Methylation in Developmental Processes 227
    13.2.1 Aging 227
    13.2.2 Flowering 229
    13.2.3 Stress and Transposons 231
13.3 Environment-Induced Epigenetic Variations 232
13.4 Epigenetic Memory 233
13.5 Conclusions 234
Acknowledgments 234
References 235

13.1 INTRODUCTION: DNA METHYLATION IN PLANTS

DNA methylation, one of the most characterized epigenetic modifications in eukaryotic organisms, plays a fundamental role in gene regulation during development (Finnegan et al., 2000; Meehan, 2003). In animals, specific DNA methylation patterns are essential for embryo development, with hypomethylation related to spontaneous abortions in mice (Li et al., 1992). DNA methylation in animals is also a determinant factor implicated in imprinting phenomena (Razin and Cegar, 1994), X chromosome inactivation (Surani, 2001), and silencing of exogenous agents, such as transposons, retrotransposons, or retroviruses (Yoder et al., 1997). In plants, it has been ascribed an important role in at least two different cellular processes. The first role of methylation refers to the control of the expression of single-copy genes, such as the FWA transcription factor, the BALL pathogen-resistance gene (BAL), and the phosphoribosylanthranilate isomerase (PAI) family of tryptophan biosynthesis genes (reviewed in Chan et al., 2005). Although it is often said that "methylation blocks gene expression," this is an oversimplification. Methylation changes the interactions between proteins and DNA, which leads to alterations in chromatin structure and either a decrease or an increase in the transcription rate (Jones and Takai, 2001). Furthermore, methylation of cytosines is a mechanism that serves as a host genome defense system. It has been proposed that methylation of cytosine residues is involved in suppressing transpositions (Miura et al., 2001, Okamoto and Hirochika, 2001), posttranscriptional gene silencing (PTGS) of transgene sequences (Mallory et al., 2001), and regulation of gene silencing that occurs in allotetraploids (Shaked et al., 2001, Madlung et al., 2002). Plant DNA methylation is also involved in imprinting phenomena, and there is also evidence that methylation is associated with silencing of imprinted alleles in some late-acting genes in the endosperm of various species (Vinkenoog et al., 2000). In plants, DNA methylation is also implicated in the correct transition between developmental phases, but, in contrast with animal behavior, demethylation does not result in lethal consequences. Three recent pioneering reports (Zhang et al., 2006; Vaughn et al., 2007; Zilberman et al., 2007) defining the DNA methylome of *Arabidopsis thaliana* have reinforced the significance of the two aforementioned tasks of DNA methylation as an epigenetic marker in this model.

It is interesting to note that plants have several peculiarities in their methylation machinery with respect to animals and, perhaps, the most significant is the presence of an additional *methylation code*. Methylation in animal genomic DNA occurs predominantly at the cytosine residues of sequences such as dinucleotide cytosine–guanine (CpG). In plants and filamentous fungi, DNA methylation is more common in the trinucleotide cytosine–any nucleotide–guanine (CpNpG), although it can also be detected in CpG sequences (Gruenbaum et al., 1981). Plants have three classes of DNA methyltransferases (DNMTs) that differ in protein structure and function and with different preference on cytosines in CpG and/or CpNpG contexts: (1) the METI family, which acts as the mammalian homolog of maintenance Dnmtl, (2) the chromomethylases (CMTs), which may preferentially methylate cytosines in CpNpG sequences, and (3) the remaining class, with similarity to Dnmt3 methyltransferases of mammals, putative de novo methyltransferases. Additional proteins, for example, DDM1 (Decrease of DNA Methylation 1), a member of the SNF2/SWI2 family of chromatin remodeling proteins, are also required for methylation of plant DNA.

From a biochemical point of view, differential behavior also could be found within the mammalian and plant DNMT proteins. For example, the amino-terminal domain of plant MET1 lacks the zinc-binding domain of mammalian Dnmtl (Bestor, 1992). In contrast, plants possess an acidic region in the N-terminal domain composed of at least 50% of glutamic and aspartic acid. Although the role of this region is not yet well defined, the fact that it remains preserved in all MET1 enzymes suggests that it could adopt a biological significance specific for plants (Finnegan and Kovac, 2000). In addition, plants showed a deletion of approximately 40 to 41 amino acid residues within the target recognition domain with respect to mammals (Genger et al., 1999). These domains are much less conserved among plants, suggesting that there are differences of preference for the substrate or that this region does not confer substrate specificity (Finnegan and Kovac, 2000). In addition, more DNMTs have been identified in plants than animals (Jeltsch, 2002).

The abundance of DNMTs in plants, the presence of two different codes of methylation (CpG and CpNpG contexts), and the existence of the family of CMT enzymes, among other peculiarities, may be related to the fact that plants require specific interaction between developmental programs and signaling pathways from external stimuli that must be coordinated at the level of chromatin organization. Plants are sessile organisms that require a fine and reversible adaptation of their genetic program to the microenvironment of each habitat. Because these adjustments must be reversible, the epigenetic mechanisms constitute an effective regulation system that could adapt gene expression to specific environmental situations.

Cell differentiation, development, and aging are all processes controlled through the temporal and spatial activation and silencing of specific genes. The question arises as to what turns the genes on and off. In this chapter, several aspects related to the involvement of DNA methylation in the aforementioned developmental processes are discussed.

13.2 IMPLICATION OF DNA METHYLATION IN DEVELOPMENTAL PROCESSES

13.2.1 Aging

Aging implies continuous changes in metabolite translocation and transcriptional activity associated with alteration of metabolic routes and repair activity of cells. In some eukaryotic organisms, especially woody species, aging and maturation induce reduction of cellular competence, limiting the possibility of domestication. Tools can help identify high-quality genotypes at the embryonal or juvenile stages, but even these are not very efficient, and cloning of adult trees, when they have expressed their potential, is difficult or impossible.

In animals, among other processes, aging involves epigenetics changes, like promoter hypermethylation of some CpG islands associated with gene silencing and whole DNA hypomethylation of CpGs dispersed throughout repetitive sequences as well as of transcriptionally relevant regions of some genes. DNA hypomethylation promotes genomic instability, amplification of oncogenes, and also silencing of the genes through an RNAi mechanism (Burzynski, 2005; Fraga et al., 2005). Despite the well-known differences between animals and plants, certain similarities such as oxidative stress damage and telomere shortening are present among organisms. However, the knowledge about epigenetic events concomitant with plants' aging is in infancy, maybe because their mode of development is more complex. While the outcome of animal embryogenesis is a mini-edition of the adult animal, with all organs being at least initiated, plant embryogenesis results in a bipolar structure. All other organs of the mature plant are formed postembryonically. These distinctive developmental strategies of plants and animals concur with different tasks of their stem cells. Whereas the major task of an animal stem cell is to replenish highly specialized body cells with a limited lifespan, plant stem cells (apical cells of meristems) provide the material for the continued formation of new organs or in same cases organisms (Bäurle and Laux, 2003).

At whole-methylation level, along with age-specific levels, there are species-, tissue-, and organ-specific levels in both plants and animals. In the former, the age-dependent demethylation of whole DNA is evident, and some investigators are inclined even to consider degree of DNA methylation as a sort of biological clock that measures the age and forecasts lifespan. Distortions in DNA methylation may lead to premature aging (Wilson et al., 1987; Richardson, 2003; Vanyushin, 2005). Until the last decade, little was known about whole-DNA methylation in plants' aging and development and information was not conclusive (Burn et al., 1993; Diaz-Sala et al., 1995; Kakutani, 1997; Lambé et al., 1997; Finnegan et al., 1998). In general, animals have their major epigenetic switches during early developmental stages. Plant cells still have the potency to undergo major changes at later stages, including not only transition from vegetative to floral organs but also root formation from stem and leaves or shoot formation from differentiated organs like roots (Fransz and de Jong, 2002).

Since 1989, research efforts of the group EPIPHYSAGE (www.uniovi.es/epiphysage/) have been focused on: (a) woody plant morphogenesis manipulation (*Corylus avellana, Olea europaea, Juglans regia, Pinus radiata, Eucalyptus globulus, Castanea sativa*) and (b) the definition of the molecular bases of differentiation–redifferentation processes. Results obtained allow validation of two molecular markers of aging, maturation, and phase change referred to endogenous polyamine contents and whole-DNA methylation level.

Fraga et al. (2002a,b) found differences in the extent of DNA methylation between meristematic areas of juvenile and mature *Pinus radiata* D. Don. trees, whereas differences between differentiated tissues were small. Another species of interest is *Castanea sativa,* both the juvenile and mature phases of which may occur on the same mature tree. The upper parts of a tree exhibiting determinate growth are chronologically younger and often exhibit mature characteristics, whereas the lower, physiologically older parts may retain juvenile attributes. In this system, ontogenic development involves an increase of whole-DNA methylation in shoot apex during active growth, but the differences are reset with dormancy (Hasbún et al., 2007).

The aging of plants is related to the changes in cell and organ interactions, rather than to the changes of the meristematic cells themselves. The capacity for self-maintenance is considered one of the common features of stem cells. Animal stem cells are preserved in most adult tissues, and stem cells in plants are preserved in the apex during long-term growth of the axial organs. However, preservation of stem cells in plants is determined by the organization of cell interactions in the meristem. There is not population of stem cells in plants, which would be preserved during the entire life. Stem cells could exist during one or more growth periods of a given axis (shoot or root). When the apex growth is suppressed, a new apex arises from the descendant's cells. These were originated from the apical stem cells, maintained quiescent, and activated by suppression of apical dominance. The change of stem cells often takes place in the site of axis growth, when the outwardly monopodial growth is actually hidden sympodial. The change of axes is related to rejuvenation and reset of the cell division count. Each axis has a limited size due to complex organ interactions at large sizes. However, the change of axes can lead to infinite growth, just as in vegetatively propagating plants (Ivanov, 2003).

The easy appearance of a new axis and entire plant from the already differentiated plant cells that preserve their totipotency is a striking capacity of plants providing for abundant growth and easy vegetative propagation of many species. This may be due to specific features of the plant genome, hidden polyploidy, and hybridogenic origin of many species (Ivanov, 1978). In fact, basal sprouting is a universal attribute of temperate angiosperm trees but is much less common among gymnosperms.

Maturational changes are particularly persistent and difficult to reverse in conifers, as has been demonstrated in several species (Greenwood, 1995; Del Tredici, 2001).

Attempts to reverse maturation and restore the regenerative competence associated with the juvenile state have met with varying degrees of success. Although regenerative competence has been temporarily restored by a variety of cultural treatments, many maturational characteristics appear to change independently of one another. Also, a temporary increase in vigor (reinvigoration) may be a response to changed cultural conditions rather than true rejuvenation (Poethig, 1990; Greenwood, 1995). In meristematic areas of *P. radiata* and *C. sativa*, there is a gradual decrease in the extent of DNA methylation as the degree of reinvigoration increased (Fraga et al., 2002c; Hasbún et al., 2007).

The observed changes in extent of DNA methylation during aging and reinvigoration indicate that reinvigoration could be a consequence of epigenetic modifications opposite in direction to those that occur during aging. But it is not clear how increases in methylation of vegetative tissues are compatible with the fact that in reproductive tissues some specific genes needed for the reproductive transition become activated. However, in chestnut floral the differentiation period converges with a gentle and temporal decrease of genomic DNA methylation (Hasbún et al., 2007).

13.2.2 FLOWERING

Recent research has demonstrated that DNA methylation participates actively in regulating timing of flowering in *Arabidopsis* and other species.

The floral transition is regulated by a complex genetic network that monitors the developmental state of the plants as well as environmental conditions such as light and temperature. The photoperiod and vernalization (low temperature) pathways are the main floral inductive signals in *Arabidopsis*. A primary response to both signals is the transcriptional activation of floral-meristem-identity genes LEAFY (LFY), APETALA1 (AP1), and CAULIFLOWER (CAL) at the shoot apex.

Vernalization is found in many plant species endemic to regions at high altitudes or high latitudes. It is thought that growth through the cold of winter ensures that flowering and seed development occur in the most favorable weather conditions. FLOWERING LOCUS C (FLC) is a key component of the vernalization response pathway: the low-temperature treatment modulates expression of FLC, and removal of this floral repressor allows the plant to flower when it is developmentally competent to do so. It is important to note that diverse authors have observed a correlation between the methylation DNA status of the plant and FLC expression (Sheldon et al., 2000; Genger et al., 2003; Finnegan et al., 2005). FLC expression is decreased in plants that have low levels of DNA methylation, supporting the idea that a reduction in DNA methylation mediates vernalization. The finding that demethylation decreases FLC expression was unexpected, as reduced methylation generally enhances, rather than represses, gene expression. In this case, methylation could regulate FLC expression directly or indirectly by controlling the transcription of a regulator of FLC. Control of FLC by methylation could be direct because methylation of the FLC promoter may facilitate the binding of a transcription factor that requires methylation of its recognition site or may prevent the binding of a repressor of transcription. Methylation is removed by vernalization, blocking transcription. Alternatively, methylation could regulate FLC expression indirectly, by regulating the expression of a repressor of FLC. In unvernalized plants, transcription of the repressor is blocked by methylation; following vernalization, the methylation block is removed, allowing transcription of the repressor, which blocks FLC transcription (Finnegan et al., 2000).

Flowering is induced by more than one pathway. Recent data suggest that methylation is important in controlling the expression of the FWA gene, which acts independently of FLC. DNA methylation affects flowering time by regulating the expression of two repressors of flowering. Genome-wide demethylation of DNA downregulates FLC, promoting flowering, while at the same time demethylation activates transcription of FWA. Demethylation has opposing effects on flowering time. But the activation of FWA does not delay flowering in all ecotypes; C24 FWA protein seems to be nonfunctional even when its transcription is activated by demethylation (Genger et al., 2003).

An important step in investigations on the mechanism of vernalization was observation that in some plants treatment with 5-azacytidine (5-azaC) led to earlier flowering or to reduction in the cold requirement. 5-azaC is a demethylation agent that reduces the level of DNA methylation. Treatment of plant and animal cells with 5-azaC results in the demethylation of DNA directly by incorporation of the analog in place of cytosine during DNA replication and indirectly by inhibition of the action of the methyltransferase enzyme (Burn et al., 1993; Horváth et al., 2003)

Zluvova et al. (2001) studied global changes in DNA methylation using an indirect immunohistochemical approach, during shoot apical meristem development in *Silene latifolia*. The central zone of shoot apical meristem remains highly methylated during the whole period of vegetative growth, and in this region, only low cell division activity was found. However, upon the transition of the shoot apical meristem to the floral bud, the meristem decreased its high methylation status and its cells started to divide. Shoot apical meristems represent a permanent pluripotent cell line in which a balance of proliferation- and differentiation-promoting genes leads to both self-maintenance of meristematic cells and morphogenesis of plant body. Reproductive organs are formed by a transition of shoot meristem to floral buds. Often a hidden epigenetic change accompanies this transition. At a first moment happen a rapid demethylation process related to reprogramming in stem cell differentiation (Wolf et al., 2001); specific demethylation events in differentiated tissues could then to further changes in gene expression as needed, later remethylation takes places accompanied by proliferation and tissues differentiation. Work by Zapater et al. (2005) confirms these results, and they suggest the existence of a progressive DNA methylation along with plant development.

We have already shown that the FLC acts as the major floral repressor in *Arabidopsis* and the photoperiodic floral regulatory and vernalization signals antagonize this FLC-mediated floral repressive activity. But in *Arabidopsis*, an autonomous floral-promotion pathway promotes flowering independently of the photoperiod and vernalization pathways by repressing FLC. He et al. (2003) reported that FLOWERING LOCUS D (FLD), one of six genes in the autonomous pathway, encodes a plant homolog of a protein found in histone deacetylase complexes in mammals. Lesions in FLD result in hyperacetylation of histones in FLC chromatin, upregulation of FLC expression, and extremely delayed flowering. Thus, the autonomous pathway regulates flowering in part by histone deacetylation, indicating that multiple means exist by which this pathway represses FLC expression. More links between DNA methylation and histone hypoacetylation are accumulating, but the example of FLC repression and flowering is clearest in plants.

A functional interrelationship between DNA methylation and chromatin components has emerged recently. Methyl-CpG-binding proteins and DNMTs have been detected in protein complexes together with chromatin-remodeling enzymes and histone deacetylase (HDAC) (Lusser, 2002). This indicates cooperation between different epigenetic processes, DNA methylation, and histone acetylation directed toward both gene silencing and cell differentiation.

Recent studies of FLC regulation show other important chromatin modifications that are used to activate FLC expression; thus trimethylation of histone 3 (H3) at lysine 4 (K4) is associated with the chromatin of active gene. When the *Arabidopsis* PAF1 protein complex associates with RNA polymerase II (Pol II), an H3-K4 methyltransferase is recruited to FLC chromatin by Pol-II-PAF1 complex, resulting in K4 trimethylation in the 5' portion of transcribed FLC chromatin; then FLC chromatin is remodeled to an active state, thus facilitating FLC transcription by Pol II. Vernalization converts active FLC chromatin into a heterochromatin-like state so that the PAF1 complex is not able to access the FLC locus (He and Amasino 2005).

13.2.3 STRESS AND TRANSPOSONS

DNA methylation has an important role in genetic stability, since it controls transposon and other mobile elements jumping, being an effective mechanism for its long-term silencing. These elements are characterized for having CpNpG islands across its sequence and when they are unmethylated, they can move across the genome and in this way they may interrupt gene sequences and silence them, being a source of variability (Kubis ct al., 2003).

It is known that transposon expression is due to biotic and abiotic stresses as well as to several physiological changes like pathogen attack, environmental cues, in vitro culture, germination, interspecies introgression, and tissue irradiation (Hirochika et al., 1996; Grandbastien, 1998; Kalendar et al., 2000; Maekawa et al., 2003; Liu et al., 2004; de Diego et al., 2006; Smýkal et al., 2007). It has been demonstrated that in normal conditions transposons are methylated in maize (Rabinowicz et al., 2003), but a decrease of global methylation under tissue culture, due to triggers or stress conditions, is tightly related to expression of transposable elements controlled by the methylation degree of its sequence (Han et al., 2004; Liu et al., 2004). In plants, silent transposable elements, like MULEs (Mutator Like Element) or the CACTA family, are methylated and can be reactivated in methylation-defective mutants DDM1. DDM1 is also required for methylation of tandem repeats at the centromere and at the nucleolar organizer (Miura et al., 2001; Singer et al., 2001). The expression of the two best-characterized plant retrotransposons is thus induced by different biotic or abiotic factors that can elicit plant defense responses. It has been shown that *N. tabacum* Tnt1A expression is related to the early steps of defense gene activation pathways (Grandbastien et al., 1997). It is known that under stress conditions demethylation occurs not only in transposable elements specifically, but also in genic sequences, inducing an alteration of gene expression by changing chromatin structure (Steward et al., 2002).

In addition to variations in methylation levels regulating expression of transposable elements, transposition of these elements has an influence on DNA methylation. When transposition takes place both in animals and plants, insertion of LTR elements may cause changes in the epigenetic state of the flanking sequences, modifying its expression (Whitelaw and Martin 2001; Kashkush et al., 2003). Moreover, in animals it has been shown that integration of foreign DNA not only causes methylation alterations in adjacent host sequences but also causes alterations in remote sequences from the insertion sites (Remus et al., 1999; Müller et al., 2001).

13.3   ENVIRONMENT-INDUCED EPIGENETIC VARIATIONS

Sensing environmental changes and initiating a gene expression response is important for plants. The response can be short term (e.g., synthesis of protectant molecules against stress) or long term (e.g., phenotypic plasticity, with alteration of developmental programs). Epigenetic systems must be part of a relay from sensing a change in the environment to a change in gene expression. Their ability to alter rapidly and reversibly, yet with the potential to keep a stable memory through many cell divisions, is key to the flexibility of plant responses to the environment (Grant-Downton and Dickinson, 2006).

Surprisingly, not much is known about epigenetics and plant responses to environmental changes, with one major exception: the regulation of flowering time in *A. thaliana.*

In *Arabidopsis*, one of the major environmental controls of flowering time is temperature: a period of low temperature accelerates flowering. This process, known as vernalization, is found in many plant species. In *Arabidopsis,* the flowering repressor FLC (FLOWERING LOCUS C) has been identified as a key component of the vernalization response pathway. The low-temperature treatment modulates expression of FLC and DNA methylation in such a way that demethylation of DNA decreases FLC expression (Sheldon et al., 2000; Genger et al., 2003; Finnegan et al., 2005).

Recent studies of flowering-time control shows other epigenetic mechanisms, chromatin modification, that regulate the transition flowering. Genetic and molecular studies have revealed three systems of FLC regulation that influence the state of FLC chromatin: vernalization, FRI, and the autonomous pathway (He and Amasino, 2005).

Different kinds of environmental stress can influence epigenetic mechanisms. It is important to emphasize that the epigenetic modifications on DNA and chromatin constitute the link between the genotype and the phenotype. The plant genome's response to environmental and genetic stress generates novel genetic and epigenetic methylation polymorphisms. DNA methylation can generate heritable phenotypic variation by influencing gene expression. DNA polymorphisms in *cis* or *trans* elements that trigger cytosine methylation can generate methylation polymorphisms. Alternatively, identical alleles may take on different methylation states.

Environmental and genetic perturbations induce genetic and epigenetic changes that trigger methylation (Lukens and Zhan, 2007).

In tree research, dormancy is most frequently referred to as absence of visible growth in any plant structure containing a meristem provoked by limitations in environmental factors. In seeds, the stage of dormancy is defined as the failure of a viable intact seed to complete germination under favorable conditions. Both situations are controlled by environmental conditions, and different authors associate these stages with different levels of DNA methylation and histone acetylation (Law and Suttle, 2004). During tuber meristem dormancy break transient demethylation of DNA precedes increases in cell division and the resumption of meristem growth. In addiction, DNA demethylation is linked to histone multiacetylation during potato tuber dormancy emergence in such a way that DNA demethylation is followed by increased H3 and H4 histone acetylation and ultimately tuber meristem reactivation (Law and Suttle, 2004).

13.4   EPIGENETIC MEMORY

Plants do not have dedicated germlines segregated and maintained from early developmental stages. Instead, germline cells are formed de novo late in the development by differentiation from somatic tissues. This fact allows the possibility of transmission of any stable epigenetic information acquired during development (Grant-Downton and Dickinson, 2005). Several epialleles that occur naturally with regard to flower symmetry (Cubas et al., 1999), pigmentation levels (Chandler et al., 2000), pathogen resistance (Stokes et al., 2002), and trait evolution across generations (Kakutani et al., 1999) have been described.

Seed plants are characterized by double fertilization, which requires two and three post-meiotic mitotic divisions to develop male and female gametophytes, respectively (for a complete description of sexual reproduction of higher plants, see Boavida et al., 2005). To ensure transgenerational inheritance of epigenetic marks in plants, the information must be maintained during gamete formation, fertilization, embryogenesis, and somatic growth and differentiation of the progeny (Takeda and Paszkowski, 2006).

Oakeley et al. (1997) reported the global reduction of DNA methylation during male gametogenesis development in *Nicotiana* after using immunocytological detection. DNA methylation seemed to reduce about 20%, compared with vegetative nucleolus, just before pollen germination. However, these results can be masked by the fact that generative nucleus is more compact than vegetative, which can interfere with antibody penetration. These classes of experiments have not been repeated and confirmed for other plant species. An updated study using antibodies against 5-mdC and histone marks would be extremely informative.

A study of the evolution of DNA methylation with generations in mutant DDM1 *Arabidopsis* plants demonstrated that ddm1 mutation leads to a hypomethylation of the plants (Vongs et al., 1993; Kakutani et al., 1999). F1 DDM1/ddm1 heterozygotes produced by backcrossing of ddm1 homozygotes to wild type showed a 5-mdC-level intermediate between parents. Testing backcrossed lines and progenies, researchers have demonstrated that the hypomethylated status originated in ddm1/ddm1 individuals can be inherited in a stable way during gametogenesis and somatic mitosis even with the presence of DDM1 activity.

Recent studies (Mathieu et al., 2007) in different *Arabidopsis* met1-3 mutants showed that the maintenance loss of CG methylation triggers genome-wide activation of alternate epigenetic mechanisms. These mechanisms involve the inhibition of the expression of DNA demethylases, retargeting of histone H3K9 methylation, and small RNA-directed DNA methylation acting stochastically. New methylation patterns are progressively formed over generations in absence of CG methylation. Non-CG methylation, mediated by small RNA, is preferentially associated with methylated transposable elements but not with methylated genes, indicating that most genic methylation is not guided by small RNA (Vaughn et al., 2007). These patterns are necessary to rescue CG methylation, which orchestrates the distribution of non-CG and H3K9 methylation, because these processes are essential for stable transgenerational inheritance of epigenetic information.

Epigenetic memory mechanisms with regard to epialleles will influence evolution in wild plant populations through their effects on both phenotypic trait distribution and fitness (Kalisz and Purugganan, 2004) in two different ways: first, epiallelic phenotypes can be less extreme than mutations in genes that cause loss of function (Cronk, 2001); second, epiallele segregation could benefit the progeny of adapted individuals during periods of rapid change or even modulate phenotypic effects of genes that in other environments could be deleterious (Cronk, 2001; Riddle and Richards, 2002). Data to support this speculation about population-level effects of DNA methylation are currently lacking.

13.5   CONCLUSIONS

DNA methylation is implicated in the correct transition between developmental stages (i.e., flowering, dormancy, etc.). Plants have several peculiarities in their methylation machinery with respect to animals, having three different classes of DNA methyltransferases: the MET1 family, which acts as the mammalian homolog DNMT1; the chromomethylase family, which preferentially methylates CpNpG sequences; and de novo methyltransferases, which are a homolog to the mammalian DNMT3 family. Methylation is stable during development, albeit not static, as demonstrated in several experimental situations, being the basis of cellular plasticity. Cellular differentiation processes are associated with an initial demethylation stage, which permits later cellular reprogramming processes. Specific demethylation issues relate to flowering, stem cell differentiation, etc. The formation of germline cells late in development permits the transmission of stable epigenetic information acquired during. This process involves many different processes regarding DNA methylation, histone modifications, and siRNA.

ACKNOWLEDGMENTS

The financial support needed to guarantee our progress into the insight of aging, phase change, and reinvigoration, as well as the progress being made in the research for quality markers has come from EU Projects FAIR3-CT96-1445, INCO 10063, and MCT-AGL2000-2126, AGL 2004-00810/FOR, AGL2007-62907/FOR Spanish National Projects. The Spanish M.E.C.D. supported fellowships of all young researchers.

REFERENCES

Amasino, R. 2004. Vernalization, competence and the epigenetic memory of winter. *The Plant Cell*, 16: 2553–2559

Bäurle, I., and Laux, T. 2003. Apical meristems: The plant's fountain of youth. *Bioessays*, 25: 961–970.

Bestor, T.H. 1992. Activation of mammalian DNA methyltransferases by cleavage of a Zn binding regulatory domain. *EMBO Journal*, 11: 2611–2617.

Boavida, L.C., Becker, J.D., and Feijó, J.A. 2005. The making of gametes in higher plants. *The International Journal of Developmental Biology*, 49: 595–614.

Burn, J., Bagnall, D., Metzger, J., Dennis, E., and Peacock, W. 1993. DNA methylation, vernalization, the initiation of flowering. *Proceedings of the National Academy of Sciences USA*, 90: 287–291.

Burzynski, S.R., 2005. Aging: gene silencing or gene activation? *Medical Hypotheses*, 64: 201–208.

Chan, S.W., Henderson, I.R., and Jacobsen, S.E. 2005. Gardening the genome: DNA methylation in *Arabidopsis thaliana*. *Nature Reviews: Genetics*, 6: 351–360.

Chandler V.L., Eggleston, W.B., and Dorweiler, J.E. 2000. Paramutation in maize. *Plant Molecular Biology*, 43: 121–125.

Cronk, Q.B. 2001. Plant evolution and development in a post-genomic context. *Nature Reviews: Genetics*, 2: 607–609.

Cubas, P., Vincent, C., and Coen, E. 1999. An epigenetic mutation responsible for natural variation in floral symmetry. *Nature*, 401: 157–161.

de Diego, J.G., Rodríguez, F.D., Rodríguez Lorenzo, J.L., Cervantes, E., and Grappin, P. 2006. cDNA-AFLP analysis of seed germination in *Arabidopsis thaliana* identifies transposons and new genomic sequences. *Journal of Plant Physiology*, 163: 452–462.

Del Tredici, P. 2001. Sprouting in temperate trees: a morphological and ecological review. *Botanical Review*, 67: 121–140.

Diaz-Sala, C., Rey, M., Boronat, A., Besford, R., and Rodríguez, R. 1995. Variations in the DNA methylation and polypeptide patterns of adult hazel (Corylus avellana L.) associated with sequential in vitro subcultures. *Plant Cell Reports*, 15: 218–221.

Finnegan, E., Genger, R., Peacock, W., and Dennis, E. 1998. DNA Methylation in plants. *Plant Molelcular Biology*, 49: 223–247.

Finnegan, E.J., Peacock, W.J., and Dennis, E.S. 2000. DNA methylation, a key regulator of plant development and other processes. *Current Opinion in Genetics & Development*, 10: 217–223.

Finnegan, E.J., Kovac, K.A., Jaligot, E., Sheldon, C.C., Peacock, W.J., and Dennis, E.S. 2005. The downregulation of flowering locus C (FLC) expression in plants with low levels of DNA methylation and vernalization occurs by distinct mechanisms. *The Plant Journal*, 44: 420–432.

Fraga, M.F., and Esteller, M. 2002. DNA methylation: a profile of methods and applications. *BioTechniques*, 33: 632–649.

Fraga, M.F., Rodríguez, R., and Cañal, M.J. 2000. Rapid quantification of DNA methylation by high performance capillary electrophoresis. *Electrophoresis*, 21: 2990–2994.

Fraga, M.F., Uriol, E., Diego, B.L., Berdasco, M., Esteller, M., Cañal, M.J., and Rodríguez, R. 2002. High-performance capillary electrophoretic method for the quantification of 5-methyl 2´-deoxycytidine in genomic DNA: application to plant, animal and human cancer tissues. *Electrophoresis*, 23: 1677–1681.

Fraga, M.F., Canal, M.J., and Rodriquez, R. 2002a. In vitro morphogenic potential of differently aged Pinus radiata D. Don. *Plant Cell Tissue and Organ Culture*, 70: 139–145.

Fraga, M.F., Canal, M.J., and Rodriquez, R. 2002b. Phase-change related epigenetic and physiological changes in Pinus radiata D. Don. *Planta*, 215: 672–678.

Fraga, M.F., Rodriguez, R., and Canal, M.J. 2002c. Genomic DNA methylation-demethylation during ageing-invigoration of Pinus radiata. *Tree Physiology*, 22: 813–816.

Fraga, M.F., Ballestar, E., Paz, M.F., Ropero, S., Setien, F., Ballestar, M.L., Heine-Suner, D., Cigudosa, J.C., Urioste, M., Benitez, J., Boix-Chornet, M., Sanchez-Aguilera, A., Ling, C., Carlsson, E., Poulsen, P., Vaag, A., Stephan, Z., Spector, T.D., Wu, Y.Z., Plass, C., and Esteller, M. 2005. Epigenetic differences arise during the lifetime of monozygotic twins. *Proceedings of the National Academy of Sciences USA*, 102: 10604–10609.

Fransz, P., and de Jong, J. 2002. Chromatin dynamics in plants. *Current Opinion in Plant Biology*, 5: 560–567.

Frommer, M., McDonald, L. E., Millar, D. S., Collis, C. M., Watt, F., Grigg, G. W., Molloy, P. L., and Paul, C. L. 1991. A genomic sequencing protocol that yields a positive display of 5-methylcytosine residues in individual DNA strands. *Proceedings of the National Academy of Sciences USA*, 89: 1827–1831.

Genger, R.K., Kovac, K.A., Dennis, E.S., Peacock, W.J., and Finnegan, E.J. 1999. Multiple DNA methyltransferases genes in *Arabidopsis thaliana*. *Plant Molecular Biology*, 41: 269–278.

Genger, R.K., Peacock, W.J., Dennis, E.S., and Finnegan, E.J. 2003. Opposing effects of reduced DNA methylation on flowering time. *Planta*, 216: 461–466.

Grandbastien, M. 1998. Activation of plant retrotransposons under stress conditions. *Trends in Plant Science*, 3: 181–187.

Grandbastien, M., Lucas, H., Morel, J.B., Mhiri, C., Vernhettes, S., and Casacuberta, J.M. 1997. The expression of the tobacco Tnt1 retrotransposon is linked to the plant defense responses. *Genetica*, 100: 241–252.

Grant-Downton, R.T., and Dickinson, H.G. 2005. Epigenetics and its implications for plant biology. 1. The Epigenetic network in plants. *Annals of Botany*, 96: 1143–1164.

Grant-Downton, R.T., and Dickinson, H.G. 2006. Epigenetics and its implications for plant biology. 2. "The epigenetic epiphany": epigenetics, evolution and beyond. *Annals of Botany*, 97: 11–27.

Greenwood, M. 1995. Juvenility and maturation in conifers: current concepts. *Tree Physiology*, 15: 433–438.

Gruenbaum, Y., Naveh-Many, T., Cedar, H., and Razin, A. 1981. Sequence specificity of methylation in higher plant DNA. *Nature*, 292: 860–862.

Han, F.P., Liu, Z.L., Tan, M.M., Hao, S., Fedak, G., and Liu, B. 2004. Mobilized retrotransposon Tos17 of rice by alien DNA introgression transposes into genes and causes structural and methylation alterations of a flanking genomic region. *Hereditas*, 141: 243–251.

Hasbún, R., Valledor, L., Santamaría, E., Cañal, M.J., and Rodríguez, R. 2007. Dynamics of DNA methylation in chestnut trees development. *Acta Horticulturae*, 760: 563–566.

He, Y., and Amasino, R.M. 2005. Role of chromatin modification in flowering-time control. *Trends in Plant Science*, 10: 30–35.

He, Y., Michaels, S.D., and Amasino, R.M. 2003. Regulation of flowering time by histone acetylation in *Arabidopsis*. *Science*, 302: 1751–1754.

Hirochika, H., Sugimoto, K., Otsuki, Y., Tsugawa, H., and Kanda, M. 1996. Retrotransposons of rice involved in mutations induced by tissue culture. *Proceedings of the National Academy of Sciences USA*, 93: 7783–7788.

Horváth, E., Szalai, G., Janda, T., Páldi, E., Rácz, I., and Lásztty, D. 2003. Effect of vernalization and 5-azacytidine on the methylation level of DNA wheat (Triticum aestivum L. cv. Martonvásár 15). *Plant Science*, 165: 689–692.

Ivanov, V. 1978. DNA content in the nucleus and rate of plant of development. *Ontogenez*, 9: 39–53.

Ivanov, V. 2003. The problem of stem cells in plants. *Russian Journal of Development Biology*, 34: 205–212.

Jeltsch, A. 2002. Beyond Watson and Crick: DNA methylation and molecular enzymology of DNA methyltransferases. *ChemBioChem*, 3: 274–293.

Johnston, J. W., Harding, K., Bremner, D. H., Souch, G., Green, J., Lynch, P. T., Grout, B., and Benson, E. E. 2005. HPLC analysis of plant DNA methylation: a study of critical methodological factors. *Plant Physiology and Biochemistry*, 43: 844–853.

Jones, P.A., and Takai, D. 2001. The role of DNA methylation in mammalian epigenetics. *Science*, 293: 1068–1070.

Kakutani, T. 1997. Genetic characterization of late-flowering traits induced by DNA hypomethylation mutation in *Arabidopsis thaliana*. *Plant Journal*, 12: 1447–1451.

Kakutani, T., Munakata, K., Richards, E.J., and Hirochika, H. 1999. Meiotically and mitotically stable inheritance of DNA hypomethylation induced by ddm1 mutation of *Arabidopsis thaliana*. *Genetics*, 151: 831–838.

Kalendar, R., Tanskanen, J., Immonen, S., Nevo, E., and Schulman, A.H. 2000. Genome evolution in wild barley (Hordeum spontaneum) by BARE-1 retrotransposon dynamics in response to sharp microclimatic divergence. *Proceedings of the National Academy of Sciences USA*, 97: 6603–6607.

Kalisz, S., and Purugganan, M.D. 2004. Epialleles via DNA methylation: consequences for plant evolution. *TRENDS in Ecology and Evolution*, 19: 309–314.

Kaminsky, Z. A., Assadzadeh, A., Flanagan, J., and Petronis, A. 2005. Single nucleotide extension technology for quantitative site-specific evaluation of [met]C/C in GC-rich regions. *Nucleic Acids Research*, 33: e95.

Kashkush, K., Feldman, M., and Levy, A.A. 2003. Transcriptional activation of retrotransposons alters the expression of adjacent genes in wheat. *Nature Genetics*, 33: 102–106.

Kubis, S. E., Castilho, A., Vershinin, A., and Seymour, J. 2003. Retroelements, transposons and methylation status in the genome of oil palm (Elaeis guineensis) and the relationship to somaclonal variation. *Plant Molecular Biology*, 52: 69–79.

Lambé, P., Mutambel, H., Fouché, J., Deltour, R., Foidart, J., and Gaspar, T. 1997. DNA methylation as a key process in regulation of organogenic totipotency and plant neoplastic progression? *In Vitro Cellular and Developmental Biology*, 33: 155–162.

Law, R.D., and Suttle, J.C. 2004. Changes in histone H3 and H4 multi-acetylation during natural and forced dormancy break in potato tubers. *Physiologia Plantarum*, 120: 642–649.

Lewin, J., Schmitt, A.O., Adorján, P., Hildmann, T., and Piepenbrock, C. 2004. Quantitative DNA methylation analysis based on four-dye trace data from direct sequencing of PCR amplificates. *Bioinformatics*, 20: 3005–3012.

Li, E., Bestor, T.H., and Jaenisch, R. 1992. Targeted mutation of the DNA methyltransferase genes results in embryogenic lethality. *Cell*, 69: 915–926.

Liu, Z.L., Han, F.P., Tan, M., Shan, X.H., Dong, Y.Z., Wang, X.Z., Fedak, G., Hao, S., and Liu, B. 2004. Activation of a rice endogenous retrotransposon Tos17 in tissue culture is accompanied by cytosine demethylation and causes heritable alteration in methylation pattern of flanking genomic regions. *Theoretical and Applied Genetics*, 109: 200–209.

Lukens, L.N., and Zhan, S. 2007. The plant genome's methylation status and response to stress: implications for plant improvement. *Current Opinion in Plant Biology*, 10: 317–322.

Lusser, A. 2002. Acetylated, methylation, remodeled: chromatin states for gene regulation. *Current Opinion in Plant Biology*, 5: 437–443.

Madlung, A., Masuelli, R.W., Watson, B., Reynolds, S.H., Davison, J., and Comai, L. 2002. Remodeling of DNA methylation and phenotypic and transcriptional changes in synthetic *Arabidopsis* allotetraploids. *Plant Physiology*, 129: 733–746.

Maekawa, M., Hase, Y., Shikazono, N., and Tanaka, A. 2003. Induction of somatic instability in stable yellow leaf Mutant of rice by ion beam irradiation. *Nuclear Instruments and Methods in Physics Research B*, 206: 579–585.

Mallory, A.C., Ely, L., Smith, T.H., Marathe, R., Anandalakshmi, R., Fagard, M., Vaucheret, H., Pruss, G., Bowman, L., and Vance, V.B. 2001. HC-Pro suppression of transgene silencing eliminates the small RNAs but not transgene methylation of the mobile signal. *Plant Cell*, 13: 571–583.

Mathieu, O., Reinders, J., Caikovski, M., Smathajitt, C., and Paszkowski, J. 2007. Transgenerational stability of the *Arabidopsis* epigenome is coordinated by CG methylation. *Cell*, 130: 851–862.

Meehan, R.R. 2003. DNA methylation in animal development. *Seminars in Cell and Developmental Biology*, 14: 53–65.

Miura, A., Yonebayashi, S., Watanabe, K., Toyama, T., Shimada, H., and Kakutani, T. 2001. Mobilization of transposons by a mutation abolishing full DNA methylation in *Arabidopsis*. *Nature*, 411: 212–214.

Müller, K., Heller, H., and Doerfler, W. 2001. Foreign DNA integration. *The Journal of Biological Chemistry*, 276: 14271–14278.

Oakeley, E.J., Podestà, A., and Jost, J.P. 1997. Developmental changes in DNA methylation of the two tobacco pollen nuclei during maturation. *Proceedings of the National Academy of Sciences USA*, 94: 11721–11725.

Okamoto, H., and Hirochika, H. 2001. Silencing of transposable elements in plants. *Trends in Plant Science*, 6(11): 527–534.

Poethig, R. 1990. Phase change and the regulation of shoot morphogenesis in plants. *Science*, 250: 923–930.

Rabinowicz, P.D., Palmer, L.E., May, B.P., Hemann, M.T., Lowe, S.W., McCombie, W.R., and Martienssen R.A. 2003. Genes and transposons are differentially methylated in plants, but not in mammals. *Genome Research*, 13: 2658–2664.

Ramsahoye, B.H. 2002. Measurement of genome wide DNA methylation by reversed-phase high-performance liquid chromatography. *Methods*, 27: 156–161.

Razin, A., and Cegar, H. 1994. DNA methylation and genomic imprinting. *Cell*, 77: 473–476.

Remus, R., Kämmer, C., Heller, H., Schmitz, B., Schell, G., and Doerfler, W. 1999. Insertion of foreign DNA into an established mammalian genome can alter the methylation of cellular DNA sequences. *Journal of Virology*, 73: 1010–1022.

Richardson, B. 2003. Impact of aging on DNA methylation. *Aging Research Reviews*, 2: 245–261.

Riddle, N.C., and Richards, E.J. 2002. The control of natural variation in cytosine methylation in *Arabidopsis*. *Genetics*, 162: 355–363.

Ruiz-Garcia, L., Cervera, M.T., and Martinez-Zapater, J.M. 2005. DNA methylation increases throughout *Arabidopsis* development. *Planta*, 222: 301–306.

San Miguel, P., Gaut, B.S., Tikhonov, A., Nakajima Y., and Bennetzen J. L. 1998. The paleontology of intergene retrotransposons of maize. *Nature Genetics*, 20: 43–45.

Shaked, R., Kashkush, K., Ozkan, H., Feldman, M., and Levy, A.A. 2001. Sequence elimination and cytosine methylation are rapid and reproducible responses of the genome to wide hybridization and allopolyploidy in wheat. *Plant Cell*, 13: 1749–1759.

Sheldon, C.C., Finnegan, E.J., Rouse, D.T., Tadege, M., Bagnall, D.J., Helliwell, C.A., Peacock, W.J., and Dennis, E.S. 2000. The control of flowering by vernalization. *Current Opinion in Plant Biology*, 3: 418–422.

Singer, T., Yordan, C., and Martienssen, R.A. 2001. Robertson's mutator transposons in *A. thaliana* are regulated by the chromatin-re-modelling gene Decrease in DNA Methylation (DDM1). *Genes and Development*, 15: 591–602.

Smýkal, P., Valledor, L., Rodríguez, R., and Griga, M. 2007. Assesment of genetic and epigenetic stability in long term in vitro shoot culture of pea (*Pisum sativum* L.). *Plant Cell Reports*, 26: 1985–1998.

Steward, N., Ito, M., Ymaguchi, Y., Koizumi, N., and Sano, H. 2002. Periodic DNA methylation in maize nucleosomes and demethylation by environmental stress. *The Journal of Biological Chemistry*, 277: 37741–37746.

Stokes, T.L., Kunkel, B.N., and Richards, E.J. 2002. Epigenetic variation in *Arabidopsis* disease resistance. *Genes and Development*, 16: 171–182.

Surani, M.A. 2001. Reprogramming of genome function through epigenetic inheritance. *Nature*, 414: 122–128.

Takeda, S., and Paszkowski, J. 2006. DNA Methylation and epigenetic inheritance during plant gametogenesis. *Chromosoma*, 115: 27–35.

Tost, J., Dunker, J., and Gut, I.G. 2003. Analysis and quantification of multiple methylation variable positions in CpG islands by Pyrosequencing™. *Molecular Diagnosis Techniques*, 35: 152–156.

Tost, J., Schatz, P., Schuster, M., Berlin, K., and Gut, I.G. 2003a. Analysis and accurate quantification of CpG methylation by MALDI mass spectrometry. *Nucleic Acids Research*, 31: e50.

Vanyushin, B. 2005. Enzymatic DNA methylation is an epigenetic control for genetic functions of the cell. *Biochemistry* (Moscow), 70: 597–603.

Vaughn, M.W., Tanurd, M., Lippman, Z., Jiang, H., Carrasquillo, R., Rabinowicz, P.D., Dedhia, N., McCombie, W.R., Agier, N., Bulski, A., Colot, V., Doerge, R.W., and Martienssen, R.A. 2007. Epigenetic natural variation in *Arabidopsis thaliana*. *PLoS Biology*, e174.

Vinkenoog, R., Spielman, M., Adams, S., Fischer, R.L., Dickinson, H.G., and Scott, R.J. 2000. Hypomethylation promotes autonomous endosperm development and rescues postfertilization lethality in fie mutants. *Plant Cell*, 12: 2271–2282.

Vongs, A., Kakutani, T., Martienssen, R.A., and Richards, E.J. 1993. *Arabidopsis thaliana* DNA methylation mutants. *Science*, 260: 1926–1928.

Whitelaw, E., and Martin, D.I.K. 2001. Retrotransposons as epigenetic mediators of phenotypic variation in mammals. *Nature Genetics*, 27: 361–365.

Wilson, V.L., Smith, R.A., Longoria, J., Liotta, M.A., Harper, C.M., and Harris, C.C. 1987. Chemical carcinogen-induced decreases in genomic 5-methyldeoxycytidine content of normal bronchial epithelial cells. *Proceedings of the National Academy of Sciences USA*, 84: 3298–3301.

Wolf, R., Wendy, D., and Walter, J. 2001. Epigenetic reprogramming in mammalian development. *Science*, 203: 1089–1093.

Yoder, J.A., Walsh, C.P., and Bestor, T.H. 1997. Cytosine methylation and the ecology of intragenomic parasites. *Trends in Genetics*, 13: 335–340.

Zhang, X., Yazaki, J., Sundaresan, A., Cokus, S., Chan, S.W.L., Chen, H., Henderson, I.R., Shinn, P., Pellegrini, M., Jacobsen, S.E., and Ecker, J.R. 2006. Genome-wide high-resolution mapping and functional analysis of DNA methylation in *Arabidopsis*. *Cell*, 126: 1189–1201.

Zilberman, D., Gehring, M., Tran, R.K., Ballinger, T., and Henikoff, S. 2007. Genome-wide analysis of *Arabidopsis thaliana* DNA methylation uncovers an interdependence between methylation and transcription. *Nature Genetics*, 39: 61–69.

Zluvova, J., Janousek, B., and Vyskot, B. 2001. Immunohistochemical study of DNA methylation dynamics during plant development. *Journal of Experimental Botany*, 365: 2265–2273.

14 Epigenetics, Environment, and Evolution

Douglas M. Ruden, Parsa Rasouli, and Xiangyi Lu

CONTENTS

14.1   Summary 242
14.2   Historical Introduction of Epigenetics 242
     14.2.1   Jean-Baptiste Lamarck 242
     14.2.2   Charles Darwin 243
     14.2.3   Gregor Mendel 244
     14.2.4   The Modern Synthesis 244
     14.2.5   James Baldwin 244
     14.2.6   Ivan Schmalhausen 245
     14.2.7   Conrad Waddington 245
     14.2.8   John Cairns and Facilitated Genetic Variation 245
     14.2.9   Hsp90 Is a Capacitor for Morphological Evolution 247
14.3   Epigenetic Inheritance in *Drosophila* 247
     14.3.1   Epigenetic Transmission of PRE Occupancy through Meiosis 247
     14.3.2   Epigenetic Assimilation Experiments in *Drosophila* 248
     14.3.3   Genetic versus Epigenetic Capacitor Models 248
14.4   Transgenerational Epigenetics in Mammals 250
     14.4.1   Nutrition and DNA Methylation of Retrotransposon Mutations ... 250
     14.4.2   Endocrine Disruptors and Transgenerational
             Epigenetic Effects 251
14.5   Epigenetics, Environment, and Cancer 251
     14.5.1   Transgenerational Epigenetics and Cancer 252
     14.5.2   Clonal versus Polyclonal Models for Cancer Initiation 252
14.6   How Epigenetic Changes Can Lead to Genetic Changes 252
     14.6.1   Deamination of Methylated Cytosine to Thymidine 253
     14.6.2   Epigenetic Regulation of Trinucleotide Repeat
             Expansion and Contraction 253
14.7   Future Studies in the Role of Epigenetics in Evolution 256
Acknowledgments 256
References 257

14.1   SUMMARY

In 1859, Darwin proposed *survival of the fittest* as a means for a species to adapt to a particular environment. In 1883, in an attempt to explain transmission of increased fitness to offspring, Darwin proposed a Lamarckian-type process whereby characteristics acquired during an organism's lifetime are heritable. In 1896, Baldwin proposed that new phenotypes can be induced in a population by a stressful environment because developmental processes are highly plastic. In subsequent generations, in what is called the *Baldwin effect*, he proposed that new mutations can be selected that stabilize the new phenotype even in the absence of the stress that generated it. Since 1900, the notion of inheritance of acquired characteristics has been out of favor, but recent advances in epigenetics show that some environmentally induced alterations in the phenotype are in fact heritable. Novel epigenetic phenotypes can be induced both within an animal's lifetime, as in cancer progression, and in the next generation, as by maternal diet. Since epigenetic alterations themselves are mutagenic, such as by deamination of methylcytosine to thymidine, epigenetic alterations can direct mutations in precisely the genes needed to stabilize the new phenotype. We call this process an *epigenetic Baldwin effect* because it provides a possible epigenetic mechanism for generating stabilizing mutations in cancer and evolution.

14.2   HISTORICAL INTRODUCTION OF EPIGENETICS

To modern biologists, the term *epigenetics* is broadly defined as "nongenetic inheritance."[1] The person who originally coined the word was Conrad Waddington, but he meant it mainly to describe the development of an organism and the effects of the environment on its development.[2] In modern terms, epigenetic marks are generally considered to be DNA methylation at cytosines, particularly at CpG dinucleotides, but histone modifications such as methylation and acetylation are also important.[1] Approximately half of mammalian genes have so-called CpG islands in promoter regions that can be differentially methylated, although differential methylation of genes that do not have CpG islands might be more significant.[3]

Some animals, such as *Drosophila*, have apparently lost CpG methylation in their evolutionary lineage. In these organisms, histone modifications are presumably the most important epigenetic marks.[4] Because epigenetic marks are acquired characteristics established in an organism's lifetime, and because these marks are affected by the environment, the inheritance of epigenetic marks in the offspring resembles Lamarckian inheritance. To further understand the significance of the transmission of epigenetic marks on evolution, we give a brief historical introduction to the study of various forms of Lamarckian and Mendelian inheritance.

14.2.1   JEAN-BAPTISTE LAMARCK

In 1809, the year of Darwin's birth, Lamarck (1744–1829) published an influential book titled *Zoological Philosophy,* in which he proposed two laws of inheritance, the latter of which is known as the "Inheritance of Acquired Characteristics."[5] According to Lamarck's first law, which states a common and noncontroversial observation about use and disuse, "more frequent and sustained use of an organ strengthens that organ ... while the constant disuse of an organ impeccably weakens it."[5] In modern parlance, this can be summarized as, "use it or lose it."

In Lamarck's controversial second law, characteristics acquired during an organism's lifetime are transmitted to its progeny. He said, "Everything that nature has caused individuals to acquire or lose by the influence of the circumstances, it preserves by heredity and passes on to the new individuals descended from it."[5] The ubiquitous example of inheritance of acquired characteristics, which was proposed most likely to ridicule Lamarck's second law rather than Lamarck himself, is that giraffes evolved their long necks because successive generations stretched their necks to reach treetops.[6] In other words, evolution is driven by need rather than by selection. While this is clearly not happening, experiments by Waddington and others in the 20th century showed that the inheritance of some traits is affected by the parent's environment.

14.2.2 CHARLES DARWIN

In 1859, Darwin (1809–1882) published *On the Origin of Species* and presented a model of evolution that is best summarized as the survival of the fittest.[7] Darwin proposed that a population exhibits a random variation of phenotypes and that the fittest organisms survived and the least fit did not survive long enough to breed. Darwin did not have a modern genetic understanding of what might make an animal fit or unfit. Instead, he proposed that fitness is acquired during an animal's lifetime and that the fittest animals are able to transmit their desirable traits to their offspring. For example, in his 1883 book, published posthumously, *Variation in Animals and Plants under Domestication*, Darwin stated, "the view which has been suggested that the drooping is due to disuse of the muscles of the ear, from the animal being seldom much alarmed."[8] In other words, Darwin suggested that a dog has floppy ears because of disuse in its ancestors (i.e., Lamarck's second law).

While Lamarck did not propose a mechanism for the inheritance of acquired characteristics, Darwin presented a theory of inheritance that he called pangenesis.[8] Pangenesis was an ad hoc model that was meant to answer the following: "How can the use or disuse of a particular limb or part of the brain affect a small aggregate of reproductive cells, seated in a distant part of the body, in such a manner that the being developed from these cells inherits the characteristics of one or both of the parents?"[8] Darwin proposed in his pangenesis model the existence of small "elemental particles" that circulate in the blood and affect the germ cells.[8]

Darwin's pangenesis model was later discredited by such investigators as August Weismann, who showed in 1895 that the germline and the soma have different lineages. Weismann argued that, because of this separate lineage, the egg and the sperm cannot receive any information from the environment.[9] Weismann dismissed Darwin's pangenesis model and said of his predictions, "They are to a certain extent a mere paraphrase of the facts ... based on speculative assumptions."[9] However, recent experiments in epigenetics, described below, suggest that the environment might in fact be able to transmit such elemental particles to the germline and that these particles involve methylation of the DNA.

14.2.3 GREGOR MENDEL

In 1865, Mendel (1822–1884) published his pea genetic experiments in his classic paper, *Experiments in Plant Hybridisation.*[10] However, Mendel's laws of genetic inheritance were not well known until 1900, when they were rediscovered.[11] Mendel's first law says that two members of a gene pair segregate from each other into the gametes. His second law, which describes the interactions between two or more genes, says that gene pairs assort independently in gamete formation.[10] Chromosomes were later discovered and linked genes were identified, but otherwise Mendel's laws are still sound and they revolutionized biology and evolutionary theories. They were thought by most scientists to be the final nail in Lamarck's coffin, but this proved to be premature.

14.2.4 THE MODERN SYNTHESIS

In 1940 evolutionary biologists developed a modern Darwinian theory, called the Modern Synthesis.[11] This theory combines Darwinian theories of evolution with Mendelian models of inheritance. The Modern Synthesis dismissed other models, such as those that include the inheritance of acquired characteristics.[11] The main contribution of the Modern Synthesis was to apply sophisticated statistical analyses and models for the inheritance of complex traits in populations of organisms. However, in many aspects, the Modern Synthesis was too restrictive. For example, it eliminated any role for the environment in generating phenotypic variation. In current models of evolution, the importance of environmentally acquired characteristics is coming back in vogue.

14.2.5 JAMES BALDWIN

Baldwin (1861–1934) was an experimental physiologist who proposed a role for the environment in evolutionary selection. He proposed, for instance, that an organism has a broad range of adaptive mechanisms that allow it to survive in many different environments. Initially, when an organism is placed in a stressful environment, it is able to adapt just enough using preexisting processes to reproduce minimally. In subsequent generations, heritable changes (i.e., new mutations) arise in a few members of the population that strengthen the somatic adaptation.[12]

In Baldwin's model of evolution, these new mutations further relieve the stress and allow the organism to reproduce more optimally. The heritable changes replace the environmentally induced phenotypic changes such that the new phenotype is present even in the absence of the stress that induced the phenotype.[12] Such predictions of developmental plasticity to the environment followed by mutation to stabilize the phenotype are called the Baldwin effect.[11] In most current usage, the Baldwin effect refers to learned behaviors that later become instincts by subsequent mutations, but Baldwin meant it to include the inheritance of any phenotype.[12]

14.2.6 IVAN SCHMALHAUSEN

In 1943, Ivan Schmalhausen (1884–1963) elaborated on Baldwin's proposals in his book *Factors of Evolution.*[13] He proposed that a "norm of reaction" exists in an organism for a particular environment, and that different norms exist in different environments. The norms in his model have two characteristics—some happen to be adaptive to environmental stresses, but the majority of responses are random and nonadaptive. He calls these nonadaptive responses "morphoses" and cites the example of *Drosophila* eyes getting larger under heat shock conditions.[13] According to this theory, the existence of the morphoses provides a pool of random phenotypic variation that can be used later to adapt to different environmental stresses.

An important concept elaborated by Schmalhausen was the idea that the processes needed to initially adapt to a new environment are present as existing genetic variation in the population. He proposed that only "small" regulatory changes derived from genetic variation in the population are needed to stabilize the change in phenotype. He called the process of selecting the small regulatory changes "stabilizing selection."[13]

14.2.7 CONRAD WADDINGTON

Waddington (1905–1975), utilizing the theories of Baldwin and Schmalhausen, performed experiments in *Drosophila* that verified many aspects of their theories. As mentioned above, he coined the term *epigenetics* to describe developmental processes and the adaptation of an organism to an environment.[1] He also coined the term *genetic assimilation* to describe stabilizing selection,[2] and this term is still in modern usage. Examples of genetic assimilation experiments performed by Waddington were selection for high salt tolerance by exposing flies to higher and higher doses of salt and selecting for survivors.[14] Waddington also confirmed the existence of morphoses, which are random and nonadaptive phenotypic variations induced by stress. He showed that an extra pair of wings can be induced by exposing flies to ether during embryogenesis; extra wings do not protect flies against ether. Similarly, he blocked the development of crossveins in wings by heat shocking larvae. Crossveins are small trachea that connect two large veins in a fly wing, and their absence has nothing to do with protection against heat shock. In all of these genetic assimilation experiments, he repeated the treatments and selections for 20–25 generations.[14]

Because of their Lamarckian-sounding conclusions, Waddington's experiments were controversial and were mostly dismissed during his lifetime.[15] However, Waddington is now referred to as the "father of epigenetics" (despite his slightly different meaning for the word). As illustrated below, his experimental approach and theories are currently thought to be visionary.[15,16]

14.2.8 JOHN CAIRNS AND FACILITATED GENETIC VARIATION

In the 1950s, Cairns (1922–), working with *Escherichia coli*, asked whether mutations occur more frequently in genes that are needed to allow survival in a particular environment.[17] To the surprise of his contemporaries, he found that reversion mutations occur more frequently in mutant genes when the bacteria are grown in media

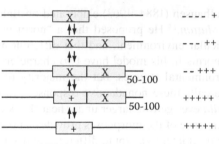

FIGURE 14.1 Amplification model for facilitated mutation. A gene with a mutation that allows it to have partial function (----+) is shown on the top (× in box). If the gene is required for growth, then duplication of the mutant gene will increase the growth rate in the selectable media (---++). Futher amplification to 50–100 copies allows nearly wild-type growth (--+++), but such amplifications are unstable (arrows). Under starvation conditions, the mutation frequency increases. Since the mutated gene is amplified, it has a 50–100 higher rate of mutation than the nonamplified genes in the genome. Occasionally, a wild-type revertant mutation restores normal function to the gene (+ in box). Finally, no longer under selective pressure, the amplified mutant genes contract and a single wild-type gene is present (bottom).

that require the wild-type gene product to efficiently grow in that media.[17] Such "directed mutations" resemble Lamarckian-type evolutionary process because the bacteria apparently evolve in response to need, much like the giraffe has a long neck in response to need. However, with further analyses, it was found that the overall mutation rate increased in response to stress and, consequently, the majority opinion was that Cairns misinterpreted his findings.

Nevertheless, a small group of scientists including John Roth felt that the increase in the overall mutagenesis rate was insufficient to explain Cairns' observations.[18–20] Roth and colleagues proposed that the genome undergoes random amplifications and contractions of large regions continuously. If the mutated gene has partial activity, its amplification has survival advantage (Figure 14.1). According to this model, there is an increase in the mutation frequency of the mutated gene compared with the mutation frequency of the rest of the genome, not because mutations are directed to mutated gene, but rather because it is amplified as much as 50- to 100-fold. When a reversion mutation restores full activity to the mutated gene, then there is no longer selective pressure for the "amplicon" and the unstable repeats contract to a single restored gene.

Roth's model, although some aspects are still being tested, nicely provides a modern molecular explanation for what appears to be Lamarckian-type facilitated mutation. Recently, Sebat and colleagues have shown that the human genome also undergoes a high frequency of whole-gene amplification and contraction (although not as dramatic) and that this type of "spontaneous mutation" can explain diseases such as autism.[21] Other examples of spontaneous and induced expansions and contractions are described in a later section.

14.2.9 Hsp90 Is a Capacitor for Morphological Evolution

In the 1990s, Susan Lindquist extended Waddington's experiments by showing that Hsp90 is a protein "capacitor" for morphological evolution and phenotypic variation in both *Drosophila* and *Arabidopsis*.[22,23] Electricity capacitors store electric charge in circuits that can be released later. A protein capacitor, such as Hsp90, stores phenotypic variation that can be released later in times of stress. Lindquist and colleagues showed that genetic or environmental inactivation of Hsp90, such as by head shock, generates a number of aberrant phenotypes in flies and plants.

Lindquist and colleagues performed genetic assimilation experiments on the novel phenotypes and, like Waddington, showed that after several generations of selection, the phenotypes remain even when the activity of Hsp90 is restored.[22,23] In the experiments described below, we provided evidence that Hsp90 might also have an epigenetic role in morphological evolution and phenotypic evolution.[15,24]

14.3 EPIGENETIC INHERITANCE IN *DROSOPHILA*

The development of any multicellular organism from one undifferentiated cell to hundreds or thousands of cell types is an epigenetic phenomenon. For example, in *Drosophila*, it has been established that cell-type-specific transcription factors establish identity, but it is chromatin-remodeling proteins that maintain the long-term effects of the transcription factors.[25] However, it has only recently been shown that epigenetic marks can be heritable (i.e., transmitted through meiosis).

14.3.1 Epigenetic Transmission of PRE Occupancy through Meiosis

In 1999, Cavalli and Paro showed that the Polycomb response element (PRE), which epigenetically maintains the expression or repression of a gene during development, can also be transmitted through meiosis.[26,27] They placed a PRE upstream of a transgene containing GAL4 binding sites driving the expression of lacZ and a linked mini-w+ gene.[26,27] When they induced the expression of lacZ and mini-w+ by heat shocking flies also containing an hsp70-Gal4 transgene, the flies turned blue (in the presence of the lacZ substrast XGAL) and the eyes turned red (because w+ is also expressed).

Surprisingly, when they segregated lacZ away from hsp70-GAL4, the lacZ and mini-w+ genes were still expressed in a majority of the offspring, even in the absence of the *trans*-acting factor GAL4. This result suggested that the activation complex at the PRE was transmitted through meiosis. They confirmed that PRE occupancy is epigenetically maintained by analyzing Polycomb binding to the PRE in salivary gland cells, and thereby solidified the conclusion that epigenetic phenotypes can be transmitted through meiosis in *Drosophila*.[26,27]

Evidently, despite Weismann's remonstration to Darwin,[9] PRE occupancy induced in the soma must also be occurring in germline cells. In retrospect, this is not that surprising because the modern understanding of germline cells is that they are the ultimate pluripotent stem cell, and stem cells have been shown in numerous studies to be epigenetically plastic.[28–30]

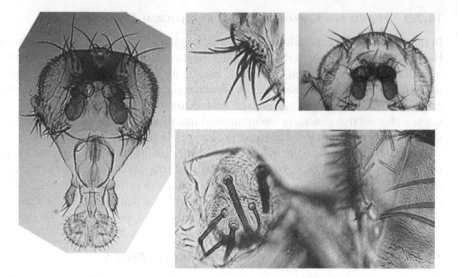

FIGURE 14.2   Mosaic of manifestations of the ELBO phenotype. Ectopic large-bristle out-growths (ELBOs) in the eyes are generated epigenetically and are transgenerationally inherited (see text).

14.3.2   EPIGENETIC ASSIMILATION EXPERIMENTS IN *DROSOPHILA*

Recently, in experiments to elaborate on Lindquist's Hsp90 experiments, we have found that maternal reduction of Hsp90 activity causes severe eye defects in *Drosophila* with the genotype Kr^{If-1} (Krueppel[Incomplete facets-1]) (Figure 14.2). We called the eye deformities ELBOs (ectopic large-bristle outgrowths) because of the numerous large bristles growing out of one or both eyes (they also sometimes look like proximal appendages—i.e., "elbows").[26,27] Even in the absence of genetic variation or further Hsp90 inactivation, we found that we could select for ELBOs and that the frequency of the ELBO phenotype increases in subsequent generations (Figure 14.3). We call this "epigenetic assimilation" to contrast it with Waddington's term, "genetic assimilation."[26,27]

In our epigenetic assimilation experiments, we proposed that Hsp90 inactivation "epigenetically destabilizes" random genes in the germline. In subsequent generations, heritable epigenetic modifications in specific genes are selected, thereby enhancing the phenotype. What is important in the epigenetic assimilation experiments is that the ELBO phenotype was selected in the absence of genetic variation, yet the frequency of the phenotype increased in each subsequent generation until it reached a plateau, when approximately 60% of the offspring had ELBOs (Figure 14.3).[24] Our epigenetic assimilation experiments suggest that both genetic and epigenetic mechanisms are available to provide a reservoir for a population to survive a stressful environment.

14.3.3   GENETIC VERSUS EPIGENETIC CAPACITOR MODELS

As discussed above, Rutherford and Lindquist borrowed the electronic term *capacitor* to describe the function of Hsp90 in regulating morphological evolution.[22]

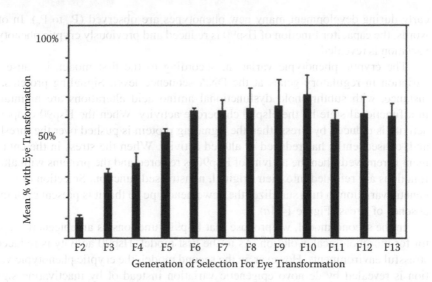

FIGURE 14.3 Epigenetic selection of the ELBO phenotype in the absence of genetic variation. The ELBO phenotype was induced by feeding the P_1 generation the Hsp90 inhibitor geldanamycin at a dose that generates ~1% ELBOs in the F_1. Selection of progeny in subsequent generations showed an increase in the percentage of offspring with the ELBO phenotype until the F_6 generation, when ~60% had the epigenetic phenotype. All of the flies in this experiment were isogenetic (i.e., iso-Kr^{If-1}) to eliminate the possibility that genetic variation was being selected.[24]

Here, we attempt to describe two capacitor models with electronic symbolism (Figure 14.4). In both models, P_0 is the ground-state phenotype, which is uniform across a population because of "canalization" or stabilization of the phenotype.[2] However, in stressful environments, Hsp90 is functionally inactivated, and if the stress occurs

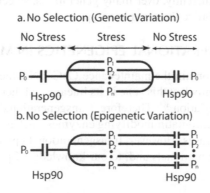

FIGURE 14.4 Genetic versus epigenetic capacitor models for morphological evolution. (*a*) Hsp90 as a genetic capacitor for morphological evolution. The function of Hsp90 is symbolized as a capacitor (short parallel lines). P_0 is the ground-state phenotype. If the stress occurs early during development, many new phenotypes are observed (P_0 to P_n). (*b*) Hsp90 as an epigenetic capacitor for morphological evolution. Note that new phenotypes (P_0 to P_n) remain even when the stress is removed.

early during development many new phenotypes are observed (P_0 to P_n). In other words, the capacitor function of Hsp90 is reduced and previously cryptic phenotypic variation is revealed.

The cryptic phenotypic variation, according to the first model, is caused by variation in regulatory genes at the DNA sequence level. Signaling proteins, for instance, with subthreshold dysfunctional amino acid alterations are maintained in a functional state by the Hsp90 chaperone activity. When the Hsp90 chaperone activity is reduced by stress, then the signaling protein is pushed over the threshold and, consequently, has reduced or altered activity. When the stress in the environment is removed, then the activity of Hsp90 is restored and the proteins with altered functions are refolded into their original, nonstressed function. Selection of cryptic genetic variation in turn stabilizes the new phenotype so that it is present even in the absence of stress (Figure 14.4a).

In the second model, we propose that Hsp90 functions as an epigenetic capacitor for morphological evolution. As in the first model, Hsp90 activity is reduced in stressful environments. However, in the second model, the cryptic phenotypic variation is revealed by de novo epigenetic variation instead of by inactivating signaling molecules with normally subthreshold missense mutations. Since epigenetically induced phenotypic variation can be selected transgenerationally (Figure 14.3), the consequence is that each of the new phenotypes (P_0 to P_n) remains even when the stress is removed (Figure 14.4). This epigenetic model greatly speeds up the rate of morphological evolution. It also eliminates the need for multigenerational stress, which is required in the genetic model for morphological evolution.

As discussed further below, we believe that both models are correct but that they act at different stages of evolutionary adaptation. We propose that epigenetic alterations can induce new phenotypes in the epigenetic assimilation experiments. Later, an epigenetic Baldwin effect occurs because the selected epigenetic changes direct new mutations in the epigenetically modified genes, some of which can stabilize the new phenotype. Concurrently, over many generations, selection of existing genetic variation can also stabilize the new phenotype.

14.4 TRANSGENERATIONAL EPIGENETICS IN MAMMALS

While some insects contain the maintenance CpG DNA methyltransferase Dnmt1, *Drosophila* has evidently lost this gene in its lineage and therefore probably does not have CpG DNA methylation.[4,31] Therefore, transgenerational epigenetic phenomena in flies likely involve chromatin-modifying enzymes, such as histone deacetylases or histone methyltransferases. However, transgenerational epigenetic phenomena have also been observed in mammals and, to our knowledge, CpG methylation has been invoked in all cases.

14.4.1 NUTRITION AND DNA METHYLATION OF RETROTRANSPOSON MUTATIONS

The laboratories of Jirtle, Waterland, and Whitelaw have shown that mice with the Agouti[variegated yellow] (A^{vy}) and Axin[fused] (A^{fu}) mutations show transgenerational epigenetic effects.[32–37] Both of these spontaneous mutations were caused by the insertion of retrotransposons in transcriptional regulatory regions. In the A^{vy} mice, when the retrotransposon is methylated, the mice have a wild-type brown coat color. However, when the retrotransposon is unmethylated, the Agouti gene is ectopically expressed in all of the tissues from the now-active long-terminal repeat (LTR) promoters and enhancers. Consequently, the mice are yellow, obese, and prone to tumors. Waterland and colleagues have shown that a maternal diet rich in methyl donors (such as folic acid) causes the progeny to have more wild-type coat colors because they have a more heavily methylated retrotransposon in the A^{vy} locus.[35] However, whether maternal diet has a global methylation effect on the genome is not yet known.

Whitelaw and colleagues have extended this further by showing that, even under normal dietary conditions, brown A^{vy} mice had more brown A^{vy} progeny, presumably because the methylated retrotransposon at the A^{vy} locus was stably inherited through meiosis. Similar findings have been made with the Ax^{fu} mutation, which causes a variable kinky-tail phenotype in mice. Mothers with diets high in methyl donors had straighter tails, and mothers with methyl-donor-poor diets had kinkier tails.[34]

These transgenerational experiments with A^{vy} and Ax^{fu} are the closest mammalian example of our epigenetic assimilation experiments in *Drosophila*. It will be interesting to determine whether such heritable epigenetic states are present on "normal" genes that do not have retrotransposon insertions. Methylation of CpG islands in the promoter regions of human genes are likely substrates for epigenetic regulation.[3]

14.4.2 ENDOCRINE DISRUPTORS AND TRANSGENERATIONAL EPIGENETIC EFFECTS

Skinner and colleagues showed that endocrine disruptors can have transgenerational epigenetic effects.[38] Exposure of a pregnant rat to the endocrine disruptors vinclozolin (an antiandrogenic compound) or methoxychlor (an estrogenic compound) decreased sperm cell number and viability and, consequently, increased the incidence of male infertility in the F_1 generation. Remarkably, these effects were transferred through the male germline to nearly all males of the F_4 generation. In all generations studied, an increase in DNA hypomethylation is observed.[38] Similar observations have been made with mothers who took the estrogenic compound diethylstilbestrol (DES).[39]

The transgenerational epigenetic ability of an endocrine disruptor to cause a disease in multiple generations has numerous implications in evolution. Transgenerational effects also need to be considered in the governmental regulation of these compounds. Endocrine disruptor contamination is ubiquitous in our environment and also contributes to cancer and obesity.[40] Epigenetic consequences of environmental contamination, dietary components, and physical considerations such as obesity need to be understood better to ensure the health of children.[39,41,42]

14.5 EPIGENETICS, ENVIRONMENT, AND CANCER

It is now widely accepted that cancer is in part an epigenetic disease, although epigenetic alterations are still viewed largely as a surrogate of genetic alterations. Studies of DNA methylation in tumor tissues have revealed at least as many epigenetic as genetic alterations for a given gene, but this is likely to be just the tip of the iceberg. We propose an epigenetic progenitor model in which cancer involves epigenetic disruption of progenitor cells, an initiating mutation, and genetic and epigenetic plasticity.[43]

14.5.1 Transgenerational Epigenetics and Cancer

In the above quote from Feinberg et al.,[43] the authors argue that an epigenetically altered gene, perhaps inherited from a parent, might induce a clone of cells that are sensitized for subsequent mutations to induce cancer. Consistent with this idea, Hitchens and colleagues have shown that an epimutation in MLH1 can be inherited in people.[44] They have shown that people with a hypermethylation of one allele of MLH1 in somatic cells throughout the body have a predisposition for the development of hereditary nonpolyposis colorectal cancer (HNCC). They found evidence that the *epimutation* (hypermethylation of MSH1) was transmitted from a mother to her son. These findings demonstrate transgenerational epigenetic inheritance of cancer susceptibility and suggest that this phenomenon might be of great importance in the development of cancer and other epigenetically influenced diseases.[44]

14.5.2 Clonal versus Polyclonal Models for Cancer Initiation

The conventional model for the origin of human cancer is the *clonal* genetic model.[43] In this model, cancer arises stepwise from a series of mutations in a single cell, and this cell eventually gives rise to a metastatic tumor. In this model epigenetic changes are important, but they are merely a *surrogate* of genetic mutations.[43] As quoted above, Feinberg and colleagues recently proposed a new model for the origin of human cancer that they called the "epigenetic progenitor theory."[43] This theory, which is a modern variation of Paget's "soil and seed" hypothesis from 1889,[45–48] proposed that stress "epigenetically destabilizes" stem cells, which proliferate to form the "soil" that can be later "seeded" by mutations or epigenetic alterations in oncogenes or tumor suppressor genes.[43] If correct, this model promises a paradigm shift in the understanding of the origins of human cancer because, instead of being clonal in origin, this model suggests that cancer may in fact be polyclonal in origin.

The epigenetic progenitor theory also has profound implications in the treatment of cancer because, for instance, treatment of benign tumors with radiation or chemotherapy, a common practice, could potentially increase the rate of later metastatic cancers.[43] Five major lines of evidence were provided to support this theory.[43] First, classical studies have shown that tumor growth properties are reversible, which suggests an epigenetic origin.[49] Second, most, if not all, tumors show global changes in DNA methylation.[50,51] Third, mice have been cloned from a mouse melanoma nucleus, indicating that most properties of cancer are reversible, and therefore epigenetic.[52–54] Fourth, neoplastic clones can be maintained by selecting stem cells in serial passage experiments in mice.[55] Finally, and most convincingly, recent data has shown that loss of imprinting (LOI) of IGF2 occurs throughout the apparently normal colonic epithelia in colorectal cancer.[56–59]

14.6 HOW EPIGENETIC CHANGES CAN LEAD TO GENETIC CHANGES

As described in the previous sections, Waddington and others showed that genetic assimilation of existing genetic polymorphisms in a population can occur over multiple generations to generate a new phenotype in the absence of additional new mutations.

We and others have shown that epigenetic assimilation of a new phenotype can occur by selecting for epigenetic variation in a population with no genetic variation. However, as discussed in this section, epigenetic alterations have a unique power because they can potentially direct new mutations that can stabilize the phenotype. As mentioned above, we call this an epigenetic Baldwin effect because it proposes a mechanism for directed mutation in precisely the genes needed to stabilize a new phenotype.

14.6.1 DEAMINATION OF METHYLATED CYTOSINE TO THYMIDINE

In a survey of ~14,000 genes in several human breast cancer and colon cancer cell lines, more than 100 genes were shown to be significantly mutated.[60] The vast majority of mutations identified were missense mutations, and 59% of the 696 colorectal cancer mutations were C:G-to-T:A transition mutations, whereas only 7% were C:G-to-G:C transversion mutations.[60] This C>T mutation profile suggests that the mutations are epigenetically derived because [5me]C is highly mutagenic; spontaneous deamination of [5me]C to T is more mutagenic than deamination of C to U because DNA repair enzymes are more likely to target the non-DNA base uracil than the normal base thymidine.

In other words, the C>T mutation profile in colon cancer suggests that a majority of the genes mutated in colon cancer were initially epigenetically altered, via DNA methylation, before they were permanently altered by mutation. Methylation of cytosines in a gene generally leads to inactivation of the gene by recruiting methylcytosine-binding proteins such as MeCP2,[61,62] so it is likely that the gene was dramatically reduced in expression epigenetically before it was inactivated genetically.

The maintenance DNA methyltransferase Dnmt1 methylates hemimethylated DNA at CpG sites during replication. There is also rare methylation of cytosines at non-CpG sites by the de novo DNA methyltransferase Dnmt3, but the CpG methylation is much more significant and probably has the most biological and evolutionary relevance. Consistent with the idea that CpG methylation is most relevant, a large fraction (44%) of the mutations in colorectal cancers were at 5'-CpG-3' dinucleotide sites. This 5'-CpG-3' preference usually led to changes of arginine residues in colorectal cancers because four of the six arginine codons contain a CpG site (i.e., the four codons CGX all code for arginine).

Several other amino acids contain the CpG dinucleotide (i.e., 5'-XCG-3'); however, they are underrepresented in the human coding sequences presumably because of their high intrinsic mutagenicity when they are methylated. Note that the 5'-XCG-3' codons contain the C at the third *wobble* position. Consequently, other nucleotides at this position will generally code for the same amino acid. Junctional CpG dinucleotides (i.e., 5'-XXC.GXX-3') are also likely important for epigenetic regulation, particularly in repeated amino acids (see below).[63]

14.6.2 EPIGENETIC REGULATION OF TRINUCLEOTIDE REPEAT EXPANSION AND CONTRACTION

Fondon and Garner[64] have shown that tandem repeat expansions and contractions in the protein-coding regions of developmental genes are a major source for rapid

TABLE 14.1

Repeat Polymorphisms in Dog Developmental Genes Preferentially Use CpG Codons

Repeat Locus	Repeat Unit	Possible Repeat Sequence	Actual Repeat Sequence	Codon Bias in Nonrepeats*	Number of Alleles	CpG in Repeat?**
Alx-4	PQ	CCX CA[G/A]	CCG (16/17)	CCT (29%) CCC (32%) CCA (28%) CCG (11%)	4***	Yes
Bmp-11/GDF-11	A	GCX	GCC (9/10)	GCT (27%) GCC (40%) GCA (23%) GCG (10%)	3	Yes
Dlx-2	G	GGX	GGC (12/13)	GGT (16.5%) GGC (34.0%) GGA (25.3%) GGG (24.2%)	5	Yes
Runx-2	Q	CA[G/A]	CAG (17/19)	CAA (26%) CAG (74%)	6***	No
Sox-9	P	CCX	CCG (8/8)	CCT (29%) CCC (32%) CCA (28%) CCG (11%)	2	Yes
Zic-2	H	CA[C/T]	CAC (8/8)	CAT (42%) CAC (58%)	2	No

* Codon bias in humans, from Zeeberg.[69] For multiple amino acid repeats, only the codon biases with CpG sites are shown. The codons with CpG sites in the codon or in the border with the repeat codon are underlined.

** CpG methylation cannot explain methylation of glutamine (Q) and histidine (H) repeats.

***Includes distinct alleles of equal length, but has variable differences in the coding sequence that are consistent with expansion and contraction events.

morphological evolution in dogs. Research in prokaryotes has shown that repeat expansions and contractions can occur at rates up to 100,000 times higher than point mutations, presumably by unequal crossing over during meiotic recombination.[65,66] Therefore, repeat expansions and contractions potentially have a much greater effect than point mutations on morphological evolution.

In a previous review, we presented arguments that repeat expansions and contractions might be epigenetically upregulated during times of stress.[63] In support of this idea, in a small survey of genes, we found that the incidence of CpG dinucleotides is much higher in vertebrate trinucleotide repeats than in other protein-coding regions, thus suggesting that CpG methylation is under stabilizing selection (Table 14.1). In contrast, in *Drosophila*, which presumably does not have CpG methylation, there is a bias against CpG sequences in repeats.[63] In Figure 14.5, we propose a model

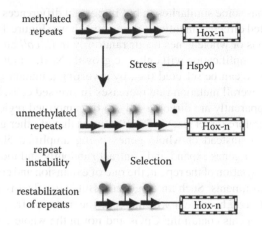

FIGURE 14.5 Combined genetic and epigenetic model for morphological evolution. Model for how stress-induced epigenetic destabilization can lead to enhanced genetic variation. Methylated CpGs are indicated (circles). Under stressful conditions, Hsp90 is functionally inactivated and, through some unknown mechanism, the CpGs become unmethylated. Unmethylated repeats are unstable and expand and contract at a higher frequency. Eventually, the gene has a new activity that decreases the environmental stress and the repeat becomes remethylated (bottom).

whereby environmental regulation of CpG methylation in repetitive sequences in germ cells could be a powerful means to increase the rate of morphological variation, and thereby the rate of morphological evolution, during times of stress.

According to our model, repeats are methylated at CpG sites in the absence of stress in germline precursor cells (i.e., in germline stem cells). Under stressful conditions, Hsp90 becomes functionally inactivated and this leads, through some unknown mechanism, to the hypomethylation of the repeats. Next, we propose that the unmethylated repeats have an increase in the rate of expansion and contraction. We base this part of the model on the observation that mice with mutations in Dnmt1, the maintenance CpG methyltransferase, have microsatellite instability.[67]

Microsatellites are long repeats of two or more bases, such as ATATAT repeated 100 times. Therefore, a polyalanine repeat with the sequence 5'-CCG.CCG.CCG.CCG.CCG-3' would be an example of a very short microsatellite repeat. It is also possible that the repeats are normally nonmethylated and stress causes their methylation. Either way, this would cause a bimodal switch that is responsive to the environment. If the rate of expansion and contraction is different in methylated versus nonmethylated repeats, then this might explain the bias for CpG dinucleotides in dog repeats.[63]

According to our model, if a mother or father were undergoing stress, such as starvation, then his or her offspring would have an increase in the frequency of repeat expansions and contractions. Occassionally, a repeat expansion or contraction will generate a new phenotype that has selective advantage in the offspring. If this occurs, the organism is no longer under stress and the repeat gets remethylated in his or her germline stem cells (Figure 14.5).

Our model has some similarites to but important differences from Roth's model of repeat-mediated directed mutations, described above (Figure 14.1). Roth proposed that amplifications of whole genes occur randomly in *E. coli* and that in a selective environment the amplification will enhance growth. Next, a mutation that stabilizes the new phenotype can be selected (i.e., by reverting a mutant gene to a wild-type gene). Since the overall mutation rate increases in stressed cells, Roth proposed that the mutations apparently are directed only to the amplified region because its copy number (and therefore target size) is 50–100 times that of other genes.

In our model, instead of whole genes being amplified 50–100 times, small repeated regions of genes expand and contract randomly and modestly. We propose that, by hypomethylation of the repeat, the rate of expansion and contraction increases in stressful environments. Such an epigenetic Baldwin effect is more efficient than what occurs in *E. coli* because it will increase the mutation frequency only of those genes that have repeats containing CpGs and not in the whole genome. Analyses of genes that have repeated amino acids show that they are significantly enriched in developmental proteins and transcription factors.[64] Consequently, directed mutation of these genes is more likely to affect morphological evolution. Much work needs to be done to verify or modify aspects of this model.

14.7   FUTURE STUDIES IN THE ROLE OF EPIGENETICS IN EVOLUTION

This is an exciting time to study the role of epigenetics in evolution. One can no longer regard the genome as a fixed structure that is faithfully transmitted to the offspring (with only a few occasional errors). We are beginning to understand how maternal and paternal diet and environmental exposure can affect the health and phenotype of the offspring via global epigenetic alterations.[33,34,68] Studies of rapid morphological evolution in dogs show that coding repeat expansions and contractions are often involved,[64] but the degree and rate of such mutations in humans is not known. We propose that expansions and contractions are under epigenetic control by the environment, thereby suggesting a further mechanism to drive morphological evolution.

Studies of spontaneous cases of autism in humans show that a significant frequency of these cases are caused by copy number changes of so-called autism genes.[21] It is possible that the great stress that humans are under in our increasingly toxic environment epigenetically contributes to these copy number changes as humans strive to evolve to adapt to this new environment. Alas, autism and many other diseases that increase generationally in frequency could be undesired *morphisms* of a stressful environment.

ACKNOWLEDGMENTS

This work was supported by NIEHS and NCI R01 grants (ES012933 and CA105349) to D.M.R.

REFERENCES

1. Haig, D., The (dual) origin of epigenetics, *Cold Spring Harb Symp Quant Biol* 69, 67–70, 2004.
2. Waddington, C. H., Canalization of development and the inheritance of acquired characters, *Nature* 150, 563–65, 1942.
3. Weber, M., Hellmann, I., Stadler, M. B., Ramos, L., Paabo, S., Rebhan, M., and Schubeler, D., Distribution, silencing potential and evolutionary impact of promoter DNA methylation in the human genome, *Nat Genet* 39 (4), 442–3, 2007.
4. Lyko, F., Ramsahoye, B. H., and Jaenisch, R., DNA methylation in *Drosophila melanogaster*, *Nature* 408 (6812), 538–40, 2000.
5. Lamarck, J. B. P., *Zoological philosophy*, Chicago Press, Chicago, 1809.
6. Shanahan, T., Chance as an explanatory factor in evolutionary biology, *Hist Philos Life Sci* 13 (2), 24968, 1991.
7. Darwin, C., *On the origin of species*, Random House, 1859.
8. Darwin, C., *Variation in animals and plants under domestication*, Appleton, New York, 1883.
9. Weismann, A., Parker, W. N., and R*onníeldt, H., *The germ-plasm: a theory of heredity*, C. Scribner's Sons, New York, 1893.
10. Mendel, G., *Experiments in plant hybridisation*, Harvard University Press, Boston, 1865.
11. Kirschner, M. and Gerhart, J., *The plausibility of life: resolving Darwin's dilemma*, Yale University Press, New Haven, CT, 2005.
12. Baldwin, J. A. M., A new factor in evolution, *Am Naturalist* 30, 441, 1896.
13. Schmalhausen, I. I., *Factors of evolution: the theory of stabilizing selection*, Blakiston, Philadelphia, 1949.
14. Waddington, C. H., Genetic assimilation of an acquired character, *Evolution* 7, 118–126, 1953.
15. Ruden, D. M., Garfinkel, M. D., Sollars, V. E., and Lu, X., Waddington's widget: Hsp90 and the inheritance of acquired characters, *Semin Cell Dev Biol* 14 (5), 301–10, 2003.
16. Pigliucci, M., Epigenetics is back! Hsp90 and phenotypic variation, *Cell Cycle* 2 (1), 34–35, 2003.
17. Cairns, J., Overbaugh, J., and Miller, S., The origin of mutants, *Nature* 335 (6186), 142–5, 1988.
18. Andersson, D. I., Slechta, E. S., and Roth, J. R., Evidence that gene amplification underlies adaptive mutability of the bacterial lac operon, *Science* 282 (5391), 1133–5, 1998.
19. Hendrickson, H., Slechta, E. S., Bergthorsson, U., Andersson, D. I., and Roth, J. R., Amplification-mutagenesis: evidence that "directed" adaptive mutation and general hypermutability result from growth with a selected gene amplification, *Proc Natl Acad Sci U S A* 99 (4), 2164–9, 2002.
20. Roth, J. R., Kofoid, E., Roth, F. P., Berg, O. G., Seger, J., and Andersson, D. I., Regulating general mutation rates: examination of the hypermutable state model for Cairnsian adaptive mutation, *Genetics* 163 (4), 1483–96, 2003.
21. Sebat, J., Lakshmi, B., Malhotra, D., Troge, J., Lese-Martin, C., Walsh, T., Yamrom, B., Yamrom, B., Yoon, S., Krasnitz, A., Kendall, J., Leotta, A., Pai, D., Zhang, R., Lee, Y. H., Hicks, J., Spence, S. J., Lee, A. T., Puura, K., Lehtimaki, T., Ledbetter, D., Gregersen, P. K., Bregman, J., Sutcliffe, J. S., Jobanputra, V., Chung, W., Warburton, D., King, M. C., Skuse, D., Geschwind, D. H., Gilliam, T. C., Ye, K., and Wigler, M., Strong association of de novo copy number mutations with autism, *Science* 2007.
22. Rutherford, S. L. and Lindquist, S., Hsp90 as a capacitor for morphological evolution, *Nature* 396 (6709), 336–42, 1998.
23. Queitsch, C., Sangster, T. A., and Lindquist, S., Hsp90 as a capacitor of phenotypic variation, *Nature* 417 (6889), 618–24, 2002.

24. Sollars, V., Lu, X., Xiao, L., Wang, X., Garfinkel, M. D., and Ruden, D. M., Evidence for an epigenetic mechanism by which Hsp90 acts as a capacitor for morphological evolution, *Nat Genet* 33 (1), 70–4, 2003.

25. Lawrence, P. A., *The making of a fly: the genetics of animal design*, Blackwell Science, Oxford, 1992.

26. Cavalli, G. and Paro, R., The *Drosophila* Fab-7 chromosomal element conveys epigenetic inheritance during mitosis and meiosis, *Cell* 93 (4), 505–18, 1998.

27. Cavalli, G. and Paro, R., Epigenetic inheritance of active chromatin after removal of the main transactivator, *Science* 286 (5441), 955–8, 1999.

28. Cairns, J., Cancer and the immortal strand hypothesis, *Genetics* 174 (3), 1069–72, 2006.

29. Dell, H., Developmental biology: marked from the start, *Nature* 445 (7124), 157, 2007.

30. Mitalipov, S. M., Genomic imprinting in primate embryos and embryonic stem cells, *Reprod Fertil Dev* 18 (8), 817–21, 2006.

31. Lyko, F., DNA methylation learns to fly, *Trends Genet* 17 (4), 169–72, 2001.

32. Rakyan, V. K., Blewitt, M. E., Druker, R., Preis, J. I., and Whitelaw, E., Metastable epialleles in mammals, *Trends Genet* 18 (7), 348–51, 2002.

33. Waterland, R. A., Do maternal methyl supplements in mice affect DNA methylation of offspring? [comment], *J Nutrition* 133 (1), 238; author reply 239, 2003.

34. Waterland, R. A., Dolinoy, D. C., Lin, J. R., Smith, C. A., Shi, X., and Tahiliani, K. G., Maternal methyl supplements increase offspring DNA methylation at Axin Fused, *Genesis* 44 (9), 401–6, 2006.

35. Waterland, R. A. and Jirtle, R. L., Transposable elements: targets for early nutritional effects on epigenetic gene regulation, *Mol Cell Biol* 23 (15), 5293–300, 2003.

36. Waterland, R. A. and Jirtle, R. L., Early nutrition, epigenetic changes at transposons and imprinted genes, and enhanced susceptibility to adult chronic diseases, *Nutrition* 20 (1), 63–8, 2004.

37. Waterland, R. A., Lin, J. R., Smith, C. A., and Jirtle, R. L., Post-weaning diet affects genomic imprinting at the insulin-like growth factor 2 (Igf2) locus, *Hum Mol Genet* 15 (5), 705–16, 2006.

38. Anway, M. D., Cupp, A. S., Uzumcu, M., and Skinner, M. K., Epigenetic transgenerational actions of endocrine disruptors and male fertility [see comment], *Science* 308 (5727), 1466–9, 2005.

39. Ruden, D. M., Xiao, L., Garfinkel, M. D., and Lu, X., Hsp90 and environmental impacts on epigenetic states: a model for the trans-generational effects of diethylstilbesterol (DES) on uterine development and cancer, *Hum Mol Genet* 14 (1), R147–R155, 2005.

40. Keith, S. W., Redden, D. T., Katzmarzyk, P. T., Boggiano, M. M., Hanlon, E. C., Benca, R. M., Ruden, D., Pietrobelli, A., Barger, J. L., Fontaine, K. R., Wang, C., Aronne, L. J., Wright, S. M., Baskin, M., Dhurandhar, N. V., Lijoi, M. C., Grilo, C. M., DeLuca, M., Westfall, A. O., and Allison, D. B., Putative contributors to the secular increase in obesity: exploring the roads less traveled, *Int J Obes* (London) 30 (11), 1585–94, 2006.

41. Ruden, D. M., Cui, X., Loraine, A. E., Ye, J., Bynum, K., Kim, N. C., De Luca, M., Garfinkel, M. D., and Lu, X., Methods for nutrigenomics and longevity studies in *Drosophila*: effects of diets high in sucrose, palmitic acid, soy, or beef, *Methods Mol Biol* 371, 111–41, 2007.

42. Ruden, D. M., De Luca, M., Garfinkel, M. D., Bynum, K., and Lu, X., *Drosophila* nutrigenomics can provide clues to human gene-nutrient interactions, *Ann Rev Nutrition* 25, 21.1–21.24, 2005.

43. Feinberg, A. P., Ohlsson, R., and Henikoff, S., The epigenetic progenitor origin of human cancer, *Nat Rev Genet* 7, 21–33, 2006.

44. Whitelaw, N. C. and Whitelaw, E., How lifetimes shape epigenotype within and across generations, *Hum Mol Genet* 15 Spec No 2, R131–7, 2006.

45. Paget, S., The distribution of secondary growths in cancer of the breast. 1889, *Cancer Metastasis Rev* 8 (2), 98–101, 1989.
46. Weber, M. H., Goltzman, D., Kostenuik, P., Rabbani, S., Singh, G., Duivenvoorden, W. C., and Orr, F. W., Mechanisms of tumor metastasis to bone, *Crit Rev Eukaryotic Gene Expression* 10 (3-4), 281–302, 2000.
47. Fidler, I. J., The pathogenesis of cancer metastasis: the 'seed and soil' hypothesis revisited, Nature Reviews, *Cancer* 3 (6), 453–8, 2003.
48. Mueller, M. M. and Fusenig, N. E., Friends or foes—bipolar effects of the tumour stroma in cancer, Nature Reviews, *Cancer* 4 (11), 839–49, 2004.
49. Lotem, J. and Sachs, L., Epigenetics wins over genetics: induction of differentiation in tumor cells, *Seminars Cancer Biol* 12 (5), 339–46, 2002.
50. Feinberg, A. P. and Vogelstein, B., Hypomethylation distinguishes genes of some human cancers from their normal counterparts, *Nature* 301 (5895), 89–92, 1983.
51. Feinberg, A. P. and Vogelstein, B., Hypomethylation of ras oncogenes in primary human cancers, *Biochem Biophys Res Comm* 111 (1), 47–54, 1983.
52. Hochedlinger, K., Blelloch, R., Brennan, C., Yamada, Y., Kim, M., Chin, L., and Jaenisch, R., Reprogramming of a melanoma genome by nuclear transplantation, *Genes Devel* 18 (15), 1875–85, 2004.
53. Jaenisch, R., Hochedlinger, K., and Eggan, K., Nuclear cloning, epigenetic reprogramming and cellular differentiation, *Novartis Foundation Symposium* 265, 107-18; discussion 118–28, 2005.
54. Jaenisch, R., Hochedlinger, K., Blelloch, R., Yamada, Y., Baldwin, K., and Eggan, K., Nuclear cloning, epigenetic reprogramming, and cellular differentiation, *Cold Spring Harbor Symposia on Quantitative Biology* 69, 19–27, 2004.
55. Singh, S. K., Hawkins, C., Clarke, I. D., Squire, J. A., Bayani, J., Hide, T., Henkelman, R. M., Cusimano, M. D., and Dirks, P. B., Identification of human brain tumour initiating cells [see comment], *Nature* 432 (7015), 396–401, 2004.
56. Cui, H., Cruz-Correa, M., Giardiello, F. M., Hutcheon, D. F., Kafonek, D. R., Brandenburg, S., Wu, Y., He, X., Powe, N. R., and Feinberg, A. P., Loss of IGF2 imprinting: a potential marker of colorectal cancer risk [see comment], *Science* 299 (5613), 1753–5, 2003.
57. Cui, H., Horon, I. L., Ohlsson, R., Hamilton, S. R., and Feinberg, A. P., Loss of imprinting in normal tissue of colorectal cancer patients with microsatellite instability [see comment], *Nature* Medicine 4 (11), 1276–80, 1998.
58. Cui, H., Niemitz, E. L., Ravenel, J. D., Onyango, P., Brandenburg, S. A., Lobanenkov, V. V., and Feinberg, A. P., Loss of imprinting of insulin-like growth factor-II in Wilms' tumor commonly involves altered methylation but not mutations of CTCF or its binding site, *Cancer Res* 61 (13), 4947–50, 2001.
59. Cui, H., Onyango, P., Brandenburg, S., Wu, Y., Hsieh, C. L., and Feinberg, A. P., Loss of imprinting in colorectal cancer linked to hypomethylation of H19 and IGF2, *Cancer Res* 62 (22), 6442–6, 2002.
60. Sjoblom, T., Jones, S., Wood, L. D., Parsons, D. W., Lin, J., Barber, T. D., Mandelker, D., Leary, R. J., Ptak, J., Silliman, N., Szabo, S., Buckhaults, P., Farrell, C., Meeh, P., Markowitz, S. D., Willis, J., Dawson, D., Willson, J. K., Gazdar, A. F., Hartigan, J., Wu, L., Liu, C., Parmigiani, G., Park, B. H., Bachman, K. E., Papadopoulos, N., Vogelstein, B., Kinzler, K. W., and Velculescu, V. E., The consensus coding sequences of human breast and colorectal cancers, *Science* 314 (5797), 268–74, 2006.
61. Adams, V. H., McBryant, S. J., Wade, P. A., Woodcock, C. L., and Hansen, J. C., Intrinsic disorder and autonomous domain function in the multifunctional nuclear protein, MeCP2, *J Biol Chem* 282 (20), 15057–64, 2007.
62. Klose, R. and Bird, A., Molecular biology. MeCP2 repression goes nonglobal, *Science* 302 (5646), 793–5, 2003.

63. Ruden, D. M., Lu, X., and Garfinkel, M. D., Epigenetic regulation of trinucleotide repeat expansions and contractions and the "biased embryos" hypothesis for rapid morphological evolution, *Curr Genomics* 6, 145–55, 2005.

64. Fondon, J. W. and Garner, H. R., Molecular origins of rapid and continuous morphological evolution, *Proc Natl Acad Sci U S A* 101 (52), 18058–63, 2004.

65. Ellegren, H., Microsatellite mutations in the germline: implications for evolutionary inference, *Trends Genet* 16 (12), 551–8, 2000.

66. Ellegren, H., Microsatellites: simple sequences with complex evolution, *Nature Rev Genet* 5 (6), 435–45, 2004.

67. Kim, M., Trinh, B. N., Long, T. I., Oghamian, S., and Laird, P. W., Dnmt1 deficiency leads to enhanced microsatellite instability in mouse embryonic stem cells, *Nucleic Acids Res* 32 (19), 5742–9, 2004.

68. Waterland, R. A. and Jirtle, R. L., Transposable elements: targets for early nutritional effects on epigenetic gene regulation, *Mol Cell Biol* 23 (15), 5293–300, 2003.

69. Zeeberg, B., Shannon information theoretic computation of synonymous codon usage biases in coding regions of human and mouse genomes, *Genome Res* 12 (6), 944–55, 2002.

15 Epigenetics and Epigenomics

José Ignacio Martín-Subero and Reiner Siebert

CONTENTS

15.1 Introduction 261
15.2 Technical Approaches for Studying the Epigenome 262
    15.2.1 A Historical Perspective 262
    15.2.2 Genome-Wide Detection of DNA Methylation Changes 263
        15.2.2.1 Microarray-Based Analysis of DNA Methylation 264
    15.2.3 Genome-Wide Detection of Histone Modifications 268
    15.2.4 High-Throughput Sequencers: A New Promise for
        Epigenomic Studies 268
15.3 Epigenomics in Biomedical Research 269
    15.3.1 Characterizing the Epigenome of Normal Cells 269
        15.3.1.1 DNA Methylation 270
        15.3.1.2 Histone Modifications 272
    15.3.2 Characterizing the Cancer Cell Epigenome 273
15.4 Future Directions 276
References 277

15.1 INTRODUCTION

Sequencing of the human genome has been one of the most important achievements in the history of science.[1,2] However, scientists worldwide are starting to realize that knowing genetic information is not sufficient to understand phenotypic manifestations. The genome encodes for potential information, but the way the DNA sequence is translated into function does not directly depend on the sequence itself, but rather on the interaction with environmental factors. And here is where the science of epigenetics comes into play, because it integrates all the different chemical languages that genome and environment use to communicate with each other.[3,4] Epigenetics literally means "upon genetics" and bridges DNA information and function by regulating gene expression without modifying the DNA sequence itself. However, a more inclusive definition of epigenetic events was recently proposed as "the structural adaptation of chromosomal regions so as to register, signal or perpetuate altered activity states."[5]

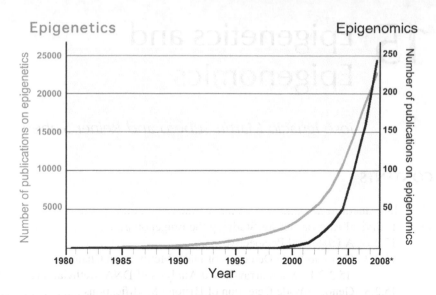

Epigenetics   Epigenomics

FIGURE 15.1 Impact of epigenetics (*gray line*) and epigenomics (*black line*) in biology and medicine, measured as the number of publications on these topics in the PubMed database (http://www.ncbi.nlm.nih.gov/sites/entrez).

The most widely studied epigenetic changes are DNA methylation of cytosines within CpG dinucleotides and a growing number of chemical modifications at different amino-acid residues of histone tails (so far, more than 100 have been identified); for example, acetylation, methylation, phosphorylation, and ubiquitination.[6] Although the present review mainly focuses on DNA methylation and histone modifications, other epigenetic factors like nuclear positioning, noncoding RNAs, and microRNAs are also associated with gene regulation and chromatin structure.[7-9] If epigenetics aims at studying epigenetic changes of specific sequences, epigenomics refers to the delineation of such potentially heritable changes across the complete genome.[10-15]

The pace of discovery of novel scientific findings goes hand in hand with technical developments. In the epigenetics field, the past few years have witnessed the introduction of new techniques, allowing the characterization of the epigenome.[11-13,16-19] Since the development of microarray-based approaches for epigenomics, the number of studies aimed at characterizing the epigenome under normal and altered conditions has been growing exponentially (Figure 15.1).

The goal of this review is to present an outline of the currently available techniques for studying the epigenome and their applications in biology and medicine.

15.2 TECHNICAL APPROACHES FOR STUDYING THE EPIGENOME

15.2.1 A HISTORICAL PERSPECTIVE

Over the past three decades, a large number of different methods have been developed to study DNA methylation changes.[17,18,20-22] The initial efforts in the 1970s were focused on the measurement of global DNA methylation content and the study of particular sequences by Southern blot analyses using methylation-sensitive restriction endonucleases. The limitations of the latter method (e.g., large amounts of high-quality DNA, sequence biases, and problems with incomplete digestions) made the study of specific sequences time consuming and not widely applicable. It was not until 1992, with the introduction of the sodium bisulfite conversion technique, that DNA methylation analyses took a revolutionary step forward.[23] Sodium bisulfite has the property of converting unmethylated cytosine into uracil, whereas methylated cytosine remains unmodified. The combination of this chemical modification with genomic sequencing and methylation-specific PCR (MSP) made the study of DNA methylation changes widely available, and a large number of studies were published from the late 1990s on (Figure 15.1).[24] However, these PCR-based approaches are restricted to the study of a few candidate genes and are not suitable as screening techniques to identify novel markers. To overcome this, techniques like amplification of intermethylated sites (AIMS) and restriction landmark genomic scanning (RLGS), which combine the use of methylation-sensitive restriction endonucleases with 1D or 2D electrophoresis, have been established.[25,26] These techniques are time consuming, and every new fragment identified as differentially methylated between a control and a test sample has to be cloned and sequenced. A step forward has been made in the recent years with the introduction of the microarray technology,[27] which has provided the basis for a new revolution in epigenetics. Exploiting the resources made available through the Human Genome Project combined with the microarray technology and classical epigenetic methods, the simultaneous study of epigenetic changes of thousands of known sequences is now possible.

The following sections provide a summary of the currently available methods for epigenomics.

15.2.2 Genome-Wide Detection of DNA Methylation Changes

The detection of global DNA methylation levels can be studied from two different perspectives. On the one hand, the whole-genome content of DNA methylation or histone modifications can be measured by different methods; for example, high-performance liquid chromatography (HPLC), high-performance capillary electrophoresis (HPCE), or luminometric methylation assay (LUMA), although this method is applicable only for DNA methylation.[18,28–30] For instance, in cancer cells, the DNA methylation content was measured in the early 1980s, and a deregulation of the epigenome was detected in the form of global hypomethylation,[31,32] which has later been associated with chromosomal instability.[33,34] A more recent study has also shown that cancer cells are additionally characterized by global loss of acetylation at lysine 16 and trimethylation at lysine 20 of H4.[35] On the other hand, genome-wide DNA methylation changes of particular sequences can be detected by microarray-based epigenomics. This technology represents a powerful approach for epigenomic profiling and the detection of novel epigenetic markers. A more detailed description of this technology is provided here.

15.2.2.1 Microarray-Based Analysis of DNA Methylation

15.2.2.1.1 Sample Preparation

Several strategies have been described to differentiate methylated and unmethylated cytosines in the context of microarray-based epigenomics (Table 15.1). These are based either on the enrichment of methylated DNA or on the chemical modification of the DNA by sodium bisulfite.

One possible technique to isolate methylated DNA sequences is by digesting the DNA with methylation-sensitive or -insensitive restriction endonucleases, and subsequent purification via, e.g., biotin labeling or adapter ligation, followed by PCR

TABLE 15.1

Techniques Used for Genome-Wide DNA Methylation Analyses

Method	Principle	Ref.
RLGS	Methylation-sensitive restriction digestion + 2D electrophoresis	26
MCA	Methylation-sensitive restriction digestion + printed membranes/dot-blot analyisis or microarray hybridization	106, 165
DMH	Methylation-sensitive restriction digestion + microarray hybridization	37
AIMS	Methylation-sensitive restriction digestion + 1D electrophoresis	25
MSO microarray	Bisulfite conversion + PCR + bead array hybridization	46
ChIP-on-chip	Chromatin immunoprecipitation with antibodies against MBDs + microarray hybridization	43
NotI digestion coupled to BAC array	Methylation-sensitive restriction digestion + microarray hybridization	36
MeDIP-on-chip	Isolation by 5-methylcytosine antibody + microarray hybridization	41
MCIp-on-chip	Isolation by MBD-Fc beads + microarray hybridization	40
HELP	Methylation-sensitive restriction digestion + microarray hybridization	38
Methylation-specific bead arrays	Bisulfite conversion + allele-specific primer extension + bead array hybridization	45
MSNP	Methylation-sensitive restriction digestion + SNP-chip hybridization	51
MMASS	Combinations of methylation-sensitive restriction digestions + microarray hybridization	166
MIRA-assisted microarray analysis	Isolation of methylated DNA by affinity to the MBD2/ MBD3L1 complex + microarray hybridization	115
MSDK	Methylation-sensitive restriction digestion + SAGE	60
aPRIMES	Differential restriction and competitive hybridization of methylated and unmethylated DNA	119
Expression profiling after demethylation	Treatment with demethylating agents + expression microarray in cells with and without treatment	133

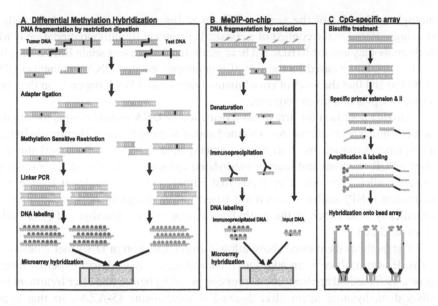

FIGURE 15.2 Graphical display of different methods used for detecting DNA methylation changes by microarrays (adapted from Nieländer et al.[167]). (A) Differential methylation hybridization based on digestion of the DNA with methylation-sensitive restriction enzymes.[37] (B) MeDIP-on-chip based on the enrichment of differentially methylated DNA fragments via immunoprecipitation of methylated DNA.[41] (C) CpG-specific microarray based on bisulfite treatment of the DNA.[45]

amplification. Examples of methods exploiting this approach are differential methylation hybridization (DMH, Figure 15.2A), HpaII tiny fragment enrichment by ligation-mediated PCR (HELP), and NotI digestion coupled with BAC arrays.[36–38] The use of methylation-sensitive enzymes is biased by the fact that not all CpG islands contain enzyme recognition sites. Therefore, not all the CpG-islands in the genome can be interrogated. To partially overcome this limitation, authors of a recent technical report applied different combinations of four methylation-sensitive enzymes (i.e., HpaII, Hin6I, AciI, and HpyCH4IV), which cover approximately 41% of all CpGs across the genome. This report also suggested that using the unmethylated DNA fraction improves the chances of detecting differential methylation between two samples.[39]

An alternative to this approach is the isolation of methylated DNA sequences by applying antibodies binding to methylated cytosines, which is less biased toward specific sequences than those methods based on methylation-sensitive endonucleases. Different strategies using this approach have been successfully applied for the detection of DNA methylation changes. Two of them use DNA as input material (e.g., methylated DNA immunoprecipitation [MeDIP] and methyl-CpG immunoprecipitation [MCIp]).[40,41] MeDIP is an adaptation of the chromatin immunoprecipitation (ChIP) protocol for DNA and uses an antibody against 5-methylcytidine to immunoprecipitate methylated DNA (Figure 15.2B).[41] MCIp uses a recombinant protein made of the methyl-CpG-binding domain (MBD) of the MBD2 protein and the Fc fraction of the human IgG1 to directly bind methylated DNA and isolate the methylated fraction of the genome.[40,42] An additional method uses whole cells as starting material in the context of a classical chromatin immunoprecipitation (ChIP) with antibodies against MBDs.[43] These methods also present some limitations, like the low resolution based on the size of immunoprecipitated DNA fragments (~200–1000 bp) and that the level of enrichment of methylated DNA depends on the abundance of CpGs in a given sequence.[44]

Another approach for array-based detection of DNA methylation is the application of a bisulfite treatment. As explained above, sodium bisulfite chemically induces a sequence variation by converting unmethylated cytosines into uracil (thymine after a PCR reaction) and leaving methylated cytosines unmodified. This sequence variation allows the use of methods that already exist for single-nucleotide polymorphism (SNP) analysis, which are based on the design of oligonucleotides that specifically bind either to the methylated (C) or to the unmethylated (U/T) allele (Figure 15.2C).[45,46]

The methods mentioned above allow a direct detection of DNA methylation patterns. However, there is an additional, but indirect, way to detect hypermethylated genes. This method applies gene expression profiling before and after treatment with DNA demethylating agents like 5-aza-2′-deoxycytosine (5-AZA), so that hypermethylated genes become reactivated after treatment.[47] Although this technique has allowed the detection of novel cancer-related hypermethylated genes, 5-AZA is highly toxic to the cells and can alter the expression levels of many genes regardless of their methylation status, leading to high false positive and false negative rates and a thorough and time-consuming data validation.[21] As an example, Shames and colleagues detected that 5-AZA treatment induced a fourfold overexpression of 866 loci (from a total of 47,000) in lung cancer cell lines, from which they selected 132 candidates, and 45 of them were studied by methylation analysis. Finally, a total of 31 genes were identified as differentially methylated lung cancer as compared with normal lung tissue.[48]

15.2.2.1.2 Microarray Platforms for Epigenomics

In line with the availability of various methods for the differentiation between methylated and unmethylated sequences, there is also a wide range of microarray platforms available for DNA methylation analysis, which differ in resolution, number, and type of genomic regions detected (Table 15.2). The initially applied microarrays used CpG island clones from libraries in which CpG-rich fragments had been isolated by MeCP2 columns.[49] These arrays are biased toward those clones contained in the available libraries and, therefore, are not representative for the complete genome. Additionally, regulatory regions of special interest might not be present. BAC/PAC arrays, initially developed for the detection of genomic imbalances by comparative genomic hybridization, also have been used for epigenomic studies.[36,41,50] Although tiling BAC/PAC arrays containing the complete genome are available, the resolution of BAC/PAC arrays is limited by the size of the inserts (~100–200 kb). Thus, only a global epigenetic signature for that large DNA stretch can be obtained, rendering the identification of differentially methylated gene promoters difficult.

In the last two years, there has been a shift toward microarrays made of short oligonucleotides (usually ranging from 25 to 60 bp), which can reach a very high

TABLE 15.2

Current Microarray Platforms Used for Epigenomic Studies

Microarray Platform for Epigenomics	Resolution	Number of Clones/Oligos*	Coverage
BAC/PAC clones	100–200 kb	Up to ~33.000	~ Complete genome
CpG islands	100–1000 bp	Up to ~12.000	CpG islands
Oligonucleotides			
Promoter	25–60 bp**	244000, 385000, 4.6 million	Promoter regions
CpG island	25–60 bp**	244000, 385000	CpG islands
Tiling	25–60 bp**	Up to 45 million (set of 7 arrays)	~ Complete genome
CpG-dinucleotide specific	1 bp	Up to 1536	Selected promoters

* Improved microarrays with higher resolution are constantly being developed, so the number of oligos on a single array increases as new platforms become available.

** This is the size of the oligonucleotide; the final resolution depends on the method used to enrich the DNA for methylated sequences or histone modifications.

resolution, are commercially available, and can be easily customized according to the user's needs. Available oligonucleotide arrays for epigenomics include, for example, promoter arrays and CpG island arrays, which contain a high density of oligomers in each studied region. A recent study has also combined the use of methylation-specific endonucleases with a SNP-ChIP to obtain an integrated genetic and epigenetic profiling.[51] Recently, oligonucleotide tiling arrays have also been developed that contain up to several millions of oligonucleotides and virtually cover the whole genome with high resolution. This approach has allowed researchers to obtain the first high-resolution DNA methylation profile of a living organism (*Arabidopsis thaliana*).[52] In contrast to arrays containing regulatory regions of genes, tiling arrays also allow the study of epigenetic modifications in noncoding areas, whose role in cell physiology has been gaining importance in recent years.[7,9]

One of the limitations of the methods cited above is that they provide only a blurry picture of the methylome, and it is not possible to determine the methylation status of specific CpGs. This problem can be overcome either by bisulfite sequencing of specific CpGs or with CpG-dinucleotide-specific microarrays.[45,53] For instance, the technology developed by Bibikova and collaborators is based on the combination of a bisulfite treatment of the test DNA, oligonucleotide annealing to the methylated or unmethylated specific CpG, oligonucleotide extension, PCR with universal differentially labeled primers for the methylated or unmethylated allele, and hybridization onto a random bead array (Figure 15.2C). This method allows the accurate quantification of the methylation status of up to 1536 individual CpGs located in the promoter regions of selected genes.[45,54]

In spite of the development of a wide range of microarray-based technologies to study the epigenome, they all show differences in terms of sample preparation, resolution, type of sequence studied, quantification accuracy, and complexity of the bioinformatic tools to analyze the data. Therefore, a systematic comparison and validation of epigenomic methods (i.e., different probe preparation and array platforms) is needed to determine their advantages, disadvantages, and suitability for a proper characterization of the methylome.

15.2.3 GENOME-WIDE DETECTION OF HISTONE MODIFICATIONS

The study of histone modifications at specific genomic regions is mostly based on a single technique, the so-called ChIP. This technique exploits the availability of antibodies that specifically detect certain histone modifications. The experimental procedure is based on an initial crosslink between histones and DNA by formaldehyde, followed by chromatin shearing and incubation with a highly specific antibody towards a histone modification. Then, the chromatin bound to the antibody is isolated by, for example, agarose beads coated with protein A or G, and finally the DNA is separated from the proteins by reversing the crosslinks and subsequent DNA extraction. This isolated DNA is enriched for sequences containing the histone modification of interest and can be then used for microarray-based studies.[55,56] The main limitations of ChIP are the necessity of a highly specific antibody for the histone modification of interest, the availability of fresh material (whole cells are required), and the large number of cells required (approximately $\sim10^7$). Some recent publications have also developed protocols for ChIP that use a smaller number of cells (e.g., as little as 100 cells), which are required to study histone modifications in, for example, small subpopulations of cells or clinical samples.[57–59]

In terms of microarray platforms for the study of histone modifications, most of the microarray types shown in Section 15.2.2.1.2, with the exception of those requiring a previous bisulfite treatment (e.g., CpG-specific microarray), can also be used, especially promoter-specific and tiling oligonucleotide arrays.

15.2.4 HIGH-THROUGHPUT SEQUENCERS: A NEW PROMISE FOR EPIGENOMIC STUDIES

In spite of the potential of microarrays to characterize DNA methylation and histone modifications across the genome, they are limited either by resolution, type, and number of sequences analyzed or by their quantification accuracy. In any case, the complete characterization of the human epigenome of a given sample requires the quantification of the methylation status of each of the ~55 million CpG dinucleotides per diploid cell and the distribution of histone marks of every DNA region, and today this is far from the possibilities of the current microarray platforms. One of the possibilities for sequencing-based methylation analysis is methylation-specific digital karyotyping (MSDK), a SAGE (serial analysis of gene expression)-like procedure that uses methylation-sensitive endonucleases to generate short-tagged DNA segments which are then sequenced and identified.[60] A similar procedure, called genome-wide mapping technique (GMAT), has been developed to map histone modifications by linking ChIP and SAGE.[61] However, the MSDK approach does not detect the methylation status of specific CpGs. This can be reached by high-throughput sequencing of sodium bisulfite-treated samples, but it is time consuming and expensive using classical sequencing approaches.[62,63] The development of a new generation of sequencers is now revolutionizing both genomics and epigenomics.[64–66] These new sequencing technologies are based, for example, on pyrosequencing using millions of picoliter-scale reactions, sequencing by synthesis, and sequencing by ligation,[64] and can sequence up to 2 gigabases of DNA in a single experiment. (The human genome is made up of ~3.1 gigabases.) The initial applications of these technologies in the epigenomics field have allowed researchers to profile the distribution of 20 different histone methylation marks in human CD4+ T cells[67] or the chromatin state (using six different histone marks) of mouse embryonic stem cells and lineage-committed cells like neural progenitor cells and embryonic fibroblasts.[68] High-throughput sequencing has also been applied to study interactions between DNA and regulatory elements by means of a technique called chromosome conformation capture carbon copy (5C).[69,70] This and related approaches are used to investigate chromatin territories within the three-dimensional space of the nucleus and have revealed an extensive network of communication between and within chromosomes.[8] With regard to the detection of DNA methylation changes, the reduction from 4 base pairs to 3 base pairs of unmethylated sequences (C is transformed to U and then to T in the PCR reaction) after bisulfite treatment poses a methodological problem, to identify the origin of the sequenced fragments. Therefore, current technologies do not yet allow direct sequencing of bisulfite-treated whole-genomic DNA, although this will be most likely achieved in the near future. The first report using high-throughput sequencers to measure DNA methylation levels has been published by Taylor and collaborators.[71] In a single experiment, they were able to sequence a total of 125 PCR products from 25 CpG islands after bisulfite treatment of pooled peripheral blood samples and clinical specimens of four different lymphoid hematopoietic malignancies. In order to differentiate the origin of the different samples, they added a four-nucleotide sample-specific tag to the 5' end of each primer. In contrast to classical bisulfite sequencing, in which 10 reads are usually generated per sequence to quantify methylation, the high-throughput approach used by Taylor et al. sequenced a mean of 1697 reads per amplicon in a single 5.5-hours/run, which allows accurate quantification of the methylation status of each CpG within PCR products.[71]

15.3   EPIGENOMICS IN BIOMEDICAL RESEARCH

15.3.1   CHARACTERIZING THE EPIGENOME OF NORMAL CELLS

It is known that epigenetics plays a key role in physiological processes like development, establishment of tissue identity, X chromosome inactivation, chromosomal stability, and gene transcription regulation.[11] Now, the application of epigenomic approaches is helping researchers understand these processes at a genome-wide level and obtain new insights on the mechanisms underlying genome regulation. Also, to convey the impact of epigenomic changes in disease, it is important to profile epigenetic patterns in normal tissues.

15.3.1.1 DNA Methylation

A series of recent epigenomic studies have correlated DNA methylation patterns in different normal tissues with sequence features, evolutionary conservation, and impact on gene regulation, which are summarized below.

Beck and collaborators have published the initial phase of the Human Epigenome Project, an ambitious international endeavor to identify, catalog, and interpret DNA methylation profiles of representative human tissues.[62,63] They performed a bisulfite sequencing of a total of 2524 amplicons from 873 genes in samples from 12 different tissues and essentially observed a bimodal distribution of DNA methylation; that is, different CpGs from a given locus were either methylated or unmethylated. Interestingly, they found that 5′ UTRs with CpG islands (i.e., defined as DNA stretches of at least 200 bp, more than 50% C+G, and a ratio of observed CpG versus expected of at least 0.6)[72] were mostly unmethylated (87.9%), whereas 5′ UTRs with low CpG content were frequently methylated (~50%), which is in line with other studies.[44,73–75] Also, a recently published microarray-based DNA methylation analysis showed that about 4% of the promoter-associated CpG islands are methylated in normal peripheral blood.[76] Comparing DNA methylation in males and females, and in different age groups, they could not find any significant difference, at least in the limited number of genes studied (n = 873). In contrast, experimental evidence suggests that DNA methylation changes in genes like lamin A/C and WRN are associated with age (reviewed by Fraga and Esteller[77]). With regard to tissue-specific DNA methylation, Eckardt et al.[62] identified that 22% of the amplicons studied were differentially methylated, with sperm showing the highest methylation differences (e.g., 20% as compared with fibroblasts) and CD4+ and CD8+ T lymphocytes the cells showing the lowest difference (~5%). Also, studying orthologous amplicons in humans and mice, they identified 70% conservation in the DNA methylation profiles, which is in agreement with the high conservation of histone modification profiles in humans and mice observed in an independent study.[78]

Another landmark study in epigenomic research has been recently published by Schübeler and collaborators.[44] Instead of bisulfite sequencing, they applied MeDIP-on-chip in fibroblasts and sperm and additionally measured gene activity (as RNA polymerase occupancy) and 2mK4-H3 (a chromatin mark associated with gene activation) status by ChIP-on-chip. They classified promoter regions into low, intermediate, and high CpG content and observed that low and high CpG promoters were mostly methylated and unmethylated in normal tissues, respectively, which is in line with the study mentioned above.[62] Comparing DNA methylation levels with RNA polymerase occupancy levels, they found that 66% of the genes with high CpG promoters were active, which contrasts with the 11% activity of low CpG promoters. A negative correlation was observed between DNA methylation and gene activity in genes with high and intermediate CpG promoters, as expected. Remarkably, DNA methylation and gene activity did not correlate in genes with low CpG promoters, suggesting that DNA methylation does not regulate gene activity in regions with low CpG density. Nonetheless, reports indicate that DNA methylation in genes lacking proper CpG islands, like Oct-4 and Nanog, can be associated with gene repression.[79,80] Comparing DNA methylation, gene activity, and 2mK4-H3 status, Weber et al.[44] confirmed that the presence of 2mK4-H3 correlated well with gene activity. However, high and intermediate CpG promoters from inactive genes were unexpectedly enriched for 2mK4-H3, and the authors suggest that a chromatin state might protect CpG-rich promoters from DNA methylation. Similarly to other studies, Weber et al. also detected clear differences between somatic tissues and sperm, with somatic cells characterized mainly by a large number of hypermethylated genes which were unmethylated in sperm samples.[38,62,81] Schilling and Rehli[81] also suggested the presence of differential tissue-specific methylation patterns targeting noncoding miRNA genes like miR-127, miR-142, miR-338, and miR-363. Another study, conducted by Ching and collaborators,[36] applied the NotI methylation-sensitive endonuclease onto BAC arrays in normal peripheral blood and normal astrocytes. Among other genes, they identified SHANK3, a gene involved in postsynaptic density, to be unmethylated and expressed in brain cells whereas it was methylated and repressed in blood cells.[36]

Bibikova et al.[54] studied embryonic stem cells and differentiated cells with a CpG-specific bead array (371 genes were studied). They identified a subset of 25 CpGs from 23 genes that were differentially methylated between embryonic stem cells and differentiated cells. We have applied an updated version of this CpG-specific array containing more than 800 genes to different hematopoietic tissues and fibroblasts (Martín-Subero et al., unpublished). As shown in Figure 15.3A, B cells isolated from tonsils or immortalized in vitro (B-cell lymphoblastoid cell lines) show similar DNA methylation profiles and cluster separately from whole peripheral

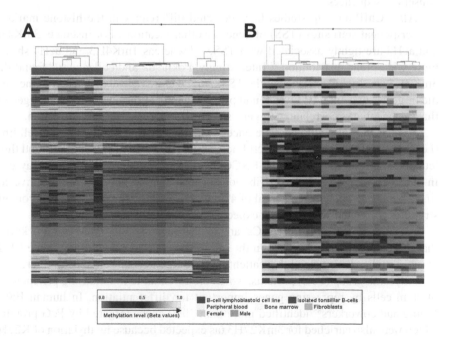

FIGURE 15.3 Heatmap from a hierarchical cluster analysis of DNA methylation data generated with the bead-array technology (Illumina Inc.) in different normal tissue samples. (a) Display of the methylation status of CpGs located in autosomal chromosomes and (b) in chromosome X. Red indicates methylated CpG loci whereas green indicates lack of DNA methylation.

blood and bone marrow samples. Fibroblasts also make a clearly distinct DNA methylation cluster. Figure 15.3B shows the DNA methylation pattern of chromosome X in normal tissues, which clearly differentiates male and female samples due to the epigenetic inactivation of one chromosome X in women.

15.3.1.2 Histone Modifications

The development of ChIP-on-chip has also enabled remarkable progress in the characterization of histone modifications at the genome-wide level.[11,55,56] Several studies using this technology are now allowing deeper insights into which histone modifications (and in which combination) are associated with a given chromatin and transcriptional state (collectively known as the histone code).[82] Also, some of these studies are allowing the characterization of the chromatin structure in different cell types, like in stem cells.[83–85]

Independent studies using different microarray platforms have established that H3 acetylation and 3mK4H3 profiles are highly concordant with each other and consistent with gene activity in humans, mice, flies, and yeast.[78,86–90] The high levels of conservation of histone marks in different species, also in orthologous regions with modest sequence conservation, indicates the universal role of histone modifications in regulating gene transcription.[78] Also, this observation strengthens the power of comparative epigenomic analyses to identify regulatory elements that lie outside conserved sequences.[11]

Also, ChIP-on-chip studies have detected differences in the histone marks at transcriptional start sites (TSS) of genes or other locations. For instance, 3mK4H3 and acH3 are tightly associated with TSS,[87,89] whereas 1mK4H3 and acH4 show a more widespread distribution.[87] Interestingly, Pokholok and colleagues[89] found that 3mK3H3 was highly enriched at the TSS, 2mK4H3 was mostly enriched in the middle of genes, and 1mK4H3 was found predominantly at the end of genes, suggesting that histone marks correlate with gene structure and transcriptional direction.

Koch et al.[87] also studied differences in histone mark profiles between cell lines (i.e., one B-cell lymphoblastoid cell line, one fetal lung fibroblast cell line, and three leukemia cell lines). Most of the ENCODE regions[91] analyzed with the array used in this study showed similar distributions of histone modifications in those five cell lines, but they also detected a total of 49 regions in which the differential chromatin structure correlated well with the expected gene expression.

As embryonic stem cells (ESCs) are able to generate any tissue and it is known that epigenetics plays a key role in the establishment of tissue identity, some ChIP-on-chip studies have turned their attention to this cell type.[92] Of special interest are the Polycomb group (PcG) proteins, which are essential to maintaining pluripotency of stem cells by repressing genes important for differentiation. In human ESCs, Young and coworkers[85] identified more than 1000 genes targeted by PcG proteins, which were also enriched for 3mK27H3 (as expected because methylation of K27H3 is catalyzed by PcG proteins). These genes were involved mostly in development.[85] This was also observed in a murine model.[93] Most interestingly, regions enriched for 3mK27H3 in ESCs, which are associated with gene repression, are also enriched for 3mK4H3, which is a mark for gene activation.[83,94] This bivalent chromatin state seems to be a hallmark of ESCs, which keep developmental genes silent. Then, upon differentiation toward a given cell lineage, required genes lose 3mK27H3 and become expressed, whereas silencing of unnecessary genes is made permanent by other epigenetic marks like 3mK9H3 and DNA methylation.[95]

15.3.2 CHARACTERIZING THE CANCER CELL EPIGENOME

Cancer cells are characterized by a disruption of the epigenomic machinery, which is reflected in multiple aberrations affecting both content and distribution of DNA methylation and histone modifications as well as alterations in nucleosome remodeling.[16,96–98] So far, DNA methylation, and especially tumor suppressor gene silencing by hypermethylation, is the best studied epigenetic modification in cancer.[99] More than 50 genes have been identified as frequently silenced in cancer by DNA methylation (e.g., P16/INK4A, P14/ARF, MLH1, or MGMT[16]), and DNA methylation patterns allow differentiation of distinct cancer entities.[100,101] However, with the advent of the science of epigenomics, a more precise and less biased delineation of the cancer cell epigenome is becoming accessible, which will produce a completely new generation of epigenetic markers in cancer. Furthermore, epigenomic studies are now comparing DNA methylation patterns with genome-wide genetic, developmental, and transcriptional patterns, which is offering new possibilities for understanding the processes underlying carcinogenesis.

Although only a few studies have characterized histone modifications in cancer cells at the genome-wide level,[16] the application of different microarray-based methods for studying DNA methylation changes in cancer has already started to yield its fruits. In one of the initial studies, Adorján and collaborators[53] studied DNA methylation profiles of different types of leukemias and solid tumors with an array covering 232 CpGs from 56 genes. Even with this small array, they detected DNA methylation marks able to differentiate different tumor types of a training set and accurately diagnose new samples from a test set. Bibikova and coworkers[45] used a bead array to measure DNA methylation levels of 1536 CpGs from 371 cancer-related genes in a panel of colon, breast, lung, and prostate cancer cell lines and healthy tissues. They identified 16 cancer-specific markers and 48 cancer subtype-specific markers that allowed a correct classification of all 24 samples under study. Furthermore, applying the methylation assay to 11 lung adenocarcinomas and 11 normal lung tissue samples, they identified 55 CpGs, which were able to predict the tumor status of further 12 lung tumors and 12 normal tissues with 100% specificity and 92% sensitivity.

In addition to the studies named above, an increasing number of microarray-based studies have focused on the detection of differentially methylated biomarkers associated with specific types of solid tumors—like, for example, breast cancer,[102–105] colorectal cancer,[106–109] prostate cancer,[110–112] lung cancer,[113–116] head and neck cell carcinoma,[117] oligodendroglioma,[118] medulloblastoma,[119] and Wilms tumors.[120] Furthermore, other microarray studies have focused on the impact of differential DNA methylation profiles in the prognosis of ovarian[121,122] and breast cancer.[123,124]

Due to the large number of different cell types of the hematopoietic system, a wide range of different leukemias and lymphomas have been identified by means of morphological, immunohistochemical, and genetic features.[125] Now, several groups are using microarray-based DNA methylation profiling to characterize the epigenome of this heterogeneous group of diseases and to identify diagnostic epigenetic marks. These studies have shown differential methylation profiles between mantle cell lymphoma (MCL) and follicular lymphoma (FL);[126] B-cell chronic lymphocytic leukemia (B-CLL), MCL, and FL;[127,128] cutaneous T-cell lymphoma and normal T cells;[129] acute lymphoblastic leukemia (ALL) and acute myeloid leukemia (AML);[130] AML and normal monocytes;[40] and ALL and normal peripheral blood.[131]

As an example, Rahmatpanah et al.[128] used the DMH technique on an 8.5-K CpG-island microarray. They detected 256 CpG islands with differential methylation among small B-cell lymphomas like B-CLL, MCL, and FL. The authors selected 10 genes for further validation with classical methods such as MSP, and a strict correlation between microarray and MSP data was missing. On the one hand, this can be explained by the higher sensitivity of the MSP in comparison to the DMH approach or to variability derived from differential hybridization efficiencies on distinct CpG-island clones.[132] This example highlights the importance of validating any novel diagnostic DNA methylation marker with classical methods and of establishing their diagnostic relevance in additional series of tumor samples.

We have recently studied a panel of 22 B-cell lymphoma cell lines and 18 normal controls from hematopoietic tissues with a bead array containing 1505 CpGs from 807 cancer-related genes (Figure 15.4; Martín-Subero et al., unpublished). The unsupervised cluster analysis shown in Figure 15.4 clearly identifies the lymphoma cell lines according to their DNA methylation profile and points to a large number of DNA hypermethylation events in lymphoma cells as compared with the controls. Studies in primary cases are currently underway to establish the diagnostic and prognostic impact of DNA methylation biomarkers in different subtypes of lymphomas.

One of the most widely used indirect strategies to detect hypermethylated loci in cancer is comparative gene expression arrays before and after treatment with DNA methyltransferase inhibitors (e.g., 5-aza-2'-deoxycytidine), which lead to expression of genes silenced by DNA methylation.[47] This approach has aided in identification of a number of candidate tumor suppressor genes, such as members of the SFRP family in colorectal cancer[133] and CEBPD in acute myeloid leukemias.[134] The advantage of this technique is that it directly identifies genes in which epigenetic changes lead to altered gene expression. However, because 5-AZA is toxic for the cells (noted in Section 15.2.2.1.1), this method leads to a large number of false-positive results and requires a throrough selection and validation of target genes.[21,48]

Most of the studies discussed that use microarrays to analyze cancer-related DNA methylation profiles have focused on the detection of differentially methylated genes. Such acquisition of differential methylation in cancer, like hypermethylation of tumor suppressor genes, is thought to provide the tumor clone with a selective (e.g., proliferative) advantage. However, recent reports have proposed that there is an instructive mechanism behind aberrant DNA methylation in cancer. Keshet and colleagues[135] performed a MeDIP-on-chip study in colon and prostate cancer and, in addition to identifying differentially methylated genes (e.g., 135 gene promoters in Caco-2, a colon cancer cell line), they studied whether these genes show distinct biological features.[135] They discovered that genes differentially methylated in

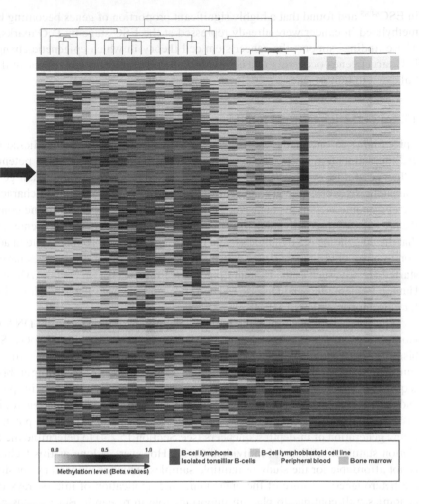

FIGURE 15.4 Heatmap from a hierarchical cluster analysis of DNA methylation data generated with the bead-array technology (Illumina Inc.) in different normal tissue samples and B-cell lymphoma cell lines. Lymphoma cell lines are characterized by a large number of hypermethylated genes in comparison with the normal controls (*arrow*). Red indicates methylated CpG loci whereas green indicates lack of DNA methylation.

cancer are enriched for functional categories (e.g., cell adhesion, cell–cell signaling, signal transduction, and ion transport) and that the expression of some of them is already repressed (or expressed at low levels) in normal cells from matched tissues. Furthermore, they detected a significant enrichment of sequence motifs and a significant clustering of such genes in chromosomal regions. In line with this finding, another study has shown that large stretches of DNA containing several genes can become hypermethylated in cancer.[136]

Three independent studies have recently provided further evidence for an instructive mechanism leading to selective methylation of certain groups of genes in cancer. These studies took advantage of ChIP-on-chip data generated using PcG antibodies in ESC[84,85] and found that a highly significant proportion of genes becoming hyper-methylated in cancer were already repressed at the ESC stage by PcG marks.[137-139] These findings support the cancer stem cell theory in which epigenetic changes of PcG target genes occurring in a cell with stem cell features might represent the initial event in tumorigenesis.[96,140,141]

15.4 FUTURE DIRECTIONS

Epigenomics is one of the most flourishing areas in biology and medicine today, and the delineation of epigenetic patterns in health and disease has the potential to influence almost every aspect of life. After the completion of the human genome, epigeneticists worldwide are now calling for an international effort to characterize the epigenome.[142-148] This endeavor is indeed a great-scale project if one considers the presence of interindividual, tissue-specific, and disease-specific epigenomes, and that the epigenome is a dynamic system that can be altered throughout life in adaptation to novel environmental cues.[149] To reach that goal, several initiatives have been started already, such as the Human Epigenome Project (HEP),[62,63,143] the Alliance for Human Epigenomics and Disease (AHEAD),[147] and the National Methylome Project for Chromosome 21 (NAME 21).[146]

The precise delineation of the human epigenome, or at least of the DNA methylome, requires sequencing of the methylation status of individual CpGs. So far, bisulfite sequencing has been performed using PCR products, which is time consuming and expensive.[62] An alternative approach is to perform a "shotgun bisulfite sequencing," which can be applied to the entire genome or a representative part of it and is easily scalable with increasing sequencing capacity.[11,150] In this regard, initiatives aiming at characterizing the epigenome are starting to exploit the potential of a new generation of in-depth sequencers (see Section 15.2.4) to determine the methylation status of millions of individual CpGs. However, as long as this technology is not affordable for the study of multiple samples, which is mandatory considering the heterogeneous nature of the epigenome, the application of microarrays in epigenomics will continue to play an important role in research. But one has also to consider that microarrays and high-throughput sequencing that provide only a consensus snapshot of the epigenome of a given tissue sample. A deeper understanding of the cellular epigenome will require the development of reliable techniques that allow measurement of the DNA methylation status of specific CpGs at the single-cell level.[151] Furthermore, high-throughput techniques for the characterization of the genome, transcriptome, and proteome are being developed simultaneously to those focusing on the epigenome. The integration of these different layers, or networks, of cell physiology into a unified cellular system will be of great importance for understanding the mechanisms underlying normal and altered physiology.[152]

The epigenomic studies reviewed here were aimed mainly at characterizing normal and cancer cells. However, as epigenomics is the language used by nature to integrate external and internal signals into the genome, it can be potentially involved in virtually all aspects of life.[4,5,153] In the last few years, a number of groundbreaking studies have shown, for example, that vernalization in plants is caused by epigenetic inactivation of the flowering locus C,[154] that phenotypic differences of genetically identical individuals can be caused by the acquisition of differential epigenetic changes throughout life,[149] that maternal supplementation with methyl donors such as folic acid is able to shift the phenotype of the offspring in Agouti mice,[155] and that maternal behavior is important in the establishment of epigenetic marks in the brain of newborn offspring.[156,157] Additionally, not only does epigenetics play a role in this life's health and disease, it might also be transmitted through generations.[158] There is statistical evidence that a person's health can be affected by the lifestyle of past generations, which might occur through the inheritance of epigenetic marks.[159-161] Experimental evidence supporting this hypothesis comes from the Agouti mouse model, in which the benefits of methyl-donor supplementation can be observed at least in two generations, or the effect of endocrine disruptors on male infertility can be observed in four generations.[162] Furthermore, a recent study has provided evidence for germline epimutations of the MLH1 gene to be associated with predisposition to nonpolyposis colorectal cancer.[163] In the near future, the science of epigenomics will certainly expand beyond the studies described herein to a genome-wide level. The identification of genomic regions susceptible to epigenetic modification by external factors will open new ways to understand how the potential of the genetic code is manifested into a phenotype. Perhaps the science of epigenomics will scientifically demonstrate the truth behind the saying, "As you sow, so shall you reap," in the sense that every physical and mental state, both under healthy and altered conditions, might be the result of a previous cause.

REFERENCES

1. Lander, E.S. et al., Initial sequencing and analysis of the human genome, *Nature*, 409, 860–921, 2001.
2. Venter, J.C. et al., The sequence of the human genome, *Science*, 291, 1304–51, 2001.
3. Bird, A., DNA methylation patterns and epigenetic memory, *Genes Dev*, 16, 6–21, 2002.
4. Jaenisch, R. & Bird, A., Epigenetic regulation of gene expression: how the genome integrates intrinsic and environmental signals, *Nat Genet*, 33 Suppl, 245–54, 2003.
5. Bird, A., Perceptions of epigenetics, *Nature*, 447, 396–8, 2007.
6. Kouzarides, T., Chromatin modifications and their function, *Cell*, 128, 693–705, 2007.
7. Chen, K. & Rajewsky, N., The evolution of gene regulation by transcription factors and microRNAs, *Nat Rev Genet*, 8, 93–103, 2007.
8. Fraser, P. & Bickmore, W., Nuclear organization of the genome and the potential for gene regulation, *Nature*, 447, 413–7, 2007.
9. Zaratiegui, M. et al., Noncoding RNAs and gene silencing, *Cell*, 128, 763–76, 2007.
10. Beck, S. et al., From genomics to epigenomics: a loftier view of life, *Nat Biotechnol*, 17, 1144, 1999.
11. Bernstein, B.E. et al., The mammalian epigenome, *Cell*, 128, 669–81, 2007.
12. Callinan, P.A. & Feinberg, A.P., The emerging science of epigenomics, *Hum Mol Genet*, 15 Suppl 1, R95–R101, 2006.
13. Fazzari, M.J. & Greally, J.M., Epigenomics: beyond CpG islands, *Nat Rev Genet*, 5, 446–55, 2004.
14. Novik, K.L. et al., Epigenomics: genome-wide study of methylation phenomena, *Curr Issues Mol Biol*, 4, 111–28, 2002.
15. Pennisi, E., Environmental epigenomics meeting. Supplements restore gene function via methylation, *Science*, 310, 1761, 2005.

16. Esteller, M., Cancer epigenomics: DNA methylomes and histone-modification maps, *Nat Rev Genet*, 8, 286–98, 2007.
17. Fraga, M.F. & Esteller, M., DNA methylation: a profile of methods and applications, *Biotechniques*, 33, 632, 634, 636–49, 2002.
18. Laird, P.W., The power and the promise of DNA methylation markers, *Nat Rev Cancer*, 3, 253–66, 2003.
19. van Steensel, B. & Henikoff, S., Epigenomic profiling using microarrays, *Biotechniques*, 35, 346–50, 352–4, 356–7, 2003.
20. Lyko, F., Novel methods for analysis of genomic DNA methylation, *Anal Bioanal Chem*, 381, 67–8, 2005.
21. Shames, D.S. et al., Methods for detecting DNA methylation in tumors: from bench to bedside, *Cancer Lett*, 251, 187–98, 2007.
22. Shen, L. & Waterland, R.A., Methods of DNA methylation analysis, *Curr Opin Clin Nutr Metab Care*, 10, 576–81, 2007.
23. Frommer, M. et al., A genomic sequencing protocol that yields a positive display of 5-methylcytosine residues in individual DNA strands, *Proc Natl Acad Sci U S A*, 89, 1827–31, 1992.
24. Herman, J.G. et al., Methylation-specific PCR: a novel PCR assay for methylation status of CpG islands, *Proc Natl Acad Sci U S A*, 93, 9821–6, 1996.
25. Frigola, J. et al., Methylome profiling of cancer cells by amplification of inter-methylated sites (AIMS), *Nucleic Acids Res*, 30, e28, 2002.
26. Kawai, J. et al., Methylation profiles of genomic DNA of mouse developmental brain detected by restriction landmark genomic scanning (RLGS) method, *Nucleic Acids Res*, 21, 5604–8, 1993.
27. Ramsay, G., DNA chips: state-of-the art, *Nat Biotechnol*, 16, 40–4, 1998.
28. Fraga, M.F. et al., High-performance capillary electrophoretic method for the quantification of 5-methyl 2′-deoxycytidine in genomic DNA: application to plant, animal and human cancer tissues, *Electrophoresis*, 23, 1677–81, 2002.
29. Gowher, H. et al., DNA of *Drosophila melanogaster* contains 5-methylcytosine, *Embo J*, 19, 6918–23, 2000.
30. Karimi, M. et al., LUMA (LUminometric Methylation Assay)—a high throughput method to the analysis of genomic DNA methylation, *Exp Cell Res*, 312, 1989–95, 2006.
31. Ehrlich, M., DNA methylation in cancer: too much, but also too little, *Oncogene*, 21, 5400–13, 2002.
32. Feinberg, A.P. & Vogelstein, B., Hypomethylation distinguishes genes of some human cancers from their normal counterparts, *Nature*, 301, 89–92, 1983.
33. Eden, A. et al., Chromosomal instability and tumors promoted by DNA hypomethylation, *Science*, 300, 455, 2003.
34. Gaudet, F. et al., Induction of tumors in mice by genomic hypomethylation, *Science*, 300, 489–92, 2003.
35. Fraga, M.F. et al., Loss of acetylation at Lys16 and trimethylation at Lys20 of histone H4 is a common hallmark of human cancer, *Nat Genet*, 37, 391–400, 2005.
36. Ching, T.T. et al., Epigenome analyses using BAC microarrays identify evolutionary conservation of tissue-specific methylation of SHANK3, *Nat Genet*, 37, 645–51, 2005.
37. Huang, T.H. et al., Methylation profiling of CpG islands in human breast cancer cells, *Hum Mol Genet*, 8, 459–70, 1999.
38. Khulan, B. et al., Comparative isoschizomer profiling of cytosine methylation: the HELP assay, *Genome Res*, 16, 1046–55, 2006.
39. Schumacher, A. et al., Microarray-based DNA methylation profiling: technology and applications, *Nucleic Acids Res*, 34, 528–42, 2006.
40. Gebhard, C. et al., Genome-wide profiling of CpG methylation identifies novel targets of aberrant hypermethylation in myeloid leukemia, *Cancer Res*, 66, 6118–28, 2006.

41. Weber, M. et al., Chromosome-wide and promoter-specific analyses identify sites of differential DNA methylation in normal and transformed human cells, *Nat Genet*, 37, 853–62, 2005.
42. Gebhard, C. et al., Rapid and sensitive detection of CpG-methylation using methyl-binding (MB)-PCR, *Nucleic Acids Res*, 34, e82, 2006.
43. Ballestar, E. et al., Methyl-CpG binding proteins identify novel sites of epigenetic inactivation in human cancer, *Embo J*, 22, 6335–45, 2003.
44. Weber, M. et al., Distribution, silencing potential and evolutionary impact of promoter DNA methylation in the human genome, *Nat Genet*, 39, 457–66, 2007.
45. Bibikova, M. et al., High-throughput DNA methylation profiling using universal bead arrays, *Genome Res*, 16, 383–93, 2006.
46. Gitan, R.S. et al., Methylation-specific oligonucleotide microarray: a new potential for high-throughput methylation analysis, *Genome Res*, 12, 158–64, 2002.
47. Karpf, A.R., Epigenomic reactivation screening to identify genes silenced by DNA hypermethylation in human cancer, *Curr Opin Mol Ther*, 9, 231–41, 2007.
48. Shames, D.S. et al., A genome-wide screen for promoter methylation in lung cancer identifies novel methylation markers for multiple malignancies, *PLoS Med*, 3, e486, 2006.
49. Yan, P.S. et al., Applications of CpG island microarrays for high-throughput analysis of DNA methylation, *J Nutr*, 132, 2430S–2434S, 2002.
50. Wilson, I.M. et al., Epigenomics: mapping the methylome, *Cell Cycle*, 5, 155–8, 2006.
51. Yuan, E. et al., A single nucleotide polymorphism chip-based method for combined genetic and epigenetic profiling: validation in decitabine therapy and tumor/normal comparisons, *Cancer Res*, 66, 3443–51, 2006.
52. Zhang, X. et al., Genome-wide high-resolution mapping and functional analysis of DNA methylation in *Arabidopsis*, *Cell*, 126, 1189–201, 2006.
53. Adorján, P. et al., Tumour class prediction and discovery by microarray-based DNA methylation analysis, *Nucleic Acids Res*, 30, e21, 2002.
54. Bibikova, M. et al., Human embryonic stem cells have a unique epigenetic signature, *Genome Res*, 16, 1075–83, 2006.
55. Bernstein, B.E. et al., The use of chromatin immunoprecipitation assays in genome-wide analyses of histone modifications, *Methods Enzymol*, 376, 349–60, 2004.
56. Huebert, D.J. et al., Genome-wide analysis of histone modifications by ChIP-on-chip, *Methods*, 40, 365–9, 2006.
57. Attema, J.L. et al., Epigenetic characterization of hematopoietic stem cell differentiation using miniChIP and bisulfite sequencing analysis, *Proc Natl Acad Sci U S A*, 104, 12371–6, 2007.
58. Kiermer, V., Embryos and biopsies on the ChIP-ing forecast, *Nat Methods*, 3, 583, 2006.
59. O'Neill, L.P. et al., Epigenetic characterization of the early embryo with a chromatin immunoprecipitation protocol applicable to small cell populations, *Nat Genet*, 38, 835–41, 2006.
60. Hu, M. et al., Methylation-specific digital karyotyping, *Nat Protoc*, 1, 1621–36, 2006.
61. Roh, T.Y. et al., High-resolution genome-wide mapping of histone modifications, *Nat Biotechnol*, 22, 1013–6, 2004.
62. Eckhardt, F. et al., DNA methylation profiling of human chromosomes 6, 20 and 22, *Nat Genet*, 38, 1378–85, 2006.
63. Rakyan, V.K. et al., DNA methylation profiling of the human major histocompatibility complex: a pilot study for the human epigenome project, *PLoS Biol*, 2, e405, 2004.
64. Fan, J.B. et al., Highly parallel genomic assays, *Nat Rev Genet*, 7, 632–44, 2006.
65. Margulies, M. et al., Genome sequencing in microfabricated high-density picolitre reactors, *Nature*, 437, 376–80, 2005.
66. Shendure, J. et al., Accurate multiplex polony sequencing of an evolved bacterial genome, *Science*, 309, 1728–32, 2005.

67. Barski, A. et al., High-resolution profiling of histone methylations in the human genome, *Cell*, 129, 823–37, 2007.
68. Mikkelsen, T.S. et al., Genome-wide maps of chromatin state in pluripotent and lineage-committed cells, *Nature*, 448, 553–60, 2007.
69. Dostie, J. et al., Chromosome Conformation Capture Carbon Copy (5C): a massively parallel solution for mapping interactions between genomic elements, *Genome Res*, 16, 1299–309, 2006.
70. Simonis, M. et al., An evaluation of 3C-based methods to capture DNA interactions, *Nat Methods*, 4, 895–901, 2007.
71. Taylor, K.H. et al., Ultradeep bisulfite sequencing analysis of DNA methylation patterns in multiple gene promoters by 454 sequencing, *Cancer Res*, 67, 8511–8, 2007.
72. Gardiner-Garden, M. & Frommer, M., CpG islands in vertebrate genomes, *J Mol Biol*, 196, 261–82, 1987.
73. Grunau, C. et al., Large-scale methylation analysis of human genomic DNA reveals tissue-specific differences between the methylation profiles of genes and pseudogenes, *Hum Mol Genet*, 9, 2651–63, 2000.
74. Smiraglia, D.J. et al., Excessive CpG island hypermethylation in cancer cell lines versus primary human malignancies, *Hum Mol Genet*, 10, 1413–9, 2001.
75. Strichman-Almashanu, L.Z. et al., A genome-wide screen for normally methylated human CpG islands that can identify novel imprinted genes, *Genome Res*, 12, 543–54, 2002.
76. Shen, L. et al., Genome-wide profiling of DNA methylation reveals a class of normally methylated CpG island promoters, *PLoS Genet*, 3, 2023–36, 2007.
77. Fraga, M.F. & Esteller, M., Epigenetics and aging: the targets and the marks, *Trends Genet*, 23, 413–8, 2007.
78. Bernstein, B.E. et al., Genomic maps and comparative analysis of histone modifications in human and mouse, *Cell*, 120, 169–81, 2005.
79. Blelloch, R. et al., Reprogramming efficiency following somatic cell nuclear transfer is influenced by the differentiation and methylation state of the donor nucleus, *Stem Cells*, 24, 2007–13, 2006.
80. Hattori, N. et al., Epigenetic control of mouse Oct-4 gene expression in embryonic stem cells and trophoblast stem cells, *J Biol Chem*, 279, 17063–9, 2004.
81. Schilling, E. & Rehli, M., Global, comparative analysis of tissue–specific promoter CpG methylation, *Genomics*, 90, 314–23, 2007.
82. Jenuwein, T. & Allis, C.D., Translating the histone code, *Science*, 293, 1074–80, 2001.
83. Bernstein, B.E. et al., A bivalent chromatin structure marks key developmental genes in embryonic stem cells, *Cell*, 125, 315–26, 2006.
84. Bracken, A.P. et al., Genome-wide mapping of Polycomb target genes unravels their roles in cell fate transitions, *Genes Dev*, 20, 1123–36, 2006.
85. Lee, T.I. et al., Control of developmental regulators by Polycomb in human embryonic stem cells, *Cell*, 125, 301–13, 2006.
86. Kim, T.H. et al., A high-resolution map of active promoters in the human genome, *Nature*, 436, 876–80, 2005.
87. Koch, C.M. et al., The landscape of histone modifications across 1% of the human genome in five human cell lines, *Genome Res*, 17, 691–707, 2007.
88. Liu, C.L. et al., Single-nucleosome mapping of histone modifications in *S. cerevisiae*, *PLoS Biol*, 3, e328, 2005.
89. Pokholok, D.K. et al., Genome-wide map of nucleosome acetylation and methylation in yeast, *Cell*, 122, 517–27, 2005.
90. Schubeler, D. et al., The histone modification pattern of active genes revealed through genome-wide chromatin analysis of a higher eukaryote, *Genes Dev*, 18, 1263–71, 2004.

91. Consortium, T.E.P., The ENCODE (ENCyclopedia Of DNA Elements) Project, *Science*, 306, 636–40, 2004.
92. Buszczak, M. & Spradling, A.C., Searching chromatin for stem cell identity, *Cell*, 125, 233–6, 2006.
93. Boyer, L.A. et al., Polycomb complexes repress developmental regulators in murine embryonic stem cells, *Nature*, 441, 349–53, 2006.
94. Azuara, V. et al., Chromatin signatures of pluripotent cell lines, *Nat Cell Biol*, 8, 532–8, 2006.
95. Sparmann, A. & van Lohuizen, M., Polycomb silencers control cell fate, development and cancer, *Nat Rev Cancer*, 6, 846–56, 2006.
96. Jones, P.A. & Baylin, S.B., The epigenomics of cancer, *Cell*, 128, 683–92, 2007.
97. Laird, P.W., Cancer epigenetics, *Hum Mol Genet*, 14 Spec No 1, R65–76, 2005.
98. Ting, A.H. et al., The cancer epigenome—components and functional correlates, *Genes Dev*, 20, 3215–31, 2006.
99. Esteller, M., Epigenetic gene silencing in cancer: the DNA hypermethylome, *Hum Mol Genet*, 16 Spec No 1, R50–9, 2007.
100. Costello, J.F. et al., Aberrant CpG-island methylation has non-random and tumour-type-specific patterns, *Nat Genet*, 24, 132–8, 2000.
101. Paz, M.F. et al., A systematic profile of DNA methylation in human cancer cell lines, *Cancer Res*, 63, 1114–21, 2003.
102. Chen, C.M. et al., Methylation target array for rapid analysis of CpG island hypermethylation in multiple tissue genomes, *Am J Pathol*, 163, 37–45, 2003.
103. Piotrowski, A. et al., Microarray-based survey of CpG islands identifies concurrent hyper- and hypomethylation patterns in tissues derived from patients with breast cancer, *Genes Chromosomes Cancer*, 45, 656–67, 2006.
104. Versmold, B. et al., Epigenetic silencing of the candidate tumor suppressor gene PROX1 in sporadic breast cancer, *Int J Cancer*, 121, 547–54, 2007.
105. Yan, P.S. et al., Dissecting complex epigenetic alterations in breast cancer using CpG island microarrays, *Cancer Res*, 61, 8375–80, 2001.
106. Estecio, M.R. et al., High-throughput methylation profiling by MCA coupled to CpG island microarray, *Genome Res*, 17, 1529–36, 2007.
107. Hayashi, H. et al., High-resolution mapping of DNA methylation in human genome using oligonucleotide tiling array, *Hum Genet*, 120, 701–11, 2007.
108. Model, F. et al., Identification and validation of colorectal neoplasia-specific methylation markers for accurate classification of disease, *Mol Cancer Res*, 5, 153–63, 2007.
109. Yan, P.S. et al., Use of CpG island microarrays to identify colorectal tumors with a high degree of concurrent methylation, *Methods*, 27, 162–9, 2002.
110. Cottrell, S. et al., Discovery and validation of 3 novel DNA methylation markers of prostate cancer prognosis, *J Urol*, 177, 1753–8, 2007.
111. Wang, Y. et al., Survey of differentially methylated promoters in prostate cancer cell lines, *Neoplasia*, 7, 748–60, 2005.
112. Yu, Y.P. et al., High throughput screening of methylation status of genes in prostate cancer using an oligonucleotide methylation array, *Carcinogenesis*, 26, 471–9, 2005.
113. Field, J.K. et al., Methylation discriminators in NSCLC identified by a microarray based approach, *Int J Oncol*, 27, 105–11, 2005.
114. Fukasawa, M. et al., Microarray analysis of promoter methylation in lung cancers, *J Hum Genet*, 51, 368–74, 2006.
115. Rauch, T. et al., MIRA-assisted microarray analysis, a new technology for the determination of DNA methylation patterns, identifies frequent methylation of homeodomain-containing genes in lung cancer cells, *Cancer Res*, 66, 7939–47, 2006.

116. Rauch, T. et al., Homeobox gene methylation in lung cancer studied by genome-wide analysis with a microarray-based methylated CpG island recovery assay, *Proc Natl Acad Sci U S A*, 104, 5527–32, 2007.
117. Adrien, L.R. et al., Classification of DNA methylation patterns in tumor cell genomes using a CpG island microarray, *Cytogenet Genome Res*, 114, 16–23, 2006.
118. Ordway, J.M. et al., Comprehensive DNA methylation profiling in a human cancer genome identifies novel epigenetic targets, *Carcinogenesis*, 27, 2409–23, 2006.
119. Pfister, S. et al., Array-based profiling of reference-independent methylation status (aPRIMES) identifies frequent promoter methylation and consecutive downregulation of ZIC2 in pediatric medulloblastoma, *Nucleic Acids Res*, 35, e51, 2007.
120. Bjornsson, H.T. et al., Epigenetic specificity of loss of imprinting of the IGF2 gene in Wilms tumors, *J Natl Cancer Inst*, 99, 1270–3, 2007.
121. Wei, S.H. et al., Prognostic DNA methylation biomarkers in ovarian cancer, *Clin Cancer Res*, 12, 2788–94, 2006.
122. Wei, S.H. et al., Methylation microarray analysis of late-stage ovarian carcinomas distinguishes progression-free survival in patients and identifies candidate epigenetic markers, *Clin Cancer Res*, 8, 2246–52, 2002.
123. Maier, S. et al., DNA-methylation of the homeodomain transcription factor PITX2 reliably predicts risk of distant disease recurrence in tamoxifen-treated, node-negative breast cancer patients—Technical and clinical validation in a multi-centre setting in collaboration with the European Organisation for Research and Treatment of Cancer (EORTC) PathoBiology group, *Eur J Cancer*, 43, 1679–86, 2007.
124. Martens, J.W. et al., Association of DNA methylation of phosphoserine aminotransferase with response to endocrine therapy in patients with recurrent breast cancer, *Cancer Res*, 65, 4101–17, 2005.
125. Harris, N.L. et al., WHO classification of tumors of haematopoietic and lymphoid tissue: Introduction, in *World Health Organization Classification of Tumors. Pathology and Genetics of Tumors of Haematopoietic and Lymphoid Tissues*, Jaffe, E.S., Harris, N.L., Stein, H. & Vardiman, J.W., Eds., IARC Press, Lyon, 2001, 12–13.
126. Shi, H. et al., Oligonucleotide-based microarray for DNA methylation analysis: principles and applications, *J Cell Biochem*, 88, 138–43, 2003.
127. Guo, J. et al., Differential DNA methylation of gene promoters in small B-cell lymphomas, *Am J Clin Pathol*, 124, 430–9, 2005.
128. Rahmatpanah, F.B. et al., Differential DNA methylation patterns of small B-cell lymphoma subclasses with different clinical behavior, *Leukemia*, 20, 1855–62, 2006.
129. van Doorn, R. et al., Epigenetic profiling of cutaneous T-cell lymphoma: promoter hypermethylation of multiple tumor suppressor genes including BCL7a, PTPRG, and p73, *J Clin Oncol*, 23, 3886–96, 2005.
130. Scholz, C. et al., Distinction of acute lymphoblastic leukemia from acute myeloid leukemia through microarray-based DNA methylation analysis, *Ann Hematol*, 84, 236–44, 2005.
131. Taylor, K.H. et al., Large-scale CpG methylation analysis identifies novel candidate genes and reveals methylation hotspots in acute lymphoblastic leukemia, *Cancer Res*, 67, 2617–25, 2007.
132. Martín-Subero, J.I. et al., Towards defining the lymphoma methylome, *Leukemia*, 20, 1658–60, 2006.
133. Suzuki, H. et al., A genomic screen for genes upregulated by demethylation and histone deacetylase inhibition in human colorectal cancer, *Nat Genet*, 31, 141–9, 2002.
134. Agrawal, S. et al., The C/EBPdelta tumor suppressor is silenced by hypermethylation in acute myeloid leukemia, *Blood*, 109, 3895–905, 2007.
135. Keshet, I. et al., Evidence for an instructive mechanism of de novo methylation in cancer cells, *Nat Genet*, 38, 149–53, 2006.

136. Frigola, J. et al., Epigenetic remodeling in colorectal cancer results in coordinate gene suppression across an entire chromosome band, *Nat Genet*, 38, 540–9, 2006.
137. Ohm, J.E. et al., A stem cell-like chromatin pattern may predispose tumor suppressor genes to DNA hypermethylation and heritable silencing, *Nat Genet*, 39, 237–42, 2007.
138. Schlesinger, Y. et al., Polycomb-mediated methylation on Lys27 of histone H3 pre-marks genes for de novo methylation in cancer, *Nat Genet*, 39, 232–6, 2007.
139. Widschwendter, M. et al., Epigenetic stem cell signature in cancer, *Nat Genet*, 39, 157–8, 2007.
140. Feinberg, A.P. et al., The epigenetic progenitor origin of human cancer, *Nat Rev Genet*, 7, 21–33, 2006.
141. Ohm, J.E. & Baylin, S.B., Stem cell chromatin patterns: an instructive mechanism for DNA hypermethylation?, *Cell Cycle*, 6, 1040–3, 2007.
142. Bradbury, J., Human epigenome project—up and running, *PLoS Biol*, 1, E82, 2003.
143. Eckhardt, F. et al., Future potential of the Human Epigenome Project, *Expert Rev Mol Diagn*, 4, 609–18, 2004.
144. Esteller, M., The necessity of a human epigenome project, *Carcinogenesis*, 27, 1121–5, 2006.
145. Garber, K., Momentum building for human epigenome project, *J Natl Cancer Inst*, 98, 84–6, 2006.
146. Jeltsch, A. et al., German human methylome project started, *Cancer Res*, 66, 7378, 2006.
147. Jones, P.A. & Martienssen, R., A blueprint for a Human Epigenome Project: the AACR Human Epigenome Workshop, *Cancer Res*, 65, 11241–6, 2005.
148. Rauscher, F.J., III, It is time for a Human Epigenome Project, *Cancer Res*, 65, 11229, 2005.
149. Fraga, M.F. et al., Epigenetic differences arise during the lifetime of monozygotic twins, *Proc Natl Acad Sci U S A*, 102, 10604–9, 2005.
150. Meissner, A. et al., Reduced representation bisulfite sequencing for comparative high-resolution DNA methylation analysis, *Nucleic Acids Res*, 33, 5868–77, 2005.
151. Nuovo, G.J. et al., In situ detection of the hypermethylation-induced inactivation of the p16 gene as an early event in oncogenesis, *Proc Natl Acad Sci U S A*, 96, 12754–9, 1999.
152. van Steensel, B., Mapping of genetic and epigenetic regulatory networks using micro-arrays, *Nat Genet*, 37 Suppl, S18–24, 2005.
153. Jirtle, R.L. & Skinner, M.K., Environmental epigenomics and disease susceptibility, *Nat Rev Genet*, 8, 253–62, 2007.
154. Bastow, R. et al., Vernalization requires epigenetic silencing of FLC by histone meth-ylation, *Nature*, 427, 164–7, 2004.
155. Waterland, R.A. & Jirtle, R.L., Early nutrition, epigenetic changes at transposons and imprinted genes, and enhanced susceptibility to adult chronic diseases, *Nutrition*, 20, 63–8, 2004.
156. Szyf, M. et al., Maternal care, the epigenome and phenotypic differences in behavior, *Reprod Toxicol*, 24, 9–19, 2007.
157. Weaver, I.C. et al., Epigenetic programming by maternal behavior, *Nat Neurosci*, 7, 847–54, 2004.
158. Pennisi, E., Environmental epigenomics meeting. Food, tobacco, and future genera-tions, *Science*, 310, 1760–1, 2005.
159. Kaati, G. et al., Transgenerational response to nutrition, early life circumstances and longevity, *Eur J Hum Genet*, 15, 784–90, 2007.
160. Pembrey, M.E., Time to take epigenetic inheritance seriously, *Eur J Hum Genet*, 10, 669–71, 2002.
161. Pembrey, M.E. et al., Sex-specific, male-line transgenerational responses in humans, *Eur J Hum Genet*, 14, 159–66, 2006.

162. Anway, M.D. et al., Epigenetic transgenerational actions of endocrine disruptors and male fertility, *Science*, 308, 1466–9, 2005.
163. Hitchins, M.P. et al., Inheritance of a cancer-associated MLH1 germ-line epimutation, *N Engl J Med*, 356, 697–705, 2007.
164. Toyota, M. et al., Identification of differentially methylated sequences in colorectal cancer by methylated CpG island amplification, *Cancer Res*, 59, 2307–12, 1999.
165. Ibrahim, A.E. et al., MMASS: an optimized array-based method for assessing CpG island methylation, *Nucleic Acids Res*, 34, 36, 2006.
166. Nieländer, I. et al., Combining array-based approaches for the identification of candidate tumor suppressor loci in mature lymphoid neoplasms, *APMIS* 115, 1107–34, 2007.

Index

A

Acetyl-CoA synthetase (ACS), 75
ACS, *see* Acetyl-CoA synthetase
Acute myeloid leukemia (AML), 29, 94
    CEBPD in, 274
    epigenomics and, 274
    median time to, 32
Acute promyelocytic leukemia (APL), 18
AD, *see* Alzheimer's disease
Advanced glycation endproducts (AGE), 219
AGE, *see* Advanced glycation endproducts
Aging, epigenetic changes in, 184–185
AIMS, *see* Amplification of intermethylated sites
Alanine aminotransferase (ALT), 33
All-trans retinoic acid (ATRA), 35, 36
ALS, *see* Amyotrophic lateral sclerosis
ALT, *see* Alanine aminotransferase
Alzheimer's disease (AD), 96, 190
AML, *see* Acute myeloid leukemia
Amplification of intermethylated sites (AIMS),
    9, 263
Amyotrophic lateral sclerosis (ALS), 96
ANAs, *see* Antinuclear antibodies
Antinuclear antibodies (ANAs), 184
Apicidin-induced apoptosis, 57
APL, *see* Acute promyelocytic leukemia
APOE, *see* Apolipoprotein E
Apolipoprotein E (APOE), 210
Aquifex aeolicus, 50
Arabidopsis thaliana, 226
Aspartate aminotransferase (AST), 33
AST, *see* Aspartate aminotransferase
Atherosclerosis
    DNA methylation patterns, 211
    risk factors for, 207
    susceptibility to, 211
ATRA, *see* All-trans retinoic acid
ATR-X syndrome, *see* X-linked α thalassemia
    mental retardation syndrome
Autism, 256
5-Azacytidine, 28–30
    administration, logistical problems with, 30
    chemistry and pharmacokinetics, 28
    clearance, 28
    clinical experience with 5-azacytidine in
        MDS and leukemia, 28–30
    efficacy of in tumor types, 34
5-Aza-2'-deoxycytidine, 30–33
    chemistry and pharmacokinetics, 30–31
    clinical experience with 5-aza-2'-
        deoxycytidine in MDS and leukemia,
        31–33

B

BAL, *see* BALL pathogen-resistance gene
Baldwin effect, 242, 244
BALL pathogen-resistance gene (BAL), 225
Base excision repair (BER) pathway, 91
BAT, *see* Brown adipose tissue
B-cell chronic lymphocytic leukemia (B-CLL),
    114
B-cell lymphomas, miRNAs and, 118
B-CLL, *see* B-cell chronic lymphocytic leukemia
BDNF, *see* Brain-derived neurotrophic factor
Beckwith–Wiedemann syndrome, 199
Benzamides, 56
BER pathway, *see* Base excision repair pathway
Bipolar disease, 196
BL, *see* Burkitt lymphoma
Brain-derived neurotrophic factor (BDNF), 162
Brown adipose tissue (BAT), 90
Burkitt lymphoma (BL), 116

C

Caenorhabditis elegans, 75, 78, 105, 107
CAGRs, *see* Cancer-associated genomic regions
CALGB, *see* Cancer and Leukemia Group B
Cambinol, 22
Cancer(s)
    aging, 94
    -associated genomic regions (CAGRs),
        111–114
    bone, 34
    breast, 92, 93, 94, 117, 190, 253, 273
    cancer stem cell theory, 276
    cell(s)
        characterization of, 273
        DNA methylation content, 263
        epigenome, 273–276
    childhood, 164
    clonal versus polyclonal models for cancer
        initiation, 252
    colon, 93, 94, 190, 253
    colorectal, 253, 274, 277

environment and, 251–252
epigenetic progenitor theory and, 252
epithelial, 92
esophageal squamous cell carcinoma, 18
gastric cancer, 18
HDACi treatment of, 56
head and neck, 273
immunosurveillance mechanisms, 60
lung, 92, 115, 266, 273
lymphoma, 52, 116, 274
malignant melanoma, 34
microsatellite instability, 22
multiple myeloma, 1118
non-small cell lung cancer, 34
ovarian, 11, 273
pancreatic, 92, 93
papillary thyroid carcinoma, 116
prostate, 18, 94, 274
-related hypermethylated genes, detection of,
     266
renal cell, 40
SirT1 and, 80, 86
sirtuins and, 92, 94
skin, 92
solid
     hypomethylating agents in, 33
     miRNA upregulation and, 115
testicular, 18
thyroid, 94
transgenerational epigenetics and cancer, 252
of unknown primary origin, 125
Cancer (DNA methylation), 3–16
     characterization of cancer, 3
     chromatin remodeling enzymes, 5
     clinical practice, 8–9
     CpG islands, 4, 10
     CpG suppression, 4
     cyclin D kinase 6, 8
     cytosine analogs, 12
     death-associated protein kinase, 10
     DNA demethylating agents, 12–13
     early event in cancer development, 5
     gene–dosage reduction, 4
     gene inactivation, 12
     genomic imprinting, 4
     HDAC inhibitors, 13
     healthy cells, 4–5
     hypermethylated genes, 7
     MeCP, 4
     methylation-specific PCR, 9
     methyl-binding protein, 4, 5
     methyl DNA immunoprecipitation, 9
     MGMT protein, 11
     miRNA silencing in human cancer, 8
     myelodysplastic syndrome, 12
     oncogenes, 3
     ovarian cancer, 11

     pattern change in cancer cells, 5–8
         aberrant gene hypermethylation in human
         cancer, 6–8
         global DNA hypomethylation in human
         cancer, 5–6
     predictive factor, 11–12
     prognostic marker, 10–11
     regulators of histone modifications, 5
     restriction landmark genomic scanning, 9
     retroviruses, 5
     tumor aggressiveness, 5
     tumor detection, 9–10
     tumor-suppressor genes, 3, 8
Cancer (histone modifications), 17–26
     aberrant gene expression, 19
     acute promyelocytic leukemia, 18
     cambinol, 22
     chromatin remodeling, 19
     chromatin silencing, 20
     dihydrocoumarin, 22
     global alterations, 18
     histone acetyltransferases, 17
     histone code, 19
     histone deacetylases, 17
     histone demethylases, 18
     histone methylation, 18
     histone methyltransferases, 20
     histone-modifying enzymes, 19–22
     indoles, 22
     posttranslational modification, 17–18
     proliferation, 22
     sirtuins, 18, 20, 22
     site-specific alterations, 19
     transcriptional coactivators, 21
     transcriptional silencing, 19
     tumor-suppressor genes, inactivation of, 21
Cancer and Leukemia Group B (CALGB), 28, 30
Cardiovascular disease (CVD), 207–223
     atherosclerosis, 208–209
     cholesterol, 209
     chromatin conformation, 214
     collagen glycation, 219
     diabetes mortality, 217
     DNA hypomethylation, 212
     DNA methylation and natural history of
         atherosclerosis, 209–214
         DNA hypermethylation, 211
         DNA hypomethylation, 209–210
         mechanisms of aberrant DNA methylation
         patterns, 211–214
     DNA methylation patterns, aberrant, 211–214
         control of chromatin structure by lipids,
         213–214
         folate and homocysteine connection, 212
         nonreplicative cells, 214
     DNA methylation and stroke, 214
     fatty acids, 213, 214

fatty streak formation, 208
future directions, 219–220
global DNA hypomethylation, 210
lipid peroxidation products, 216
lipoprotein metabolism, 209
oestrogen receptors, 211
oocyte pool maturation phase, 217
prepubertal slow growth period, 216
risk
    hit, 219
    remote control of, 215, 218
    transgenerational transmission of, 217
risk factors, 215–219
    alternatives to epigenetic mechanisms,
        218–219
    aortic coarctation, 217–218
    early non-nutritional factors, 217–218
    nutritional factors, 215–217
    tobacco smoke, 218
    very-low-density lipoproteins, 208
Castanea sativa, 228
CDK6, *see* Cyclin D kinase 6
CDK inhibitor, *see* Cyclin-dependent kinase
        inhibitor
CHARGE syndrome, 158
CHD-binding protein, *see* Chromodomain
        helicase DNA-binding protein
ChIP, *see* Chromatin immunoprecipitation
Cholesterol, 209
Chromatin
    immunoprecipitation (ChIP), 141, 162, 265
    network, 1
    remodeling, 1, 156–159
        CHARGE syndrome, 158
        Cockayne syndrome B, 158–159
        enzymes, 5
        HDACi and, 50
        histone modifications and, 19
        Schimke immuno-osseous dysplasia, 159
        X-linked α thalassemia mental
            retardation syndrome, 156–158
    structure, 262
Chromodomain helicase DNA (CHD)-binding
        protein, 158
Chronic lymphocytic leukemia (CLL), 58
Chronic myeloid leukemia (CML), 31
CLL, *see* Chronic lymphocytic leukemia
CLS, *see* Coffin–Lowry syndrome
CML, *see* Chronic myeloid leukemia
Cockayne syndrome (CS), 158
Coffin–Lowry syndrome (CLS), 163–164
Collagen glycation, 219
Corylus avellana, 228
CpG, *see* Cytosine–guanine
CpNpG, *see* Cytosine–any nucleotide–guanine
CREB-responsive gene, 122
CS, *see* Cockayne syndrome

CTLs, *see* Cytotoxic T lymphocytes
CTP, *see* Cytidine triphosphate
CVD, *see* Cardiovascular disease
Cyclic peptides, 52–56
Cyclin-dependent kinase (CDK) inhibitor, 59
Cyclin D kinase 6 (CDK6), 8
Cystic fibrosis, 189
Cytidine deaminase inhibitor, 37
Cytidine triphosphate (CTP), 28
Cytosine–any nucleotide–guanine (CpNpG), 226
Cytosine–guanine (CpG), 226
    amino acids containing, 253
    dinucleotides, 133, 242
    DNA methyltransferase, insect, 250
    island(s)
        DNA methylation and, 4
        hypermethylation, 10
        plant epigenetics and, 227
        Polycomb complexes and, 134
        promoter methylation, 19
        unmethylated state of, 134
    microarrays, epigenomics and, 267
    promoters, DNA methylation and, 270
    suppression, 4
Cytosine methylation, plant epigenetics and,
        225, 232
Cytotoxic T lymphocytes (CTLs), 123

D

DAPK, *see* Death-associated protein kinase
DC, *see* Dendritic cells
Death-associated protein kinase (DAPK), 10
Dendritic cells (DCs), 116
DES, *see* Diethylstilbestrol
Diabetes mortality, 217
Diethylstilbestrol (DES), 251
Differential methylation hybridization (DMH),
        265
Diffuse large B-cell lymphoma (DLBCL), 116
Dihydroazacytidine, 33
Dihydrocoumarin, 22
DLBCL, *see* Diffuse large B-cell lymphoma
DLT, *see* Dose-limiting toxicity
DMH, *see* Differential methylation hybridization
DNA demethylating agents, 27–48
    aberrant DNA methylation of promoter-
        associated CpG islands, 27
    acute myeloid leukemia, 29
    chronic myeloid leukemia, 31
    combination of DNA methylation inhibitors
        with histone deacetylase inhibitors,
        35–36
    cytidine deaminase inhibitor, 37
    cytidine triphosphate, 28
    dihydroazacytidine, 33
    DNA methyltransferase, 28

DNMT inhibitor, 37, 38
fazarabine, 33
gamma globin synthesis, 34
gene reexpression, 40
gene-specific DNA hypomethylation, 37
hypomethylating agents as
        immunomodulators, 34
hypomethylating agents in nonmalignant
        hematological disorders, 34–35
hypomethylating agents in solid tumors,
        33–34
incidences of hematological adverse events,
        32
in vitro effects of nucleoside analog
        hypomethylating agents, 37
in vivo effects of nucleoside analog
        hypomethylating agents, 37–38, 39
LINE test, 35, 38
methylation-specific PCR, 38
myelodysplastic syndromes, 27
non-nucleoside analogs, 38–40
nucleoside analogs, 28–33
        5-azacytidine, 28–30
        5-aza-2'-deoxycytidine, 30–33
        other nucleoside analogs, 33
optimal dosing, 41
TIMP-3 gene, 40
transient marrow hypoplasia, 31
DNA methylation, see also Cancer (DNA
        methylation)
analysis, genome-wide, 264
array-based detection of, 266
changes, lymphocyte differentiation and,
        178–179
data, hierarchical cluster analysis of, 275
establishment of, 159–160
natural history of atherosclerosis and,
        209–214
        DNA hypermethylation, 211
        DNA hypomethylation, 209–210
        mechanisms of aberrant DNA
                methylation patterns, 211–214
patterns, 157
        abnormal, Polycomb complexes and, 143
        atherosclerosis, 211
        epigenomic sequence features and, 270
        Polycomb complexes and, 143
plants, see Plant epigenetics
PRC2 products targeting, 144
reading, 160–163
stroke and, 214–215
user-friendly techniques to study, 1
DNA methyltransferase (DNMT), 4, 28, 226, 250
DNMT, see DNA methyltransferase
Dose-limiting toxicity (DLT), 34
Drosophila melanogaster, 75, 107, 131

Drugs, see DNA demethylating agents; Histone
        deacetylase inhibitors

E

EBV, see Epstein–Barr virus
E-cadherin, 11
EGF, see Epidermal growth factor
Electron transport chain (ETC), 87
Embryonic stem (ES) cells, 88, 132
        CpG identification, 271
        Dicer-deficient, 108
        differentiation of, 133
        genome-wide expression analysis of, 137
        microRNAs and, 107
        PcG products and, 132
        sirtuins and, 88
Environment, see Epigenetics, environment, and
        evolution
Epidermal growth factor (EGF), 163
Epigenetic Baldwin effect, 242
Epigenetic inheritance system, 190
Epigenetics
        definition of, 242
        introduction to, 1
Epigenetics, environment, and evolution,
        241–260
        amplicon, 246
        autism, 256
        Baldwin effect, 242, 244
        cancer, 251–252
                clonal versus polyclonal models for
                        cancer initiation, 252
                transgenerational epigenetics and cancer,
                        252
        clonal genetic model, 252
        DNA repair enzymes, 253
        ELBOs, 248, 249
        endocrine disruptor contamination, 251
        epigenetic Baldwin effect, 242
        epigenetic changes leading to genetic
                changes, 252–256
                deamination of methylated cytosine to
                        thymidine, 253
                epigenetic regulation of trinucleotide
                        repeat expansion and contraction,
                        253–256
        epigenetic inheritance in Drosophila,
                247–250
                epigenetic assimilation experiments, 248
                epigenetic transmission of PRE
                        occupancy through meiosis, 247
                genetic versus epigenetic capacitor
                        models, 248–250
        epigenetic progenitor theory, 252
        evolution driven by need, 243
        father of epigenetics, 245

future studies, 256
genetic assimilation, 245
historical introduction of epigenetics, 242–247
  Charles Darwin, 243
  Conrad Waddington, 245
  Gregor Mendel, 244
  Hsp90 as capacitor for morphological evolution, 247
  Ivan Schmalhausen, 245
  James Baldwin, 244
  Jean-Baptiste Lamarck, 242–243
  John Cairns and facilitated genetic variation, 245–246
  Modern Synthesis, 244
laws of genetic inheriatnce, 244
laws of inheritance, 242
loss of imprinting of IGF2, 252
LTR promoters, 251
model of evolution, 244
model for morphological evolution, 255
Modern Synthesis, 244
morphisms of stressful environment, 256
morphoses, 245
norm of reaction, 245
pangenesis, 243
predictions of developmental plasticity, 244
protein capacitor, 247
soil and seed hypothesis, 252
spontaneous mutation, 246
stabilizing selection, 245
surrogate of genetic mutations, 252
transgenerational epigenetic phenomena, 250
transgenerational epigenetics in mammals, 250–251
  endocrine disruptors and transgenerational epigenetic effects, 251
  nutrition and DNA methylation of retrotransposon mutations, 250–251
Epigenetics and its genetic syndromes, 155–174
ADD domain, 156, 157
ATR-X syndrome, 156
childhood cancers, 164
chromatin remodeling, 156–159
  CHARGE syndrome, 158
  Cockayne syndrome B, 158–159
  enzyme, 159
  Schimke immuno-osseous dysplasia, 159
  X-linked α thalassemia mental retardation syndrome, 156–158
Cockayne syndrome, 158
corepressor complexes, 161
CpG islands, 160
CREB-binding protein, mutations in, 164
cyclin-dependent kinase-like 5 gene, 161
defect in HAT activity, 164

diseases of chromatin, 155
epidermal growth factor, 163
establishing DNA methylation, 159–160
human disease-causing mutations, 167
ICF syndrome, 159, 167
imprinting control region, 162
loss of protein-coding genes, 166
mental retardation, 156
methylation-associated gene silencing, 162
missense mutations, 156, 164
modifying histones, 163–165
  Coffin–Lowry syndrome (CLS), 163–164
  Rubenstein–Taybi syndrome (RSTS), 164
  Sotos syndrome, 164–165
mutations in ATRX, 156
myelodysplasia, 156
NSD1 gene, 165
nucleotide excision repair pathway, 158–159
reading DNA methylation, 160–163
RSK regulated members, 163
silencing by position effect, 165–167
silent heterochromatin, 165
T-cell immunodeficiency, 159
α thalassemia, 156
X chromosomal inactivation, 161
xeroderma pigmentosa, 158
Epigenomics, 261–284
acute myeloid leukemias, 274
biomedical research, 269–276
  cancer cell epigenome, 273–276
  epigenome of normal cells, 269–273
chromatin structure, 262
CpG promoters, 270
definition of epigenetic events, 261
DNA methylation analysis, 264
ENCODE regions, 272
future directions, 276
gene regulation, 262
hematopoietic malignancies, 269
Human Epigenome Project, 276
human genome, 261, 269
limitations of ChIP, 268
microarray platforms, 266
shotgun bisulfite sequencing, 276
technical approaches for studying epigenome, 262–269
  genome-wide detection of DNA methylation changes, 263–268
  genome-wide detection of histone modifications, 268
  high-throughput sequencers, 268–269
  historical perspective, 262–263
Epstein–Barr virus (EBV), 122, 123
ERC, see Extrachromosomal rDNA circles
ES cells, see Embryonic stem cells
ESR, see Oestrogen receptors
ETC, see Electron transport chain

Eucalyptus globulus, 228
Evolution, *see* Epigenetics, environment, and
   evolution
Extrachromosomal rDNA circles (ERC), 94, 95

F

FA, *see* Fatty acids
Facioscapulohumeral dystrophy (FSHD),
   165–167
Fatty acids (FA), 213
  chromatin conformation and, 214
  nuclear receptors, 214
Fazarabine, 33
FDA, *see* U.S. Food and Drug Administration
5-Fluorouracil (5-FU), 11
FMR gene, *see* Fragile X mental retardation-1
   gene
FMRP, *see* Fragile X mental retardation protein
FOXO transcription factors, 85
Fragile X mental retardation-1 (FMR) gene, 4
Fragile X mental retardation protein (FMRP),
   120
FSHD, *see* Facioscapulohumeral dystrophy
5-FU, see5-Fluorouracil

G

GDH, *see* Glutamate dehydrogenase
Genetic syndromes, *see* Epigenetics and its
   genetic syndromes
Genome hypomethylation, 5
Genomic imprinting, major psychosis and, 194
GH, *see* Growth hormone
Glutamate dehydrogenase (GDH), 91
Graft-versus-host disease (GVHD), 60
Graft-versus-leukemia (GVL) effects, 60
Growth hormone (GH), 94
GVHD, *see* Graft-versus-host disease
GVL effects, *see* Graft-versus-leukemia effects

H

HATs, *see* Histone acetyltransferases
H-cadherin, 11
HCMV, *see* Human cytomegalovirus
HCV, *see* Hepatitis C virus
HD, *see* Huntington's disease
HDAC, *see* Histone deacetylase
HDACi, *see* Histone deacetylase inhibitors
HDL, *see* High-density lipoproteins
HDMs, *see* Histone demethylases
HEP, *see* Human Epigenome Project
Hepatitis C virus (HCV), 124

Hereditary nonpolyposis colorectal cancer
   (HNCC), 252
Heterochromatin, sirtuins and, 80
High-density lipoproteins (HDL), 208
High-performance capillary electrophoresis
   (HPCE), 263
High-performance liquid chromatography
   (HPLC), 263
Histone(s)
  code, 19
  –DNA binding alterations, 17
  modifications, 1, *see also* Cancer (histone
    modifications)
    epigenomics and, 272
    in lupus, 183
    regulators of, 5
  modifying, 163–165
    Coffin–Lowry syndrome (CLS), 163–164
    Rubenstein–Taybi syndrome (RSTS), 164
    Sotos syndrome, 164–165
Histone acetyltransferases (HATs), 17, 142
Histone deacetylase (HDAC), 17, 35, 74, 177, 231
  -containing repressor complexes, 177
  plant epigenetics and, 231
Histone deacetylase inhibitors (HDACi), 49–71
  anti-FasL blocking antibodies, 57
  antitumor effects, 61, 62
  antitumor efficacy, 50
  antitumor properties, 60
  apicidin-induced apoptosis, 57
  apoptosis, 56–60
    cell cycle arrest and differentiation,
     59–60
    role of death receptor (extrinsic) pathway,
     57
    role of mitochondrial (intrinsic) pathway,
     58–59
  biological effects, 56–63
    angiogenesis and immune responses, 60
    apoptosis, 56–60
    clinical anticancer agents, 63
    transcription-independent effects, 61–63
  cell differentiation, 59
  characteristics of HDACi classes, 52–56
    benzamides, 56
    cyclic peptides, 52–56
    hydroxamic acids, 52
    ketones, 56
    short-chain fatty acids, 56
  chemical structures, 51
  chromatin remodeling, 50
  chromosome condensation, 61
  chronic lymphocytic leukemia (CLL), 58
  classes, 50
  clinical trials, 50, 63
  cowpox virus serpin, 57
  cyclin-dependent kinase inhibitor, 59

cytotoxic effects, 64
depsipeptide-induced apoptosis in osteosarcoma cells, 57
expression of VEGF receptors, 60
FADD-DN, 57
fusion proteins, 59
immune responses, 60
nonhistone proteins, 61
proteasomal degradation, 56
ROS-regulatory proteins, 59
SAHA-induced apoptosis, 58
structural basis of enzyme inhibitory activity, 50–52
suppression of angiogenesis, 56
T cell lymphoma, 52, 63
T-lymphoblastic leukemia cells, 57
transcription-independent effects, 61–63
cell cycle checkpoints, 61
DNA damage and repair, 61
hyperacetylation of nonhistone proteins, 61–63
Histone demethylases (HDMs), 18, 140
Histone methyltransferase (HMT), 18, 84
HIV, *see* Human immunodeficiency virus
HL, *see* Hodgkin's lymphoma, 116
HMT, *see* Histone methyltransferase
HNCC, *see* Hereditary nonpolyposis colorectal cancer
Hodgkin's lymphoma (HL), 116
Hox gene downregulation, 145
HPCE, *see* High-performance capillary electrophoresis
HPLC, *see* High-performance liquid chromatography
Human cytomegalovirus (HCMV), 123
Human Epigenome Project (HEP), 276
Human immunodeficiency virus (HIV), 106
Huntington's disease (HD), 96, 189
Hydroxamic acids, 52

competitive inhibitor of DNA methyltransferase activity, 181
epigenetics in immune differentiation and immune response, 176–179
DNA methylation changes associated with lymphocyte differentiation and activation, 178–179
transcription factors and epigenetic modifications in lymphocyte differentiation, 177–178
failure to maintain epigenetic homeostasis, 175
Ikaros proteins, 177
immunologic synapse, 180
methotrexate, 183
naive lymphocytes, 176
Polycomb group proteins, 178
procainamide, 181
RUNX1 regulation, 178
scleroderma, 184
T-cell antigen receptor, 180
T-cell differentiation, 176
T lymphocyte differentiation, 185
X chromosome skewing, 184
Immunodeficiency, centromeric instability, and facial anomalies (ICF) syndrome, 159–160, 167
autoimmunity and, 179
etiology of, 160
Imprinting control region (ICR), 162
Indoles, 22
Insulin-like growth factor (IGF)
binding protein 3, 11
pathway, 78
International Working Group (IWG) criteria, 29
IRAK1 protein, 122
IWG criteria, *see* International Working Group criteria

I

ICF syndrome, *see* Immunodeficiency, centromeric instability, and facial anomalies syndrome
ICR, *see* Imprinting control region
IGF, *see* Insulin-like growth factor
Ikaros proteins, 177
Immunity, 175–188
antigen receptor genes, 176
antinuclear antibodies, 184
autoimmunity, 179–185
epigenetic changes in other autoimmune disorders, 183–185
ICF syndrome, 179
lymphocyte development, 176
systemic lupus erythematosus, 179–183

J

JmjC, *see* Jumonji C
Juglans regia, 228
Jumonji C (JmjC), 141

K

Kaposi's sarcoma-associated virus (KSHV), 123
Ketones, 56
KSHV, *see* Kaposi's sarcoma-associated virus

L

LAT, *see* Latency-associated transcript
Latency-associated transcript (LAT), 123

LDL, *see* Low-density lipoproteins
Leishmania infantum, 77
Lipid peroxidation products, 216
LOI, *see* Loss of imprinting
Long-terminal repeat (LTR) promoters, 251
Loss of imprinting (LOI), 252
Low-density lipoproteins (LDL), 208
LTR promoters, *see* Long-terminal repeat
    promoters
LUMA, *see* Luminometric methylation assay
Luminometric methylation assay (LUMA), 263
Lupus
    drug-induced, DNA methylation and, 181–182
    histone modifications in, 183
    idiopathic, 182
Lymphocyte(s)
    cytotoxic T lymphocytes, 123
    development, 176
    differentiation, epigenetic modifications in,
        177–178
    naive, 176

M

Major psychosis, etiology of, 189–206
    antipsychotic drugs, 194
    Beckwith–Wiedemann syndrome, 199
    bipolar disease, 196
    brain cell types, 198
    cerebral maldevelopment, 193
    developmental arrest, 191
    DNA sequence-based theory, 194
    epigenetic inheritance system, 190
    epigenetic metastability, 191
    epigenetic misregulation, 197
    epigenetic model, 197
    experimental considerations, 197–200
    fundamental concepts behind epigenetic
        model, 190–191
        dynamic epigenetic status, 190–191
        regulation of genomic functions, 191
        transmission of epigenetic signals, 191
    genomic imprinting, 194, 196
    non-Mendelian features of psychiatric
        disease, 195–197
    non-shared environmental effects, 195
    parent-of-origin effect, 196
    pre-epimutation, 197
    primary site of disease manifestation, 198
    recreational drugs, 195
    schizophrenia, 199
    sexual dimorphism, 190, 196
    tissue differentiation, 194, 197
    traditional theories, 192–195
        epigenetic perspective, 194–195
        genetic theory, 192–193
        neurochemical theory, 193–194

        neurodevelopmental theory, 193
Malaria, 76
MALT, *see* Mucosa-associated lymphoid tissue
    lymphoma
Mantle cell lymphoma (MCL), 274
Marginal zone lymphoma (MZL), 114
MBD, *see* Methyl-binding domain
MCL, *see* Mantle cell lymphoma
MDS, *see* Myelodysplastic syndrome
MeDip, *see* Methyl DNA immunoprecipitation
MEFs, *see* Mouse embryonic fibroblasts
Mendelian diseases, 189
Methotrexate, 183
Methylation-specific PCR (MSP), 9, 38, 263
Methyl-binding domain (MBD), 143, 160
Methyl DNA immunoprecipitation (MeDip), 9
Methylmethane sulfonate (MMS), 90
MicroRNAs in cell biology and diseases,
    105–130
    BART cluster, 122
    B-cell integration cluster, 116
    B-cell lymphomas, 118
    B-cell proliferation, 122
    BHRF1 cluster, 122
    Burkitt lymphoma, 116
    cancer, 111–120
        cancer-associated genomic regions,
            111–114
        high-throughput miRNA profiling
            methods, 114–115
        oncogenes, 115–118
        tumor suppressor genes, 118–120
        of unknown primary origin, 125
    cell biology, 107–111
        microRNAs as differentiating genes,
            107–110
        microRNAs as pro-/anti-apoptotic genes,
            110–111
    chromosomal location, 114
    chromosomal rearrangements, 119
    chromosomal translocations, 117
    CREB-responsive gene, 122
    Dicer, 106
    diffuse large B-cell lymphoma, 116
    Drosha, 106
    Epstein–Barr virus, 122, 123
    germline point mutation, 118
    GPIIB promoter, 109
    hepatitis C virus, 124
    host–virus interactions, 123
    HOXB8 expression, 121
    human cytomegalovirus, 123
    human diseases, 111–124
        cancer, 111–120
        endocrine diseases, 124
        immunity, 121–122
        neurological diseases, 120–121

viral diseases, 122–124
human immunodeficiency virus, 106
identification of founding members, 105
IFN-mediated upregulation of miR, 122
immunoglobulin heavy-chain enhancers, 118
Kaposi's sarcoma-associated virus, 123
latency-associated transcript, 123
lung cancer, 115
maintenance methylation, 120
marginal zone lymphoma, 114
mediators of inflammatory response, 116
mucosa-associated lymphoid tissue
        lymphoma, 114
multiple myeloma, 118
OCD circuit, 121
oncogenic, 112–113
oncosuppressors, 115
overexperession mechanisms, 117
physiology, 105–107
pre-miRNAs, 106
pri-miRNAs, 106
quantitative RT-PCR, 115
Ras-dependent cellular proliferation, 118
reexpression, 119
schizophrenia, 125
seed sites, 107
serum response factor, 110
stem cell differentiation, 108
stem cell fate, 110
suppressor, 112–113
T-cell activation, 116
transcription factor, 119
tumor suppressor genes, 111
viral miRNAs, 123
MM, see Multiple myeloma
MMS, see Methylmethane sulfonate
Modern Synthesis, 244
Monozygotic (MZ) twins, 190, 195
Mouse embryonic fibroblasts (MEFs), 84
MSP, see Methylation-specific PCR
Mucosa-associated lymphoid tissue lymphoma
        (MALT), 114
MULE, see Mutator Like Element
Multiple myeloma (MM), 58
Mutator Like Element (MULE), 231
Myelodysplasia, 156
Myelodysplastic syndrome (MDS), 12, 27
classification of, 29
studies of 5-aza-2′-deoxycytidine in, 32
MZL, see Marginal zone lymphoma
MZ twins, see Monozygotic twins

N

Nampt, see Nicotinamide
        phosphorybosiltransferase
Natural killer (NK) cells, 182

NBS1, see Nijmegen breakage syndrome 1
NER pathway, see Nucleotide excision repair
        pathway
NHEJ mechanism, see homologous end-joining
        mechanism
Nicotinamide phosphorybosiltransferase
        (Nampt), 90
Nijmegen breakage syndrome 1 (NBS1), 86
NK cells, see Natural killer cells
Nonhomologous end-joining (NHEJ)
        mechanism, 74
Non-small cell lung cancer (NSCLC), 34
NSCLC, see Non-small cell lung cancer
Nucleotide excision repair (NER) pathway,
        158–159

O

Obsessive-compulsive disorder (OCD), 121
OCD, see Obsessive-compulsive disorder
Oestrogen receptors (ESR), 211
Olea europaea, 228
Oncogene(s)
    identification of, 3
    miRNAs as, 115
    overexpression, sirtuins and, 86
Oncosuppressors, miRNAs as, 115

P

PAI, see Phosphoribosylanthranilate isomerase
Pangenesis, 243
Papillary thyroid carcinoma (PTC), 116
Parent-of-origin effect, 196
Parkinson's disease (PD), 96
PBMC, see Peripheral blood monocytes
PcG, see Polycomb group
PCR
    methylation-specific, 9, 38, 263
    real time, 115
PD, see Parkinson's disease
Peripheral blood monocytes (PBMC), 40
Peroxisome proliferator-activated receptors
        (PPARs), 213
PEV, see Position effect variegation
PhoRC, see Pho repressive complex
Pho repressive complex (PhoRC), 139
Phosphoribosylanthranilate isomerase (PAI), 225
Pinus radiata, 228
PKC, see Protein kinase C
Plant epigenetics, 225–239
    absence of visible growth, 233
    apiallele segregation, 234
    BALL pathogen-resistance gene, 225
    biological clock, 227
    cell differentiation, 227

chromomethylase, 234
cytosine methylation, 225, 232
DNA methylation in plants, 225–227
DNA methyltransferases, 226
DNMT proteins, 226
environment-induced epigenetic variations,
        232–233
epigenetic memory, 233–234
FLC expression, 229–230
hidden sympodial, 228
histone deacetylase, 231
implication of DNA methylation in
        developmental processes, 227–232
    aging, 227–229
    flowering, 229–231
    stress and transposons, 231–232
retroviruses, 225
seed plants, 233
target recognition domain, 226
tryptophan biosynthesis genes, 225
tuber meristem dormancy, 233
vernalization, 232
X chromosome inactivation, 225
Plasmodium, 76–77
PMBL, *see* Primary mediastinal B-cell
        lymphoma
Polycomb complexes, chromatin modifications
        by, 131–154
    abnormal DNA methylation patterns, 143
    activating marks, 132
    bivalent domains, 133
    cell cycle progression deubiquitinase, 142
    cell-type-specific expression patterns, 132
    ChIP-on-ChIP experiments, 145
    chromatin compaction activity, 140
    chromatin immunoprecipitation, 136, 141
    citosine methylation, 134
    CpG islands, 134
    CTBP repressor complex, 140
    genome-wide maps of epigenetic landmarks
            associated with protein-encoding
            genes, 132–134
        DNA methylation and gene expression,
            134
        histone modifications in promoter
            proximal regions, 132–134
    histone acetyl transferase, 142
    histone-based mechanism, 146
    histone-binding proteins, 136
    histone demethylases, 140
    hormone-inducible PSA gene model, 143
    Hox gene downregulation, 145
    Hox gene upregulation, 141
    intersection of PcG and DNA methylation
            pathways, 143–145
    limited methylation of DNA, 134
    linker histone, 137

mammalian PcG products, 135
nucleosomes, 133, 137
PcG repressive complexes, 135–140
    maintaining of repressed state, 138–139
    marking chromatin for repression,
            136–138
    other complexes, 139–140
perspectives, 145–146
Pho repressive complex, 139
Polycomb response elements, 145
PRC2 products targeting DNA methylation,
        1444
PRC3 complex, 137
protein inactivation, 144
recruiting of PcG complexes and mechanisms
        of repression, 145
reversal of PcG-dependent epigenetic marks,
        141–143
    histone H2A deubiquitinases, 142–143
    histone H3K27 demethylases, 141–142
Smad-interacting protein 1, 140
transcription factors, 137, 139
transcription initiation, 133
ubiquitylation assay, 138
X chromosome inactivation, 138
Polycomb group (PcG), 131
    -mediated gene repression, 139
    products
        cell proliferation and, 132
        isolation, 138
        mammalian, 135
    proteins, 8, 178, 272
    recruiting, 145
    regulation, developmental processes and, 145
    system, cellular diversity and, 132
    transcriptional repression, 145
Polycomb response elements (PREs), 145, 247
Position effect variegation (PEV), 75
PPARs, *see* Peroxisome proliferator-activated
        receptors
PREs, *see* Polycomb response elements
Primary mediastinal B-cell lymphoma (PMBL),
        116
Progressive systemic sclerosis (PSS), 184
Protein(s)
    ataxin-3, 96
    capacitor, morphological evolution and, 247
    chromodomain helicase DNA-binding
            protein, 158
    -coding genes, loss of, 166
    CREB-binding protein, 164
    DNMT, 226
    -encoding genes, Polycomb complexes and,
            132–134
        DNA methylation and gene expression,
            134

histone modifications in promoter proximal regions, 132–134
enzymatically inactive, 136
fusion, 8, 59
histone-binding, 136
hyperacetylation of, 62
Ikaros, 177
inactivation, Polycomb complexes and, 144
IRAK1 protein, 122
methylation-reading proteins, 84
methyl-binding, 4
methyl-CpG-binding proteins, 143
MGMT, 11
nonhistone, 61
PML-RAR, 8
Polycomb, 8, 80, 178, 272
ROS-regulatory, 59
Sir2, 20
Smad-interacting protein 1, 140
uncoupling protein 1, 90
uncoupling protein 2, 87, 90
Protein kinase C (PKC), 116
PSS, see Progressive systemic sclerosis
Psychiatric disorders, see Major psychosis, etiology of
Psychosis, see Major psychosis, etiology of
PTC, see Papillary thyroid carcinoma

R

RA, see Refractory anemia
RAEB, see Refractory anemia with excess blasts
RA with ringed sideroblasts (RARS), 29
RARS, see RA with ringed sideroblasts
Reactive oxygen species (ROS), 94
Real time (RT)-PCR, 115
Refractory anemia (RA), 29
Refractory anemia with excess blasts (RAEB), 29
Restriction landmark genomic scanning (RLGS), 9, 263
Retinoic acid, 109
Retroviruses
    DNA methylation and, 5
    plant, 225
Rett syndrome, 160–163
Rheumatoid arthritis, epigenetic changes in, 183–184
RISC, see RNA-induced silencing complex
RLGS, see Restriction landmark genomic scanning
RNAi, see RNA interference
RNA-induced silencing complex (RISC), 106
RNA interference (RNAi), 122
ROS, see Reactive oxygen species
RSTS, see Rubenstein–Taybi syndrome
RT-PCR, see Real time PCR
Rubenstein–Taybi syndrome (RSTS), 164

S

Saccharomyces cerevisiae, 74
S-adenosylhomocysteine (SAH), 212
S-adenosylmethionine (SAM), 209
SAH, see S-adenosylhomocysteine
SAHA, see Suberoylanilide hydroxamic acid
Salmonella enzyme, 75
SAM, see S-adenosylmethionine
Schimke immuno-osseous dysplasia (SIOD), 159
Schizophrenia, 192, 193, 199
Schizosaccharomyces pombe, 75
Seed sites, 107
Serum response factor (SRF), 110
SGP, see Slow growth period
Silencing RNA techniques (siRNA), 109
Silene latifolia, 230
Single-nucleotide polymorphisms (SNPs), 121, 193
SIOD, see Schimke immuno-osseous dysplasia
siRNA, see Silencing RNA techniques
Sirtuin(s), 20
    deacetylases, HDACi and, 62
    inhibitors, first known, 22
    structural homology, 18
Sirtuins in biology and disease, 73–104
    aging, 94–95
    ataxin-3, 96
    axonal degeneration, 96
    base excision repair pathway, 91
    Bax-dependent apoptosis, 85
    B-cell differentiation, 93
    binding partners, 79
    budding yeast, 76
    cancer, 92–94
    cellular stress, 90
    checkpoint factor, 86
    Drosophila and C. elegans Sir2 homologs, 77–78
    electron transport chain, 87
    embryonic stem cells, 88
    extrachromosomal rDNA circles, 94, 95
    FOXO transcription factors, 85
    general features and targets, 82–83
    glutamate-/glutamine-dependent insulin production, 91
    heterochromatin, 80, 81
    histone deacetylase, 74
    histone methyltransferase, 84
    human embryonic kidney cell line, 79
    insulin-like growth factor pathway, 78
    mammalian sirtuins, 78–92
        cellular localization, 78–79
        dual activity, 79
        H4K16Ac, 79–80
        SirT1, 80–88
        SirT2, 88–89
        SirT3, 90, 95

SirT4 and SirT5, 91
  SirT6 and SirT7, 91–92
methylation-reading proteins, 84
mitochondrial gene expression, 90
neurological diseases, 96
neuronal protection, 80
nicotinamide phosphorybosiltransferase, 90
Nijmegen breakage syndrome 1, 86
nucleolar compartments, 91
oncogene overexpression, 86
oxidative stress, 86, 92
pattern of localization, 78
polycomb factor, 77
polyQ tract expansion, 96
prokaryotic Sir2 homologs, 75
Sir2 family, 74–75
SIR genes, 74
Sir2 homologs in lower eukaryotes, 76–77
  Plasmodium, 76–77
  Trypanosoma, 77
  yeast, 76
SirT1, 80–88
  cell differentiation, 87–88
  cell survival, 85–86
  heterochromatin regulation, 80–85
  H3K9 methylation and, 84
  metabolic homeostasis, 87
SirT2, 88–89
  cytoskeleton, 89
  H4K16Ac and cell cycle regulation, 89
  overexpression, 88
tumor suppressor genes, 93
uncoupling protein 2, 87
SLE, *see* Systemic lupus erythematosus
Slow growth period (SGP), 216
SMA, *see* Spinal muscular atrophy
SMC, *see* Smooth muscle cells
Smooth muscle cells (SMC), 208
SNPs, *see* Single-nucleotide polymorphisms
Soil and seed hypothesis, 252
Solid tumors, hypomethylating agents in, 33
Sotos syndrome, 164–165
Spinal muscular atrophy (SMA), 120
SRF, *see* Serum response factor
Stabilizing selection, 245
Stroke, DNA methylation and, 214–215
Suberoylanilide hydroxamic acid (SAHA), 50
  expression of VEGF receptors, 60
  graft-versus-leukemia effects, 60
  -induced apoptosis, 58
Sulfolobus, 75
Systemic lupus erythematosus (SLE), 179–183
  DNA demethylation and autoimmunity,
    180–181
  DNA methylation and drug-induced lupus,
    181–182

DNA methylation and idiopathic lupus,
    182–183
DNA methylation and T-cell autoreactivity,
    180
histone modifications in lupus, 183

T

T-ALL cells, *see* T-lymphoblastic leukemia cells
TAR, *see* Transactivating response
T-cell
  antigen receptor (TCR), 180
  autoreactivity, 180
  differentiation, 176
  immunodeficiency, 159
  lymphoma, 52, 63
  rheumatoid arthritis, 183
TCR, *see* T-cell antigen receptor
TGF-β, *see* Transforming growth factor-β
α Thalassemia, 156
THBS-1, *see* Thrombospondin-1
Thrombospondin-1 (THBS-1), 11
T-lymphoblastic leukemia (T-ALL) cells, 57
Transactivating response (TAR), 106
Transactivating response RNA-binding protein
    (TRBP), 106
Transcriptional repression domain (TRD), 160
Transcriptional start sites (TSS), 272
Transforming growth factor-β (TGF-β), 123
Transient receptor potential melastatin-related
    channel 2 (TRPM2), 79
TRBP, *see* Transactivating response RNA-
    binding protein
TRD, *see* Transcriptional repression domain
TRPM2, *see* Transient receptor potential
    melastatin-related channel 2
Trypanosoma, 77
Trypanosomes, 77
Tryptophan biosynthesis genes, 225
TSGs, *see* Tumor suppressor genes
TSS, *see* Transcriptional start sites
Tumor aggressiveness, DNA methylation and, 5
Tumor detection, DNA methylation in, 9
Tumor suppressor genes (TSGs), 93, 111
  CpG islands of, 8
  DNA methylation and, 8
  identification of, 3, 74
  inactivation of, 21
  microRNAs and, 111, 118
  sirtuins and, 93

U

UCP1, *see* Uncoupling protein 1
UCP2, *see* Uncoupling protein 2
Uncoupling protein 1 (UCP1), 90

Uncoupling protein 2 (UCP2), 87, 90
U.S. Food and Drug Administration (FDA), 52
    SAHA approval, 52
    vorinostat approval, 63

V

Valproic acid (VPA), 35, 36, 50
VEGF receptors, SAHA expression of, 60
Very-low-density lipoproteins (VLDL),
        208, 216
Virus(es)
    Dengue virus, 124
    Epstein–Barr virus, 122, 123
    hepatitis C virus, 124
    HIV, 106
    influenza B virus, 124
    Kaposi's sarcoma-associated virus, 123
    miRNAs, 123
    PFV-1 virus, 124

RNAi defense system and, 122
VLDL, *see* Very-low-density lipoproteins
VPA, *see* Valproic acid

W

WHO, *see* World Health Organization
World Health Organization (WHO), 29

X

X chromosomal inactivation (XCI), 161, 225
X chromosome skewing in autoimmunity,
        epigenetic changes in, 184
XCI, *see* X chromosomal inactivation
Xenopus laevis, 142
Xeroderma pigmentosa, 158
X-linked α thalassemia mental retardation
        (ATR-X) syndrome, 156–158

T - #0051 - 041019 - C0 - 234/156/17 - PB - 9780367403492